ESSENTIALS OF PROJECT AND SYSTEMS ENGINEERING MANAGEMENT

Third Edition

ESSENTIALS OF PROJECT AND SYSTEMS ENGINEERING MANAGEMENT

Third Edition

HOWARD EISNER

Distinguished Research Professor and Professor, Engineering Management
and Systems Engineering Department
The George Washington University
Washington, DC

WILEY

John Wiley & Sons, Inc.

Library of Congress Cataloging-in-Publication Data:

Eisner, Howard, 1935–
    Essentials of project and systems engineering management / Howard Eisner. – 3rd ed.
        p.   cm.
    Includes bibliographical references and index.
    ISBN 978-0-470-12933-3 (cloth)
1. Systems engineering–Management.   2. Project management.   I. Title.
    TA168.E38   2008
    658.4'06–dc22   2007050287

Printed in the United States of America.

10 9 8 7 6 5 4 3 2 1

ABOUT THE AUTHOR

Since 1989, Howard Eisner has served as Distinguished Research Professor and Professor in the Engineering Management and Systems Engineering Department at The George Washington University in Washington, DC. For the prior thirty years, he held various technical and management positions in industry, including president of two systems and software engineering companies (Intercon Systems Corporation and the Atlantic Research Services Company). He also served as a board member of three high-tech companies. He is a Life Fellow of the Institute of Electrical and Electronics Engineers (IEEE) and a Fellow of the International Council on Systems Engineering (INCOSE) and a member of several engineering honor societies. Dr. Eisner has written two books on systems engineering and related topics. He has also written a book on personal and corporate reengineering and one on ways of thinking about and managing large complex systems. He holds a BEE from the City College of New York, an MS from Columbia University, and a Doctor of Science from The George Washington University.

CONTENTS

PART IV TRENDS, PERSPECTIVES, AND INTEGRATIVE MANAGEMENT

12 Systems/Software Engineering and Project Management Trends

13 Selected New Perspectives

14 Integrative Management

PREFACE

This book has two primary objectives: (1) to define and describe the essentials of project and systems engineering management, and (2) to show the relationship and interconnection between project management and systems engineering.

The subject of project management is well-trodden territory and is explored at considerable length in numerous books. Systems engineering, though, is not as well known, as measured perhaps by the literature that describes and supports it. However, this literature has clearly been on the upswing as the need for systems engineering has been increasing and expanding. Like project management, systems engineering deals with a variety of methods for designing and building a system that are largely independent of the domain itself. Slowly, but noticeably, systems engineering is finding its way into a greater number of college curricula and taking its place alongside the more traditional engineering disciplines, such as electrical engineering, mechanical engineering, chemical engineering, and the like.

More often than not, systems engineering is carried out in the real world in the context of a project. In a typical scenario, a company might set up a project whose basic purpose is to design and build some type of system. Thus, there is almost always a strong connection between project management and systems engineering, whether it is formally recognized or not. Many students have asked about this sometimes murky connection during my courses in systems engineering. They want to know more about how systems engineering fits into the structure of a project and its various management-oriented tasks and activities. These questions, directly and indirectly, have led to this book.

Indeed, this may be the first book that attempts to bring these two important subjects together.

This third edition provides new and expanded materials, including these subject areas:

- Problems in systems and software
- Errors in systems
- Numerical trade-offs with detection and false alarm probabilities
- Likelihood ratio and detection threshold setting
- Support for systems engineering within government, especially the Department of Defense (DoD) and NASA
- International Council on Systems Engineering (INCOSE)
- Investments in major large-scale systems
- Systems Engineering Management Plan (SEMP) and Systems Engineering Plan (SEP)
- Standards
- Group processes and decision making
- Test and evaluation
- Additional requirements problem areas
- Trading off of requirements
- DODAF, MoDAF, enterprise, and service-oriented architectures
- Multiple views of architectures
- System interoperability
- Modeling and simulation
- Unified Modeling Language (UML) and Systems Engineering Modeling Language (SysML)
- Systems Engineer and software engineering
- Nonconstant failure rates in systems
- Quantitative least squares fit
- Acquisition of systems (a directive and an instruction)
- Defense acquisition performance assessment
- Capability-based acquisition
- System complexity
- Integration of systems
- "Top dozen" integration list
- Thinking outside the box

As compared with the second edition, this edition adds a chapter (a new chapter 13) so that it has a total of 14 chapters and an appendix. This makes it suitable for a 15-week course in project and systems engineering management.

At the same time, Systems Engineers and Project Managers in an industrial environment, or with a government agency, will find the essentials of what they need to know under one cover.

I am pleased to dedicate this book at both a professional and a personal level. With respect to the former, I dedicate it to my graduate students and colleagues in the engineering management and systems engineering department in the School of Engineering and Applied Science, The George Washington University. At a personal level, I dedicate the book to my wife, June Linowitz, whose patience, support, and love helped make it possible.

HOWARD EISNER
Bethesda, Maryland

PART I
OVERVIEW

___1
SYSTEMS, PROJECTS, AND MANAGEMENT

1.1 INTRODUCTION

This is a book about management, with emphasis on managing the design, development, and engineering of systems. It addresses two primary questions:

1. What does the Project Manager (PM) need to know?
2. What does the Chief Systems Engineer (CSE) need to know?

The focus is therefore on the essentials of what the PM and CSE must master in order to be successful in building various types of systems and managing project teams.

This chapter is largely introductory, dealing with the preliminary definitions of systems and projects, problems encountered in building systems, the systems approach, key managerial responsibilities, and organizational matters that significantly impact the way in which systems are planned, designed, and constructed.

1.2 SYSTEMS AND PROJECTS

There are many definitions of systems, one of which is simply that "a system is any process that converts inputs to outputs" [1.1]. We look here at systems by example and, for that purpose, start by examining a radar system. This is certainly a system, performing the functions of search and tracking of objects in space, as in an air route or surveillance radar at or near an airport. A system normally has functions that it carries out (such as search and tracking), and it does so by means of its subsystems. At the same time, such airport radar systems, together with other systems (such as communications and landing systems), are part of a larger system known as an air traffic control (ATC) system. Examined from the perspective of an air traffic control system, the radar systems actually serve as subsystems of the larger system. In the same vein, the air traffic control system may be regarded as a subsystem of a larger national aviation system (NAS) that consists also of airports, air vehicles, and other relatively large systems (e.g., access/egress) in their own right.

Our view of systems, therefore, is rather broad. In the preceding context, the radars, air traffic control, and national aviation system are all systems. Such systems normally are composed of hardware, software, and human elements, all of which must interoperate efficiently for the overall system to be effective. We adopt this broad perspective in the definition of systems, drawing on examples that affect our everyday life, such as automobile systems, telephone systems, computer systems, heating and cooling systems, transit systems, and information systems.

Projects are formal enterprises that address the matter of designing and developing the various systems just cited. A project is an assemblage of people and equipment, and it is normally managed by a Project Manager (PM). Project personnel work toward satisfying a set of goals, objectives, and requirements, as set forth by a customer. Projects may also have a limited scope of work, dealing only with, for example, the design phase of a system, rather than its construction or entire life cycle. The success of a system is dependent on the skills of the people on a project and how well they are able to work together. Ultimately, the success, or lack of it, is attributed to the many skills that the PM is able to bring to bear in what is often an extremely complex situation and endeavor. The PM, in short, must not only have considerable technical skills, but must also have a deep understanding of the fine art of management.

1.2.1 Definitions of Systems Engineering

The Chief Systems Engineer (CSE) normally reports to the Project Manager and focuses upon building the system in question. The overall process that the CSE employs is known as *Systems Engineering*, a central theme in this text. We will define Systems Engineering in terms of increasing complexity and detail in various parts of this book, starting here with five relatively simple expressions, namely:

1. As developed by the International Council on Systems Engineering (INCOSE)
2. As articulated by the Department of Defense (DoD)

3. As represented in an earlier text by this author
4. As summarized by the Defense Systems Management College (DSMC)
5. As viewed by the National Aeronautics and Space Administration (NASA)

The INCOSE definition is that Systems Engineering is [1.2]:

> An interdisciplinary approach and means to enable the realization of successful systems.

This definition is rather sparse and emphasizes three aspects: "interdisciplinary," "realization," and "successful." Especially for large-scale systems, it is clearly necessary to employ several disciplines (e.g., human engineering, physics, software engineering, and management). Realization simply confirms the fact that systems engineering processes lead to the physical construction of a real-life system (i.e., it goes beyond the formulation of an idea or concept). Finally, our expectation is that by utilizing the various disciplines of systems engineering, the outcome will be a successful system, although this result is certainly not guaranteed.

A definition provided by the Department of Defense (DoD), a strong supporter as well as user of systems engineering as a critical discipline, is that Systems Engineering [1.3]:

> Involves design and management of a total system which includes hardware and software, as well as other system life-cycle elements. The systems engineering process is a structured, disciplined, and documented technical effort through which systems products and processes are simultaneously denned, developed and integrated. Systems Engineering is most effectively implemented as part of an overall integrated product and process development effort using multidisiplinary teamwork.

Key words from this definition include: "design and management," "hardware and software," "structured, disciplined and documented," and "overall integrated" effort that involves "multidisciplinary teamwork." These important notions will be reiterated and expanded upon in later parts of this book.

A third definition, formulated by this author, is that Systems Engineering is an [1.4]:

> Iterative process of top-down synthesis, development, and operation of a real-world system that satisfies, in a near-optimal manner, the full range of requirements for the system.

Here, key ideas have to do with "iterative," "synthesis," "operation," "near-optimal," and "satisfies the system requirements." Designing and building a system usually involves several loops of iteration, for example, from synthesis to analysis, from concept to development, and from architecting to detailed design. The notion of synthesis is emphasized, since the essence of systems engineering is viewed from the perspective of design rather than analysis. Design precedes analysis; if there is no coherent design, there is nothing to analyze. The term "near-optimal" suggests that large-scale systems engineering does not lead to a provably optimal design, except under very special circumstances. The normal cases all involve attempts to find an appropriate balance between a variety of desirable features. Trade-off analyses are utilized to move in the direction of a "best possible" design. Finally, in terms of the basic definition, we find a need to satisfy the full range of requirements for the system. The focus on constructing a system that is responsive to the needs of the user-customer is central to what systems engineering is all about.

The Defense Systems Management College text summarizes Systems Engineering as [1.5]:

> An interdisciplinary engineering management process to evolve and verify an integrated, life cycle balanced set of system solutions that satisfy customer needs.

Here, key words emphasize a "management process," "verification," a "balanced set of solutions," and "customer needs." This definition, therefore, tends to see systems engineering through a "management" prism, requires a balanced set of solutions as well as verification of those solutions, and the satisfaction of what the customer states as a set of needs.

The last definition cited in this chapter is that represented by NASA, namely, that Systems Engineering is [1.6]:

> A robust approach to the design, creation and operation of systems.

NASA expands this short explanation by emphasizing:

1. Identification and quantification of goals
2. Creation of alternative system design concepts
3. Performance of design trades
4. Selection and implementation of the best design
5. Verification that the design is properly built and integrated
6. Post-implementation assessment of how well the system meets the stated goals

The above five definitions of systems engineering, in the aggregate, give us a point of departure for our further exploration of systems engineering and the management thereof. We will also see other representations that tend to add further detail and structure to these short-form definitions. For example, Chapter 2 cites several standards that relate to systems engineering. Further,

Chapter 7 defines the thirty elements that this author considers to be the essence of large-scale systems engineering.

1.2.2 System Cost-Effectiveness

The project team, led by the PM and the CSE, are also in search of a *cost-effective* solution for the customer. In order that this concept have real substance, we must be in a position to ultimately quantify both the system cost as well as its effectiveness.

System cost will be approached from a life-cycle perspective. This means that a life-cycle cost model (LCCM) will eventually be constructed, with the following three major categories of cost:

1. Research, development, test, and evaluation (RDT&E)
2. Acquisition or procurement, and
3. Operations and maintenance (O&M)

The latter category, by our definition, will also include the cost of system disposal when it is necessary to do so. System cost will also be viewed as an independent variable, expressed as "cost as an independent variable" (CAIV). The Department of Defense (DoD) sees CAIV as a "strategy that entails setting aggressive yet realistic cost objectives when defining operational requirements and acquiring defense systems and managing achievement of these objectives" [1.3].

System effectiveness will also need to be calculated. One perspective regarding system effectiveness is that it is a function of three factors [1.4]:

1. Availability
2. Dependability
3. Capability

Availability is sometimes called the readiness reliability, whereas dependability is the more conventional reliability that degrades with time into the system operation. Capability is also referred to as system performance. The approach adopted here with respect to effectiveness is somewhat less restrictive, allowing the CSE's team the flexibility to select those effectiveness measures that are fundamental to the system design as well as of special importance to the customer and user.

System cost-effectiveness considerations may thus be visualized as a graph of effectiveness (ordinate) plotted against total life-cycle cost (abscissa). As such, we see that this type of graph implies that several systems can be built, each representing a "point" on such a plot. Our overall task as architects and designers of systems is to find the point design that is to be recommended to the customer from among a host of possible solutions. This further implies that the process will include the exploration of several alternatives until a preferred alternative is selected.

1.2.3 Support for Systems Engineering

Systems engineering has had major supporters over the years, in both methodology as well as applications. In this section we take a brief look at systems engineering perspectives from both the Department of Defense (DoD) as well as the National Aeronautics and Space Administration (NASA).

A few years back, the Deputy Under Secretary of Defense (DUSD) for Acquisition, Technology and Logistics (A,T&L) confirmed a policy of support for systems engineering across the department [1.7]. A key sentence is quoted next:

> All programs responding to a capabilities or requirements document, regardless of acquisition category, shall apply a robust SE (systems engineering) approach that balances total system performance and total ownership costs within the family-of-systems, systems-of-systems context.

Eight other notions emphasized in this policy document include:

1. A rigorous SE discipline is needed.
2. We are integrating increasingly complex systems.
3. Programs will formulate an SEP (Systems Engineering Plan).
4. We are attempting to institutionalize SE across all of the DoD.
5. Establish a senior-level SE forum, with participation by a flag officer.
6. Drive good SE practices back into the way we do business.
7. Make SE an important consideration during source selection.
8. Evaluate the adequacy of current policies and procedures and recommend changes where necessary.

The list is certainly a massive endorsement of SE as a key discipline with the DoD. It underscores the belief that a more widespread and rigorous application of SE will lead to better system performance, within schedule and budget. Also, with the large number of people and programs within the DoD, we can see a great need for very large-scale education and training with respect to the numerous elements of systems engineering. The aim of all of this, of course, is to provide "affordable, supportable and above all, capable solutions for the warfighter."

In 2006, NASA established a set of "Systems Engineering Processes and Requirements" [1.8], promulgated through its Office of the Chief Engineer. From a top-level perspective, and using NASA's words:

> NASA missions are becoming increasingly complex, and the challenge of engineering systems to meet the cost, schedule and performance requirements within acceptable levels of risk requires revitalizing systems engineering. . . .

The engineering of NASA systems requires the application of a systematic, disciplined, engineering approach that is quantifiable, recursive, iterative, and repeatable for the development, operation, maintenance, and disposal of systems integrated into a whole throughout the life cycle of a project or program.

Next we cite eight key points in NASA's approach to documenting a desired set of systems engineering processes [1.8]:

1. Increasing system complexity will be accompanied by the reduction of operations staff;
2. Systems are moving toward increased autonomy;
3. A robust approach is needed to meet NASA objectives;
4. Systems-level thinking is also needed;
5. Common technical processes are critical to implementing NASA products and systems;
6. A revolutionary advancement of SE is essential;
7. NASA must also deal with the implications of past failures;
8. Consistency across the administration is required to meet stated goals.

It is important to recognize that NASA expressed the need for revitalization of all processes with respect to systems engineering. The language in this NASA approach has the ring of urgency as well as determination. This NPR (NASA Procedural Requirements—7123.1) document provides a thrust in the direction of strengthening and applying well-defined processes. Indeed, NASA's overall framework for an improved SE capability includes:

a. Common technical processes
b. Tools and methods
c. Workforce considerations

The 17 items that NASA lists within the common technical processes are of special interest in this text, and are listed here:

System Design Processes

- Requirements Definition Processes
  1. Stakeholder Expectations Definition
  2. Technical Requirements Definition
- Technical Solution Definition Processes
  3. Logical Decomposition
  4. Physical Solution

Product Realization Processes

- Design Realization Processes
  5. Product Implementation
  6. Product Integration
- Evaluation Processes
  7. Product Verification
  8. Product Validation
- Product Transition Processes
  9. Product Transition

Technical Management Processes

- Technical Planning Process
  10. Technical Planning
- Technical Control Processes
  11. Requirements Management
  12. Interface Management
  13. Technical Risk Management
  14. Configuration Management
  15. Technical Data Management
- Technical Assessment Process
  16. Technical Assessment
- Technical Decision Analysis Process
  17. Decision Analysis

This articulation of key SE processes can be compared with process definitions and discussions provided in Chapters 2 and 7 (e.g., with respect to standards and elements of SE).

Of course, in addition to the DoD, NASA, and other government agencies, outstanding and continuous support has been provided by the International Council on Systems Engineering (INCOSE) [1.9]. This organization of leaders in the field has devoted time and energy to formulate, enhance, and apply the principles of systems engineering to numerous problems that arise in building and managing new and complex systems.

1.2.4 System Errors

In broad terms, all systems are said to exhibit fundamental errors known as Type I and Type II errors. These errors are related to the field of hypothesis testing whereby errors are made by (a) rejecting a hypothesis that is true (Type I error) or (b) accepting a false hypothesis (Type II error). From a systems engineering perspective, a major task of the CSE's team is to reduce such errors so as to satisfy the system requirements. Three examples of these errors are briefly discussed below.

Many of us have car alarm systems that are intended to go off when an intruder is trying to get into our car. There is an error if and when the alarm does not go off when forced entry is being attempted (Type I). At the same time, we do not wish to be awakened at 3 o'clock in the morning when the alarm goes off from the car in front of our house, without any type of intrusion (Type II error).

On a somewhat larger scale, we have radar systems that are intended to detect targets at specified ranges. When they fail to do so, an error (Type I) has been committed. On the other hand, these systems also claim, from time to time, that a target is present when no such target exists. This latter case (a Type II error) is called a false alarm. These types of errors for a search radar are explored in some detail in later chapters of this text. Specific detection and false alarm probabilities are calculated, and the relationship between them is examined.

On an even larger scale, we have situations presented by our national air transportation system. When the system fails to get you to your destination at the expected time of arrival (ETA), an error has been committed that all of us have experienced. And, if you're trying to get to New York from Washington, and wind up in Philadelphia due to bad weather, the system is delivering an unintended result.

Whether a system is large or relatively small, many times errors are the reason for failure of all or a part of the system. Therefore, often it is critical to *define and control* the most significant errors that might occur. To do so, usually an *error model* is constructed that:

a. Identifies all primary error sources, as well as their likely magnitudes.
b. Establishes mathematical relationships between these error sources.

If the errors can be shown to be independent and additive, often we can make good use of a well-known relationship from elementary probability theory: the variance of a sum is equal to the sum of the variances (of random variables). By definition, the square root of the variance is the standard deviation, which is designated as "sigma". Then we continue to work with the standard deviations to represent the errors in question. Given such an error model, we then must figure out the maximum tolerable errors and how to control (i.e., reduce) the errors to make sure the overall system error budget is not exceeded. As an example, if we are dealing with a shipboard air defense system or a spacecraft that is being placed in a precise orbit, if the error budget is exceeded, it is likely that these types of systems will fail or become severely degraded. The consequences may well be mission failure and loss of life.

One of the more important aspects of error analyses is to decide how to relate the errors defined in the system requirements document to the errors in the error model. Specifically, this question must be answered: Does the maximum error requirement correspond to the one-, two-, or three-sigma error (or greater) value? Under fairly general conditions for error modeling and analysis, designing a system for one of these three "assumptions" will result in:

1. An overall system error probability of about 32 percent (plus and minus one-sigma designation)
2. An overall system error probability of about 5 percent (plus and minus two-sigma designation)
3. An overall system error probability of about 0.27 percent (plus and minus three-sigma designation)

A simple one-sigma choice means that it is "allowable" for the overall system to fail about one third of the time. This is certainly not a recommended approach, and systems engineers must understand this issue in order to design the system properly. The issue is also directly related to the "six-sigma" notion that has been applied by numerous enterprises pursuing a specific high-quality approach to delivering products and services. More about error analyses will be presented in chapters eight and eleven, along with numerical examples.

Understanding when systems are likely to fail to do what they're supposed to do, and also do what they're not supposed to do, is often a central theme of the systems engineering activities. These, of course, can be expressed as problems that need to be solved by the management and technical personnel working on the system. At the same time, there are many problems that might be considered chronic issues when managing an engineering project. A sample of such problems is presented and discussed in the next section.

1.3   PROBLEMS IN MANAGING ENGINEERING PROJECTS

An article in the *Washington Post* [1.10] described an industry contract with the Federal Aviation Administration (FAA) in the terms "out-of-control contract" and "how a good contract goes sour." It went on to describe how a "cure letter" was sent to the contractor saying that "delays in a $4 billion contract to modernize the computers used in the nation's air traffic control system were unacceptable." Although this admonition pointed to delays and therefore could be connected to not getting the work done on time, it is likely that time delays resulted from performance issues and were also related to the cost of the program. In general terms, problems that surface on a typical project usually show themselves ultimately in terms of three main features:

1. Schedule (time)
2. Cost (as compared with the original budget)
3. Performance

These are the "big three" of project management and systems engineering management. Projects are originally planned to meet the performance requirements within the prescribed time and budget constraints.

Although there are numerous reasons why projects do not satisfy these three key aspects of a system development [1.11, 1.12, 1.13], several of the most common such reasons are:

1. Inadequate articulation of requirements
2. Poor planning
3. Inadequate technical skills and continuity
4. Lack of teamwork
5. Poor communications and coordination
6. Insufficient monitoring of progress
7. Inferior corporate support

The following discussion expands on these reasons for problems and lack of success.

1.3.1 Inadequate Articulation of Requirements

Requirements for a system are normally defined by the customer for the system and are at times referred to as "user" requirements. Such systems can be completely new or they can represent upgrades of current systems. Especially if they are new, the customer often has difficulty in expressing these requirements in a complete and consistent fashion, and in terms that can be utilized by a system developer. It is also the case that it may be simply too early to understand all system requirements. Poor requirements invariably lead to poor system design. This situation remains a problem if the requirements cannot be negotiated and modified for various contractual reasons. Both users and developers complain about requirements, but from their own perspectives. They agree that something new has to be done regarding how requirements are defined and satisfied and several proposals have been made (such as the "spiral model" for software development) to improve the situation. Flexibility is called for in contractual situations that can be quite formal and unyielding. Project Managers must keep this high on their list of potential problem areas.

1.3.2 Poor Planning

Projects normally follow a "project plan" written at the initiation of a project. The ingredients of such a plan are described in detail in Chapter 3. Such plans often are "locked in" as part of a proposal and cannot be easily modified from the point of view of the customer. Because system developments are rather dynamic, most project plans are obsolete 6 to 12 months after they have been written. They therefore need to be continually updated to reflect the current understanding and status of the system. Basing communications and future actions on outmoded or nonexistent plans can lead to large amounts of trouble.

1.3.3 Inadequate Technical Skills and Continuity

Many PMs complain that they are not able to access the necessary personnel resources in order to run their projects. In a company setting, there is obvious competition for the best people and some projects suffer simply because they cannot find or hire such people. When they are able to hire from outside the company, even if the new personnel are technically competent, it takes time for them to climb the learning curve in terms of the project itself as well as the corporate culture. Another side of the coin is the loss of key technical capabilities to other "more important" projects in a company or the possibility that various people may just get up and go to another firm. It is critically important for a PM to maintain an excellent technical staff or else face the strong possibility of inadequate technical performance, which will also show up as problems in schedule and cost.

1.3.4 Lack of Teamwork

Even with a cadre of strong technical people, if they do not operate as a team, the project is in jeopardy. The skills of the Project Manager are paramount here, as he or she must be able to forge a spirit of teamwork and cooperation. Today's systems are very complex and require day-to-day interactions of the members of the project. If these interactions do not take place, or are negative, the project suffers and loses ground. There are times, as well, that a PM must "bite the bullet" with a project person who is not able to be part of a team, preferring instead to be isolated, or act so as to represent a divisive force in a team effort. This should not be tolerated and decisive action is required to solve this type of problem. A variety of issues surrounding how to build a productive team are addressed in considerable detail in Chapter 6.

1.3.5 Poor Communications and Coordination

One of the key skills of a Project Manager, and a leader, is communication. Effective communication is critical both within the project itself and outside the project to supporting company elements (e.g., company management, accounting/finance, contracts, etc.) as well as the customer. Special efforts are required to keep necessary people continually informed about what is going on and why. Surprises as well as insufficient data and lead times can be deadly in a project situation. Project staff are especially sensitive to a PM who does not provide important information and feedback. Some staff require a special amount of "TLC" so that they can perform. Such are the facts of life in dealing with high-technology engineering projects. Many projects fail for this reason alone. A responsible PM must always be aware of the need for communication and be prepared to spend the time necessary to communicate and coordinate.

1.3.6 Insufficient Monitoring of Progress

For reasons that are not particularly clear, many Project Managers kick off their projects and then let them run "open loop" until a critical project review is scheduled. Peters and Waterman's "management by walking around (MBWA)" [1.14] is something to keep in mind in this regard. A good PM keeps in touch with people and progress every day, mostly by "walking around" and informally exploring issues, problems, and needs. Even highly competent personnel require monitoring, as long as it is done in an inobtrusive and helpful way. By careful and sensitive monitoring of progress between key milestones, one is able to keep the project on track and avoid disasters during the formal project reviews when both management and customer are present. This is especially true during the early days of a project because one "never gets a second chance to make a good first impression." Consistent and constructive monitoring and feedback from the beginning set the stage for project success.

1.3.7 Inferior Corporate Support

All organizations are expected to provide assistance and support to the projects that are often the lifeblood of these organizations. Support should be forthcoming from the PM's boss as well as the various designated support groups such as accounting and finance, contracts, graphics, production, manufacturing, and an assortment of matrixed functional elements (such as mechanical design, electrical design, software engineering, etc.). For example, accounting/finance may be expected to provide project cost reports to the PM and the project team on a periodic basis, such as monthly. If these reports are late or incorrect most of the time, the PM is operating at a distinct disadvantage. The PM should not allow this situation to continue. Although finding solutions to inadequate internal support can be a nontrivial adventure, it is usually worth the time and effort necessary to solve such a problem. However, even a good PM may have to enlist the good offices of line management to do so.

The preceding sections present just seven ways a project can go off track. There are clearly many others. If you are a Project Manager or Chief Systems Engineer, it makes sense to understand these and other problem areas so that you can find solutions before they lead to cost and schedule overruns and inadequate system performance. These key problems can be restated in terms of specific guidance to the Project Manager, as described in Exhibit 1.1:

Exhibit 1.1: Selected Ways for the PM to Avoid Problems

1. Review and analyze requirements continuously and in detail and raise problems with requirements with your management and, as necessary, with your customer.
2. Prepare the best project plan that you can and update that plan at least once a quarter; make sure that your plan is concise and readable by your project personnel.
3. Do not accept poor technical performers on your project; insist on the best technical talent who meet the highest standards of performance and creativity.
4. Build a high-energy responsive team that is able to communicate freely and solve project problems; discharge personnel that prove to be incorrigible nonteam players.
5. Maintain high standards of open and honest communication and coordination with your boss, other company people, project staff, and the customer.
6. Monitor project status and progress through informal MBWA, being sensitive to the work habits and needs of your people; establish more formal periodic status reviews.
7. Set up efficient and productive support mechanisms within your company or organization so as to maximize the effectiveness of these interactions; insist upon high standards of performance from support organizations.

Most government agencies develop systems and therefore have been struggling with these types of problems for a long time. They often try, therefore, to provide guidance internally and also to contractors as to issues and problems that they have faced in the past. For example, the National Aeronautics and Space Administration (NASA) has been building high-technology systems since its inception and attempted to head off problems by publishing a document called *Issues in NASA Program and Project Management* [1.15]. The contents of this document are as follows:

1. An Overview of the Project Cycle
2. Systems Engineering and Integration (SE&I) Management for Manned Space Flight Programs
3. Shared Experiences from NASA Programs and Project: 1975
4. Cost Control for Mariner Venus/Mercury '73
5. The Shuttle: A Balancing of Design and Politics
6. Resources for NASA Managers

Clearly, NASA is trying to learn from its history, experiences, and mistakes and have its contractors benefit from the past. A relatively new "theory" of management emphasizes "the learning organization" and proposes methods of assuring that such learning occurs [1.16]. Learning from one's own as well as another's errors is a basic rationale for this as well as other books.

1.4 THE SYSTEMS APPROACH

The "systems approach," at times difficult to define and execute, is basically a recognition that all the elements of a system must interoperate harmoniously, which, in turn, requires a systematic and repeatable process for designing, developing, and operating the system. The architecture for a system must be sound, and it must at least satisfy all the requirements for the system as set forth by the user or customer. By following a systematic and repeatable "systems" process, the developer maximizes the chances that this will be the case.

The key features and results of a systems approach may be stated as follows:

1. Follow a systematic and repeatable process.
2. Emphasize interoperability and harmonious system operations.
3. Provide a cost-effective solution to the customer's problem.
4. Assure the consideration of alternatives.
5. Use iterations as a means of refinement and convergence.
6. Satisfy all user and customer requirements.
7. Create a robust system.

Figure 1.1 provides an overview of a systems approach, the elements of which are briefly cited in what follows:

Box 1: Requirements. Requirements for the system are defined by the customer and user and become the touchstone for all design and development efforts. These are considered inviolate unless a negotiation leads to changes that should be reflected in all contractual documents. Requirements are normally provided in a formal "requirements" document. At times, a derivative document called a specification is forthcoming from the customer. The specification, however, is often written by the developer.

Box 2: Project Plan. The PM is able to develop a project plan from the statement of requirements. This is a roadmap (discussed in Chapter 3) for the important aspects of the project. If the key members of the project team have been selected, they will work with the PM in order to develop the plan. If not, they must ultimately buy into the plan as defined by the PM, or modify it appropriately.

Box 3: Functional Design of Alternatives. The architectural design of the system operates at the functional level, that is, it concentrates on the functions that the system is to perform in distinction to how these

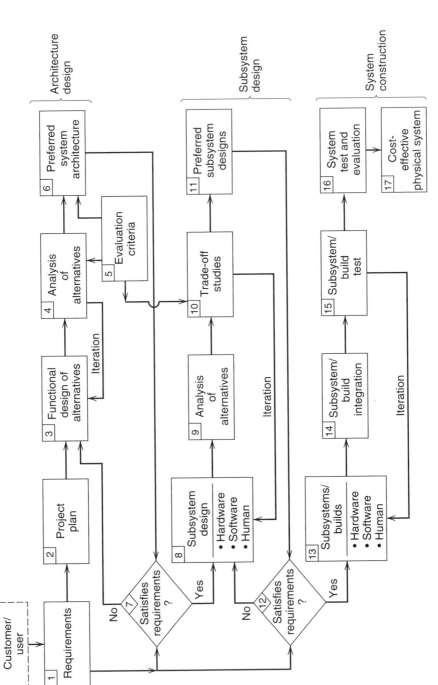

Figure 1.1. Overview of the systems approach.

functions are to be implemented in hardware, software, and human components. Several such designs are configured, each representing a feasible alternative. Often, these alternatives span concepts that range from low cost to high performance.

Box 4: Analysis of Alternatives. Each of the alternatives is analyzed in terms of cost, performance, and satisfaction of requirements. By interacting back and forth between the postulation of alternatives and their analyses, it is ultimately possible to determine the quantitative and qualitative attributes of the various viable alternatives. At the system level, two to four alternatives might be considered desirable.

Box 5: Evaluation Criteria. The analysis of alternatives could not be carried out without the clear identification of criteria against which the alternatives are evaluated. These criteria are derived from the requirements and may include such features as interoperability, growth potential, and societal risk as well as the detailed performance items listed in the requirements document. A formal evaluation framework is normally necessary in order to carry out the evaluation.

Box 6: Preferred System Architecture. This step is a selection of the system-level architecture that is most cost-effective. It represents a choice among the competing alternatives. Many projects go astray because they leap to a preferred architecture without the explicit consideration of alternatives. As an example, this may constitute the selection of time-division multiplexing as preferred over a frequency-division multiplexing approach for a communications system. System architecture is a very important part of the systems approach and the system engineering and design process and is discussed again in Chapter 9.

Box 7: Satisfies Requirements? We make this step explicit in order to emphasize the significance of assuring that the preferred system architecture meets all the designated requirements. If even one mandatory requirement is not completely met, then it is necessary to loop back and consider additional alternatives. If all the key requirements are satisfied, then and only then can the project team move on to the matter of subsystem design.

Box 8: Subsystem Design. By knowing the preferred architecture at the system level, it is then possible to move into detailed subsystem design. These subsystems involve the interplay among hardware, software, and human elements. Subsystems are naturally divided into subordinate elements, which can be called builds, configuration items (CIs), components, or other names that can be mutually understood.

Box 9: Analysis of Alternatives. Following a process similar to that utilized to develop a preferred architecture, alternatives are set forth and analyzed at the subsystem level of design. This is critical because there are numerous ways to implement a given function. Issues of timing and sizing are usually important here.

Box 10: Trade-Off Studies. A variety of trade-offs are generally considered in trying to optimize at the subsystem level. These may be power–weight–space–performance trades, attempting to find the proper balance of attributes. An iteration loop is shown explicitly to account for the possible need to postulate additional alternative subsystem designs.

Box 11: Preferred Subsystem Designs. Preferred subsystem designs flow from the previous steps, representing near-optimal choices with all relevant factors explicitly considered in the trade-off studies. At this stage of the process, one is still at the design level and the system has not, as yet, been built. There are some exceptions to this, as with the notion of rapid prototyping of subsystems in order to prove certain critical high-risk parts of a system.

Box 12: Satisfies Requirements? We again wish to make explicit the checking of the preferred subsystem designs to assure that all requirements have been met. If not, an iteration loop is shown that means we are "back to the drawing board." If so, we move on to the physical building of the system.

Box 13: Subsystems/Builds. The physical construction of the subsystems is now in order, occurring for the hardware, software, and human components, and in consonance with the subsystem designs. Builds is used here as a generic name for configuration items, components, subsubsystems, and so on. The physical construction proceeds through the various levels of indenture defined in the design process.

Box 14: Subsystem/Build Integration. After a given build (or CI) has been constructed, it must be integrated with all interoperating builds (or CIs). This is performed at all subordinate levels of the system.

Box 15: Subsystem/Build Test. Physical testing takes place as builds (CIs) are integrated to assure that they work together, are compatible, and perform as required. If integrated builds fail these tests, the process is iterated until the test leads to success. Clearly, all test plans and procedures must be based on the original or derivative requirements. Many people have suggested, especially with respect to software, that a "build a little, test a little" orientation is most likely to lead to success.

Box 16: System Test and Evaluation. A final system-level test and evaluation (T&E) step confirms that the system meets both development and operational requirements. This can be a long and protracted step, especially for systems that are to operate in a hostile field environment such as aboard a ship or aircraft. It represents an end-to-end check of the full system and a final verification that all requirements have been met.

Box 17: Cost-Effective Physical System. The result of all the previous steps, and many implicit substeps, is a cost-effective physical system.

Although these steps represent most of the elements of the systems approach, there are several that are implicit and therefore are examined in later chapters. However, this overview explains the key aspects of such an approach. It is intended to lead to a system that meets all requirements and is cost-effective and robust. These terms are examined in the chapters dealing with systems engineering management.

The Project Manager and Chief Systems Engineer are clearly key players in assuring that the systems approach is carried out with discipline and good sense. We now more formally explore their roles and responsibilities in a corporate setting.

1.5 THE PROJECT ORGANIZATION

An illustrative organization chart for a project is shown in Figure 1.2. This chart shows only the project and not the organization in which the project may be embedded, which is addressed later in this chapter.

The Project Manager (PM) is shown at the top of the chart with two other key players, the Chief Systems Engineer (CSE) and the Project Controller (PC). In this book, we strongly suggest that the chief engineer of a project be called the Chief *Systems* Engineer, stressing that the main task of the chief engineer is the systems integrity of the overall system. Some organizational structures might list the lead engineer as the chief engineer and have the systems engineer and systems engineering function in parallel with the other engineering functions such as hardware and software engineering. Some projects might be more limited in scope and therefore not require some of the functions shown. Others might indeed be larger and include additional functions such as manufacturing, production engineering, installation, operations and maintenance, and others. We will now consider the specific responsibilities of the Project Manager, Chief Systems Engineer, and the Project Controller.

1.5.1 Responsibilities of the Project Manager (PM)

Clearly, the Project Manager (PM) has responsibility for the overall project, in all its dimensions. At the top level, this focuses on the schedule, cost, and technical performance of the system. An estimate of the time that a PM might spend on each of these features might be 20% schedule, 30% cost, and 50% performance, assuming that one could divide all job-related activities into these three categories. If one includes purely administrative activities as a fourth category, the percentages might be 15% schedule, 25% cost, 35% performance, and 25% administrative. The last item would include such matters as interviewing personnel, preparing their evaluations, and similar duties.

The classical responsibilities of a PM are usually described in terms of four activities: (1) planning, (2) organizing, (3) directing, and (4) monitoring. Some people use the word "controlling" in place of this alternative of "monitoring," for which all control is subsumed within the "directing" activity.

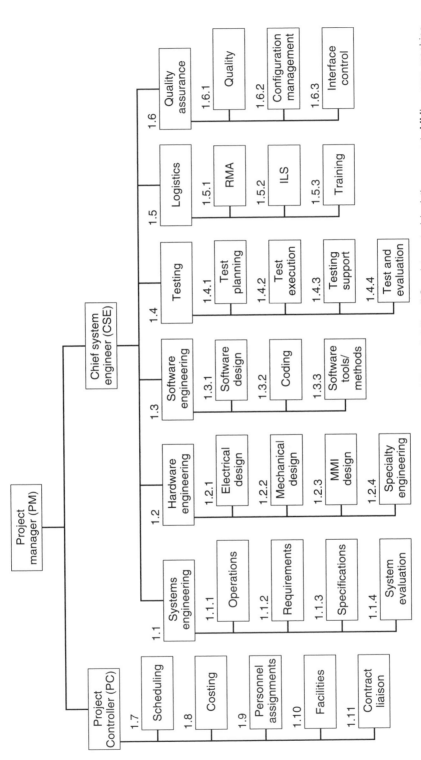

Figure 1.2. Illustrative project organization. RMA = reliability-maintainability-availability; ILS = integrated logistics support; MMI = man-machine interface.

The *planning* activity is dominant in the early stages of a project, especially with respect to the coherent preparation of a project plan. Steady-state planning involves updating this plan and thinking about and planning how to handle special problems and contingencies.

The *organizing* responsibility involves deciding how to organize the project itself (e.g., the chart of Figure 1.2), and reorganizing when and where necessary. It also means the allocation of resources to the various tasks of the project. This shows up as the preparation of initial tasking, work breakdown structures, responsibility matrices for the project, and the like.

The *directing* activity is the formal and informal day-to-day running of the project and its various meetings as well as the delineation of assignments when changes or fine-tuning is required to solve problems.

The *monitoring* duty involves the continuous reading of the status of all aspects of the project in relation to the system requirements and the project plan. If monitoring results in the discovery of problems, remedial action is taken under the directing activity.

An often frustrating factor comes into play when the PM's responsibilities and authority are not congruent. Because the PM usually has full responsibility for the success or failure of the project, it can be extremely difficult if this person cannot, for example, hire or fire, negotiate with outside vendors and subcontractors, and make final arrangements with a counterpart customer. Incommensurate authority is one of the "red flags" of most PMs. A summary list of the various responsibilities and duties of a Project Manager is provided in Exhibit 1.2.

Exhibit 1.2: Selected Duties and Responsibilities of a PM

Cost/Budget
- Confirming that the project can be completed within budget
- Reviewing periodic (e.g., monthly) cost reports
- Obtaining valid cost-to-complete estimates
- Assessing and mitigating project cost risks
- Assuring the validity of system life-cycle costs

Schedule
- Establishing an up-to-date master schedule
- Assuring that all interim milestones are met
- Determining ways to make up time when slippage occurs
- Obtaining valid time-to-complete estimates
- Scheduling internal and customer status reviews

Technical Performance
- Assuring that the system satisfies all technical requirements
- Confirming the validity of the technical approach
- Continuous tracking of technical performance status

- Installing systems and software engineering methods/practices
- Obtaining computer tools for systems and software engineering

Administrative

- Personnel interviewing, hiring, and evaluation
- Interfacing with corporate management
- Interfacing with internal project support groups
- Coaching and team building
- Assuring the availability of required facilities

1.5.2 Responsibilities of the Chief Systems Engineer (CSE)

As suggested by the organization chart of Figure 1.2, the Chief Systems Engineer (CSE) is the key manager of all the engineering work on the project. Thus, the CSE is both a technical contributor as well as a manager. Indeed, the CSE might well have twice as many direct reports as does the PM.

The CSE, under the PM, assumes primary responsibility for the technical performance of the system. In terms of time allocations, the CSE might experience 15% schedule, 15% cost, and 70% technical performance. The CSE has some administrative responsibilities, largely having to do with the management of the technical team. The CSE is definitely a *systems* engineer and should spend a great deal of energy in finding the correct technical solution for the customer.

The fact that both the PM and the CSE have, to some extent, overlapping responsibilities, suggests that it is critically important that these two people work together productively and efficiently. Friction between these key players will seriously jeopardize project success. They must communicate and share information extremely well, and understand each other's weaknesses and strengths. One-on-one meetings are standard so that potential problems are solved before they might hurt the efforts of the entire team. A summary list of the key responsibilities and duties of the Chief Systems Engineer is shown in Exhibit 1.3.

Exhibit 1.3: Ten Responsibilities and Duties of the Chief Systems Engineer (CSE)

1. Establish the overall technical approach
2. Evaluate alternative architectural system designs
3. Develop the preferred system architecture
4. Implement a repeatable systems engineering process
5. Implement a repeatable software engineering process
6. Oversee use of computer tools and aids
7. Serve as technical coach and team builder
8. Hold technical review sessions

  9. Attempt to minimize overall project time period

 10. Develop cost-effective system that satisfies requirements

1.5.3 Responsibilities of the Project Controller

The Project Controller (PC) is the third player in the project management triumvirate. The PC has no technical performance responsibilities, focusing instead on schedule, cost, personnel assignment, facilities, and contract liaison issues. Time spent on these matters is estimated as 25% schedule, 45% cost, 10% personnel, 10% facilities, and 10% contract liaison. Cost issues have to do with assuring that the PM and CSE get the cost reports that they need and also that the overall project stays within budgeted costs.

    The Project Controller is likely to be the "keeper" of the master schedule for the project, although inputs are obviously required from engineering personnel. The PC need not be an engineer, although an understanding of what engineering does is clearly a requirement. Good PCs can anticipate problems by in-depth analyses of project cost and schedule data. By examining trends and timetables, the PC may be able to spot trouble spots before they are evident to other project personnel. This person therefore can be worth his or her weight in gold, primarily to the PM. A brief citation of some of the PC's responsibilities and duties is provided in Exhibit 1.4.

> **Exhibit 1.4: Ten Responsibilities and Duties of the Project Controller (PC)**
>
>   1. Maintain overall project schedule
>   2. Assess project schedule risks
>   3. Assure validity and timeliness of project cost reports
>   4. Track special cost items (e.g., travel, subcontractors)
>   5. Develop project cost trends
>   6. Assess project cost risks
>   7. Maintain life-cycle cost model for system
>   8. Verify and maintain personnel assignments
>   9. Assure that necessary facilities are available
> 10. Maintain appropriate liaison with contracts department

1.6 ORGANIZATIONAL ENVIRONMENTS AND FACTORS

There are many who claim that the organizational environment in which a project is performed is the critical factor in the ultimate success or failure of a project [1.17]. This item was alluded to earlier under the topic of "inferior corporate support." We examine this issue here in somewhat greater detail with respect to the particular corporate entities with which the Project Manager (and the Chief Systems Engineer and Project Controller) must interact.

Interactions with project staff, in the main, are reserved for the discussions in Chapters 5 and 6.

1.6.1   Corporate Organizational Structures

Although to a large extent a project has a great deal of internal structural coherence, it exists within a given overall corporate organizational structure. That corporate structure, depending on its configuration and processes, can have major impacts on how well a project is able to function.

In general, it can be said that there are three generic types of corporate structures, as illustrated in Figure 1.3: *(a)* the functional structure, *(b)* the project structure, and (c) the matrix structure.

As shown in the figure, the functional structure is organized fundamentally by functional areas such as engineering, marketing, sales, manufacturing, production, and so forth. Projects, as such, either for internal or outside customers, are formed *within* a functional group for the duration of the project and then are dissolved. As projects come and go, the basic functional structure remains. A PM is selected from the functional group that is likely to have the most to do with the project from a functional discipline perspective. Depending upon how high up the PM is in the functional organization, as well as other factors such as the technical scope of work of the project, the PM may have to reach across functional lines to access resources for the project. This can work very well because all functional managers are in the same position of requiring resources from other groups from time to time. Projects therefore can do very well in functionally structured organizations, but only if the functional line management is supportive of project needs and requirements.

Figure 1.3 next shows the "pure" project structure, in which the entire organization consists of a set of projects. This structure is prevalent in service organizations, and especially in professional services contractors that do work for the federal government. In such cases, each contract tends to establish a project, and projects come and go as the contracts under which they are operating are completed without renewals or further work requirements. The PM usually starts a project with key personnel from a project that is phasing down or being completed. Such an overall corporate orientation is conducive to project autonomy and support because it is its only focus. Projects can flourish in that type of environment, but from time to time, they do not have ready access to specialized expertise that might reside in a functional group.

The third type of overall corporate structure shown in Figure 1.3 is the matrix structure. This might be viewed as a hybrid between the previous two forms, with the coexistence of functional groups together with the formal recognition of a project group. In principle, this structural corporate form can provide an ideal mix of the advantages of both project autonomy and functional expertise. However, real-world pressures and competition between project and functional groups can also yield a nonsupportive environment. Theories aside, much of the success of a matrix structure, in terms experienced by the Project Manager, depends on the quality of corporate management.

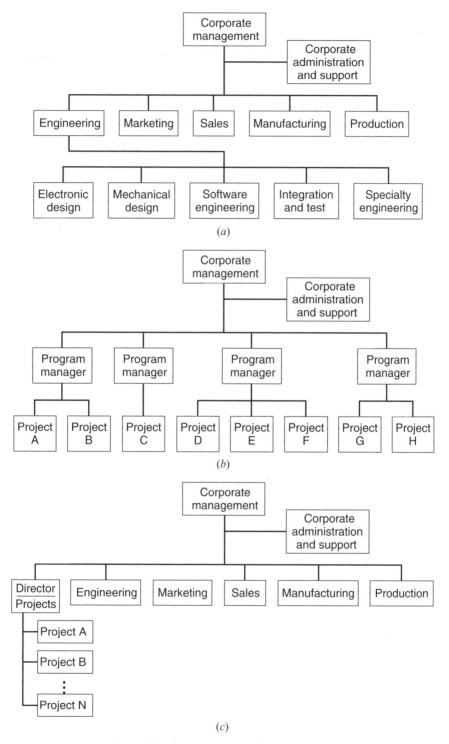

Figure 1.3. Corporate organizational structures.

1.6.2 Interactions with Management

The PM reports "upward" to management, as represented perhaps by a pro-gram manager, or a division director, or a vice president. The specific title may be less important than the nature of the relationship between the PM and the boss. A project management position may carry with it the assumption that the PM runs the project, that is, that the PM has full responsibility and authority for the project. This can be true for the former, but in real life is rarely true for the latter. That is, the PM's authority is limited, and that is a key matter that has to be negotiated between the PM and the boss. Failure to resolve this issue can lead to significant stress for both parties, which will carry over to the CSE, PC, and other members of the project team. Some organizations attempt to recognize and solve this problem through the formal use of an "authority matrix," which defines the boundaries of authority at the various levels in the organization. Such a matrix might deal with precise definitions of limits with respect to such activities as:

1. Hiring personnel and setting salaries
2. Giving raises and bonuses
3. Negotiating and signing contracts
4. Expenditures of monies for different categories
5. Signing and verifying time cards and charges
6. Negotiating with customers

In the absence of a culture that requires such matrix definitions, it falls to the two parties, the PM and the boss, to negotiate a working relationship. If you are a PM, or aspire to be one, you should seriously consider how to begin a dialogue with your boss with respect to your authority and lack of it. A good working understanding is crucial to the success of the project.

1.6.3 Interactions with Matrixed Functional Managers

Depending on the organizational structure of the enterprise at large, it may be necessary to interact with matrixed functional managers so as to obtain resources, the principal one of which is people. Especially in large organiza-tions, there are managers of software engineering, or electronic engineering, or mechanical engineering groups. If a PM needs three software engineers for the project, the corporate culture may call for requesting such persons from the head of the software engineering group or department. This involves inter-views with candidates, selections of the best persons for the job, conflicts with current assignments, and ultimately commitments of people for various spe-cific lengths of time. Depending as well on the circumstances (e.g., the project load and level of business) as well as the people and personalities involved, this interaction may be easy or it may be difficult. If a PM cannot get satisfaction in terms of obtaining the necessary commitments of the right personnel, it may be necessary to work up the chain of command and across to the functional manager chain of command at a higher level than the PMs counterpart. Such are the necessary vagaries of working "across the company" in order to secure the needed project personnel and support. Much of this can be avoided if the PM has a go-ahead to hire, but the well-run organization will almost always have an eye out for borrowing or transferring people from one group to another to maximize productivity for the enterprise at large.

1.6.4 Interactions with Accounting/Finance

Another kind of interaction occurs when an accounting/finance group has responsibility for project cost accounting, and this group does not report to the PM. This is a very common situation, calling for an early understanding of what types of reports will be provided to the project management team.

The centerpiece of such reports is likely to be project cost reports, which define the costs expended to date, and during the last reporting period (e.g., month), by various categories of cost (e.g., direct labor, overhead, and general and administrative costs). The PM may designate the Project Controller as the point of contact in obtaining the required cost reports with the desired format and frequency. A smooth interaction in this regard that works effectively is considered critical to the success of the project. No project can be run efficiently without timely and accurate cost information.

1.6.5 Interactions with Contracts

As with accounting and finance, most organizations have a contracts department that does not report to the PM. Thus, a linkage has to be established with certain contracts personnel in order to understand the precise requirements of the contract, provide all the necessary contract deliverables, and, when appropriate, negotiate modifications to the contract. Various types of contracts (e.g., cost type vs. fixed price vs. time and materials) will be handled by a PM in different ways. Contract provisions may allow certain costs to be traded without contracting officer approval, or they may not. Certain contracts have limits on expenditures by category (e.g., use of travel or consultant fees), and so forth. The PM must thoroughly understand these types of contract provisions in order to effectively manage the project. The PC may also be utilized by the PM as the primary point of contact with contracts so as to conserve the time demands on the PM but still have a solid and constructive interaction. PMs who neglect this relationship are headed for trouble.

1.6.6 Interactions with Marketing/Sales

For purposes of this discussion, we can consider two circumstances. For the first, the company is primarily focused upon developing products to sell to other businesses or to the public at large (e.g., consumer products), and for the second, the company does most of its system development under a contract with a specific customer. We call the first case "commercial" and the second "contract."

In the commercial case, marketing/sales has the task of trying to figure out what types of which products should be made, and for what classes of customers. When they have made such a determination, they then establish a requirement that engineering make the selected products, including the features that are considered most desirable. Timetables are also established and a project is up and running to meet these needs. Thus, there is a direct link between the project and sales/marketing such that the requirements for the system (product) in question are determined by the marketing/sales staff. The vagaries of the marketplace often come into play such that the PM, CSE, and PC are under enormous pressure with difficult schedules and performance requirements. In smaller companies and projects, this can lead to twelve-hour-a-day work assignments, due to the usual lack of resources. The project team thus needs to be functioning well, and also needs to stay in constant touch with the marketing/sales people.

In the contract case, as with doing work for federal, state, and local governments, marketing/sales get involved early in talking to the customer and conveying the system requirements to the project team, usually before a request for proposal (RFP) has been written or conveyed. Both the project team and marketing/sales work together in order to shape the proposal response so as to maximize the probability of a win. Once the contract is indeed won, marketing/sales usually shift their focus to the potential follow-on contract, talking to both project team and customer to make sure that the PM, CSE, and PC are considering the future contract as well as the current one. Thus, there is continual contact between the project team and the marketing/sales organization in order to be in a position to make the most competitive bids, and win as many of them as possible.

1.6.7 Interactions with Human Resources

Typically, the Human Resources Department (HRD) focuses upon at least the following:

1. Recruiting in order to satisfy project needs
2. Administering benefit programs (health insurance, etc.)
3. Managing the overall personnel review/evaluation process
4. Recommending salary and total compensation increases
5. Advising on special personnel problems

The need for new people for a project is usually established by the PM and the HRD is tasked to find these people. This is a most critical link in the relationship with the HRD, since without the right people at the right time, the project risks begin to escalate. Less interaction is required relative to the benefit programs, since they tend to be standard for all employees. Personnel performance reviews are held periodically and guidance is usually given to the PM by the HRD for consistent execution. This leads to salary and compensation increases, which tend to result in at least some unhappy people. This, in turn, might result in the HRD folks working with the PM in order to convey the correct messages to the employees in question. Typically, the PM, CSE, and PC are the most important folks in determining what the compensation will be for each member of the project team. The best results are usually obtained when there is good communication with the Human Resources people.

1.6.8 Interactions with Corporate Information Officer (CIO) Office

The Office of the CIO has, among others, the following responsibilities:

1. Identify information needs of the entire enterprise
2. Focus these needs at the project level
3. Build or acquire the systems in order to satisfy these needs
4. Operate these systems in order to provide support to the various projects
5. Reengineer the systems as the needs change

As suggested earlier in the interactions with Accounting/Finance, there is a critical need for the PM, CSE, and PC to be able to track cost, schedule, and technical performance of the project. The cost information typically comes from accounting/finance, but may have to be converted into a project management format by the CIO office. A well-run organization understands that this highly critical interface issue needs to be worked and resolved successfully. This means that the reports sent to the project team must be timely and accurate. Certain special needs may have to be satisfied, such as cost at the task and subtask levels, and being able to establish "crosswalks" to the project work breakdown structure (WBS), as an example. Schedule information may be captured in a Project Management Software package (such as Microsoft Project). In such a case, project cost information may have to be transferred into such a package in order to analyze and display cost and schedule status charts. In all cases, project people need to provide timely inputs, such as monthly cost to complete and time to complete estimates, usually at the task and activity level.

1.6.9 Interactions with Corporate Technology Officer (CTO) Office

Tasks that are typically carried out within the Office of the CTO are:

1. Identifying technologies that are needed now and into the future by the entire enterprise
2. Relating these technologies to individual programs and projects

3. Investing in and acquiring the necessary technologies
4. Training project personnel with respect to these technologies
5. Assisting project personnel with technology transfer and insertion in order to provide additional value to the customers as well as a highly competitive position

Interaction with the CTO office can ultimately have a profound effect upon the success of a project, especially the so-called high-tech projects. Such projects are continuously exploring new technologies that will result in superior performance at an affordable cost. If this can be achieved, the project will enhance its chances of success, both immediately and into the future. As we see in the next chapter, a project will often have a requirement to formulate a Systems Engineering Management Plan (SEMP), one of whose elements has to do with technology transitioning. This means that the project must consider and ultimately define how various technologies need to transition from current to future systems.

Another type of technology that is usually needed by a project team can be described by a set of computer-based tools. Computer-Aided Software Engineering (CASE) tools are an example, and they provide the software engineering team the tools that it needs to get its job done. A project team that is provided with superior tools of this type will be able to operate at higher levels of productivity, which will translate into higher efficiency and a better overall result for the customer. Technology of all types can be viewed as discriminators that lead to better solutions, which, in turn, lead to project success.

1.6.10 Interactions with Customers

Finally, but certainly not last in importance, is the matter of interactions with customers. Usually, there is a customer counterpart with whom the PM has direct and day-to-day contact. This is true whether the customer is in the same organization or is an outside client. Although more is discussed on this critical subject in other parts of this book, there is little that is more important than an honest, trusting, and effective relationship with the customer. At the same time, the relationship cannot transcend or violate the terms and conditions of the contract between the two entities. For example, the PM cannot agree to do tasks that are not called for under the scope of work of the contract. All increases or modifications in scope must be handled through formal changes in the contract itself.

Another key factor in customer interaction involves the PM's boss and his or her boss, and so forth up the organization. No PM "owns" the customer; the formal relationship is between corporate groups. A good organization has multiple points of contact up and down the organization. This can be particularly effective when problems occur that cannot be resolved between the PM and the client counterpart. Relationships up the chain of command can be brought to bear in attempts to resolve difficulties that arise and find solutions acceptable to both parties. The nature and success of a PM's interaction with the customer are affected and supported by bosses up the chain of command. This is yet one more reason for establishing a solid working relationship between a PM and the boss. For better or worse, this relationship can dominate the life of a PM, working smoothly and successfully, or with stress and possible failure.

The effective PM truly sees the Project Manager, Chief Systems Engineer, and Project Controller as a triumvirate that works together on a day-to-day basis to anticipate and respond to the myriad demands of managing a project. It is one of the most difficult jobs, with lots of stumbling blocks and hurdles. The PM must be a highly skilled and competent individual in order to stay focused on the key issues and create a team that moves forward effectively and solves the many problems that invariably arise.

1.7 LARGE-SCALE ORGANIZATION AND MANAGEMENT ISSUES

The focus of this book is upon project and systems engineering management. Some of the above discussion suggests that success or failure of a project depends significantly upon the larger organizational structure within which the project is being carried out. If it is embedded in a highly bureaucratic situation, success becomes more difficult. Examples of such situations, of course, can be found in both industry and government. Large bureaucratic organizations often chip away at problems, but tend not to be able to truly solve them. The reasons for this are varied, but they clearly can affect the PM, CSE, and PC, who are laboring in the trenches, trying to make a difficult problem more tractable.

The federal government, with its large size and tendency toward bureaucracy, is a good example of a set of large-scale organizations (i.e., the various executive departments) that have a wide variety of internal problems that are extremely difficult to solve. One can get some idea as to what these problems are by looking at reports produced on a continuing basis by the General Accountability Office (GAO), whose job it is to investigate problem areas in the executive agencies. Exhibit 1.5 provides a sample listing of some of the reports of the GAO.

Exhibit 1.5: A Sampling of General Accountability Office (GAO) Report Titles

- Defense Transportation: Process Reengineering Could Be Enhanced by Performance Measures
- Managing for Results: Strengthening Regulatory Agencies' Performance Management Practices

- Management Reform: Elements of Successful Improvement Initiatives
- Department of Energy: Need to Address Longstanding Management Weaknesses
- Executive Guide: Creating Value Through World-Class Financial Management
- Defense Acquisitions: Need to Revise Acquisition Strategy to Reduce Risk for Joint Air to Surface Standoff Missile
- Defense Acquisitions: Comprehensive Strategy Needed to Improve Ship Cruise Missile Defense
- Defense Acquisitions: Improvements Needed in Military Space Systems' Planning and Education
- Defense Acquisitions: Achieving B-2A Bomber Operational Requirements
- Air Traffic Control: FAA's Modernization Investment Management Approach Can Be Strengthened
- Combat Identification Systems: Changes Needed in Management Plans and Structure
- Defense Acquisitions: Advanced Concept Technology Demonstration Program Can Be Improved
- Defense Logistics: Actions Needed to Enhance Success of Reengineering Initiatives
- Internal Revenue Service: Custodial Financial Management Weaknesses
- Information Security: Opportunities for Improved OMB Oversight of Agency Practices
- Defense Information Resource Management (IRM): Critical Risks Facing New Materiel Management Strategy
- Department of Transportation: University Research Activities Need Greater Oversight
- Battlefield Automation: Army Needs to Determine Command and Control Priorities and Costs
- Department of Energy: Management Problems Require a Long-Term Commitment to Change
- Military Satellite Communications: Opportunity to Save Billions of Dollars
- Acquisition Reform: Contractors Can Use Technologies and Management Techniques to Reduce Costs
- Defense Management: Impediments Jeopardize Logistics Corporate Information Management
- Tactical Intelligence: Joint STARS Needs Current Cost and Operational Effectiveness Analysis
- NASA Aeronautics: Impact of Technology Transfer Activities Is Uncertain
- Financial Management: Reliability of Weapon System Cost Reports is Highly Questionable
- Drug Control: Heavy Investment in Military Surveillance Is Not Paying Off

- Simulation Training: Management Framework Improved, But Challenges Remain
- DoD Computer Contracting: Inadequate Management Wasted Millions of Dollars
- Financial Management: IRS Lacks Accountability Over its ADP Resources
- Patent and Trademark Office: Key Processes for Managing Automated Patent System Development Are Weak
- DoD Information Services: Improved Pricing and Financial Management Practices Needed for Business Area
- Information Security: Serious Weaknesses Place Critical Federal Operations and Assets at Risk
- Space Surveillance: DoD and NASA Need Consolidated Requirements and a Coordinated Plan
- Defense IRM: Strategy Needed for Logistics Information Technology Improvement Efforts
- Unmanned Aerial Vehicles: Maneuver System Schedule Includes Unnecessary Risk
- Department of State IRM: Modernization Program at Risk Absent Full Implementation of Key Best Practices
- Air Traffic Control: Complete and Enforced Architecture Needed for FAA Systems
- Tax System Modernization: Imaging System's Performance Modernization Improving But Still Falls Short of Expectations
- Air Traffic Control: Improved Cost Information Needed to Make Billion Dollar Modernization Investment Decisions
- Major Management Challenges and Program Risks: A Governmentwide Perspective

Scanning Exhibit 1.5 we see a variety of problem areas, including:

1. Overall management deficiencies
2. Risks that need to be reduced
3. Costs that are too high or not well enough known
4. Schedules that are not workable
5. Requirements difficulties
6. Need for better performance and effectiveness measurement of systems
7. Need for use of best practices
8. Investment decision issues
9. Overall financial management issues
10. Need for systems reengineering and improvements

These are all familiar themes in the worlds of project management and systems engineering. However, in the context of large-scale organizational issues, they may well be beyond the scope of what the PM, CSE, and PC are able to tackle and provide effective solutions for. Indeed, the last-cited item in Exhibit 1.5 offered solution areas for twenty individual federal government agencies, solutions that emphasized the following four areas:

1. Adopting a results orientation
2. Effectively using information technology to achieve program results
3. Establishing financial management capabilities that effectively support decision making and accountability
4. Building, maintaining, and marshaling the human capital needed to achieve results

Massive efforts will be required to address these areas for the twenty government agencies.

The last point to be made in relation to the above is the fact that we are seeing increasing amounts of software in our systems such that software itself, its development and maintenance, is fast becoming our number one "systems" problem. Exhibit 1.6 lists some of the GAO reports that highlight the various aspects of software that need to be addressed.

Exhibit 1.6: Selected GAO Reports That Focus Upon Software Issues

- Land Management Systems: Major Software Development Does Not Meet BLM's Business Needs
- Weather Forecasting: Improvements Needed in Laboratory Software Development Processes
- Defense Financial Management: Immature Software Development Processes at Indianapolis Increase Risk
- Embedded Computer Systems: Defense Does Not Know How Much It Spends on Software
- Embedded Computer Systems: F-14D Aircraft Software Is Not Reliable
- Embedded Computer Systems: Significant Software Problems on C-17 Must Be Addressed
- Embedded Computer Systems: New F/A-18 Capabilities Impact Navy's Software Development Process
- Space Station: NASA's Software Development Approach Increases Safety and Cost Risks
- Mission-Critical Systems: Defense Attempting to Address Major Software Challenges
- Software Tools: Defense Is Not Ready to Implement I-CASE Departmentwide

The ubiquitous nature of software in our systems has led this author to include a separate chapter (Chapter 10) in this book that highlights significant software issues and attempts to define approaches that are and have been taken in order to find effective solutions. The PM, CSE, and PC are all likely, in the twenty-first century, to have to deal with an increasing number of problems associated with software as critical parts of our future systems.

In addition to providing the above reports, the GAO apparently carries out an annual assessment of selected major weapon system programs. In a March 2005 report regarding these programs [1.18], the agency looked at 54 programs that represented an overall investment of some $800 billion. The GAO tends to explore cost, schedule, and performance from a knowledge-based perspective. That is, the GAO looked at critical junctures in these programs and assessed the degree to which actual knowledge at those junctures was better or worse than knowledge suggested by best practices. In other words, at these points in the programs, did we know what we should have known? If not, we were implicitly accepting higher levels of risk with respect to cost, schedule, and performance. The three specific program elements examined in some detail had to do with:

1. Technology maturity
2. Design
3. Production

This is certainly an interesting approach and perspective. The GAO concluded that, of the fifty-four programs that were examined, the majority cost more and took longer to develop than planned. The potential impacts of accepting lower levels of knowledge were cited in terms of adverse cost and schedule consequences, leading to fewer quantity buys than were originally planned.

In March 2006, the GAO examined fifty-two weapon system programs at an investment level of over $850 billion. Looking at the five-year investment numbers (from 2001 to 2006), we started at about $700 billion and ended at nearly $1.4 trillion (!). As before, a picture of shortfalls was portrayed in cost, schedule, and performance. Technology perspectives were highlighted, with these results:

> Programs that began with immature technologies have experienced average research and development cost growth of 34.9 percent; programs that began with mature technologies have only experienced cost growth of 4.8 percent.

Another quote of special interest is:

> DoD often exceeds development cost estimates by approximately 30 to 40 percent and experiences cuts in planned quantities, missed deadlines, and performance shortfalls.

A knowledge-based assessment with respect to technology, design, and production continued to be the dominant mode of analysis; actual results were compared with suggested best practices.

If we look at space system acquisitions within the DoD, another report in April 2006 cited substantial cost and schedule overruns. The impacts of these problems, over the following five years, were estimated to be a reduction of some $12 billion available for new systems or to explore new technologies. Several problem causes were articulated as well as methods for problem reduction. The latter included:

- Using practices suggested by the GAO
- Allowing the Science and Technology (S&T) community to bring the technologies to maturation
- Using an evolutionary development approach
- Improving collaboration on requirements
- Shifts in thinking about how to develop space systems
- Changes in incentives

Accepting inputs from another agency is quite a problematic undertaking, considering that all managers within the DoD operate within a definitive and well-thought-out management structure. We might infer from some of these results that being a weapon system manager within the DoD is a most stressful and difficult vocation.

QUESTIONS/EXERCISES

1.1 From your own experience or your reading, identify
  a. a project with major problems
  b. three reasons the project got into trouble
  c. what might have been done to
    - fix the problems
    - avoid the problems

1.2 For a project of your selection, discuss ways in which the systems approach
  a. was used effectively
  b. was not used, and the consequences

1.3 Critique the systems approach diagram of Figure 1.1 Are there ways that you would modify the diagram? Explain.

1.4 Discuss three advantages and disadvantages each for the following organizational structures:
  a. functional
  b. project
  c. matrix

1.5 Draw a project organization chart for a project of your own selection.

1.6 Identify three responsibilities, other than those listed in this chapter, of
  a. the Project Manager
  b. the Chief Systems Engineer
  c. the Project Controller

1.7 Locate another two definitions of systems engineering from the literature. Which of the various definitions do you find most satisfying? Why?

1.8 Define three additional areas in which systems exhibit Type I and Type II errors. How would you describe such errors? Are these errors related to one another? Explain.

1.9 The section on errors shows specific error probabilities for plus and minus one-, two-, and three-sigma situations. Verify these numbers. What assumptions were needed in order to obtain these values? What is the corresponding "four-sigma" error probability?

1.10 For a system with three additive independent errors (standard deviations) of 2, 3, and 4, what is the variance associated with the overall maximum error? What is the maximum standard deviation?

REFERENCES

1.1 Chapman, W. L., A. T. Bahill, and A. W. Wymore (1992). *Engineering Modeling and Design*. Boca Raton, FL: CRC Press.

1.2 International Council on Systems Engineering (INCOSE), 2033 Sixth Avenue, #804, Seattle, WA 98121.

1.3 Department of Defense (DoD) Website: web2.deskbook.osd.mil

1.4 Eisner, H. (1988). *Computer-Aided Systems Engineering*. Englewood Cliffs, NJ: Prentice-Hall, p. 17.

1.5 Defense Systems Management College (DSMC) (1999). *Systems Engineering Fundamentals*. Ft. Belvoir, VA: DSMC Press, p. 3.

1.6 Shishko, R. (1995). *NASA Systems Engineering Handbook*, SP-6105. Linthicum Heights, MD: National Aeronautics and Space Administration, Scientific and Technical Information Program Office, p. 4.

1.7 Wynne, Michael W. (2004). "Policy for Systems Engineering in DoD," 3010 Defense Pentagon. Washington, DC 20301–3010, February 20.

1.8 Office of the Chief Engineer (2006). *NASA Systems Engineering Processes and Requirements*." NPR 7123.1. Washington, DC: NASA, March 13.

1.9 See www.incose.org.

1.10 Burgess, J. (1993). "Out-of-Control Contract." *Washington Post*, March 8.

1.11 Archibald, R. D. (1976). *Managing High Technology Programs and Projects*. New York: John Wiley, p. 10.

1.12 Kezsbom, D. S., D. L. Schilling, and K. A. Edward (1989). *Dynamic Project Management*. New York: John Wiley.

1.13 Kerzner, H. (2000). *Project Management: A Systems Approach to Planning, Scheduling and Controlling*, 7th edition. New York: John Wiley.

1.14 Peters, T. J., and R. H. Waterman, Jr. (1982). *In Search of Excellence.* New York: Bantam Books.

1.15 Hoban, F. T., ed. (1992). *Issues in NASA Program and Project Management*, NASA SP-6101(05). Washington, DC: National Aeronautics and Space Administration, Office of Management and Facilities.

1.16 Senge, P. M. (1990). *The Fifth Discipline — The Art & Practice of The Learning Organization.* New York: Doubleday.

1.17 Frame, J. D. (1987). *Managing Projects in Organizations.* San Francisco: Jossey-Bass.

1.18 GAO Highlights. "Assessment of Selected Major Weapon Programs," March 2005 and March 2006. www.gao.gov.

2
OVERVIEW OF ESSENTIALS

2.1 INTRODUCTION

This chapter provides an overview of the essentials of project management and the systems engineering process as well as its management. This is followed by an examination of the essentials of the systems acquisition life cycle, which itself is correlated with the systems engineering process. The chapter concludes with a citation of some of the standards relevant to project management and systems engineering.

2.2 PROJECT MANAGEMENT ESSENTIALS

The previous chapter explored some of the elements of project management, including the roles of the project triumvirate—the Project Manager (PM), the Chief Systems Engineer (CSE), and the Project Controller (PC)—and the various organizational interfaces that affect how they approach and perform their jobs. Here we continue this exploration, but focus on the essentials of project management in a more step-by-step fashion. An overview of the essentials of project management is depicted in Figure 2.1. This figure shows the project management triumvirate as a team that oversees the essential project management functions of planning, organizing, directing, and monitoring.

Starting from the left side of Figure 2.1, the customer is shown as the source of a request for proposal (RFP) that might start off the project. For cases in which the project is internal to an organization, there is no formal RFP, but the requirements nevertheless should be spelled out in some type of document. In

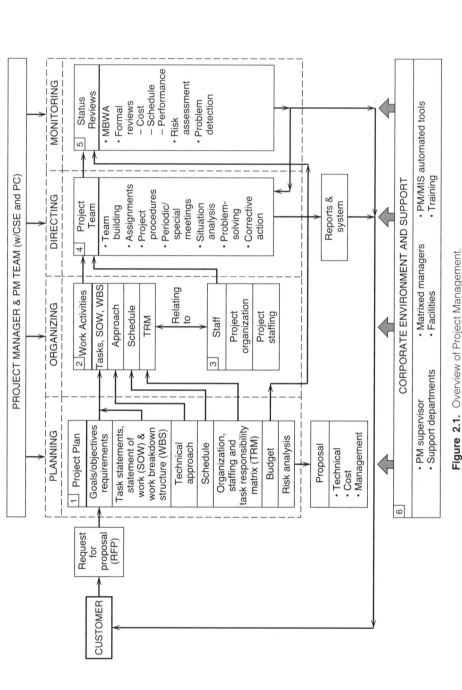

Figure 2.1. Overview of Project Management.

either situation, the planning phase begins with the development of _
plan (Box 1). Such a plan contains seven essential elements, discussed in
greater detail in the next chapter. These elements are:

1. Needs, goals, objectives, and requirements
2. Task statements, a statement of work (SOW), and a work breakdown structure (WBS)
3. The technical approach to the project
4. A project schedule
5. Organization, staffing, and a task responsibility matrix (TRM)
6. The project budget
7. Risk analysis

If an RFP response is called for, the preceding elements of the project
plan form the basis for preparing the proposal, which normally contains three
volumes:

1. A technical proposal
2. A management proposal
3. A cost proposal

This three-volume proposal is delivered to the customer for evaluation. A
feedback loop from the monitoring function, when it is necessary to update
the project plan, is not shown in the figure but is implicit in the process.

The seven project plan elements provide the input to the organizing function, as shown in Figure 2.1. This function is divided into two essential
activities, namely, work activities (Box 2) and activities having to do with the
project organization and staffing (Box 3). The work activities have four parts,
all flowing from the project plan. These are:

1. The tasks, SOW, and WBS
2. The technical approach
3. The schedule
4. The task responsibility matrix (TRM)

The last three, in particular, interact strongly with the staffing for the project
because more or less staff stretches or compresses the schedule and the ability
to perform tasks in series or in parallel. The TRM is an explicit assignment
of people to the various tasks and elements of the work breakdown structure
(WBS). The two elements of the staff activities are

1. The project organization, and
2. The project staffing

Initially, the project organization is shown without specific people filling the various organizational roles; as the project is staffed with specific individuals, names are added to the organization chart to make it clear as to precisely who fills what roles. Particular individuals may fill more than one project role. A typical project organization chart was shown in Chapter 1 in Figure 1.2. The preceding work and staffing activities constitute the essentials of the organizing function for the project.

The organizing function is normally followed by the directing function. The essential element of this function is to establish a coherent and effective project team (Box 4). Typical directing activities include:

1. Team building
2. Clarifying assignments for various team members
3. Articulating project and team procedures
4. Executing both periodic and special team meetings
5. Carrying out situation analyses, as a team
6. Problem solving when the inevitable problems occur
7. Implementing corrective actions

All of these directing activities are an ongoing part of the project and constitute the main expenditure of energy and time in order to perform the various project tasks. Project reports and systems (physical and procedural, as called for in the statement of work) flow from the work of the project team, operating individually and collectively.

This is followed by the steady-state monitoring function, which explicitly and continuously takes stock of the project status. This is executed by (Box 5):

1. MBWA (management by walking around)
2. Formal and periodic schedule, cost, and performance reviews, some of which are attended by upper management and the customer
3. Risk assessment
4. Problem detection

When a problem is discovered, there is feedback to the directing function in which corrective action is taken. Some texts consider the combination of problem detection and corrective action as a separate project function known as project control. Results of the directing and monitoring functions are provided to the customer in the feedback loop shown in the figure.

All of the previously cited project functions are carried out in the context of and support by a corporate environment (Box 6). As mentioned in

Chapter 1, this support, or lack of it, can have a major influence on the success or failure of the project. Some of the influencing factors include:

1. The supervisor (boss) of the project manager
2. The support departments (e.g., finance/accounting, contracts, and human resources)
3. The matrixed functional managers (if resources are to be obtained in a matrix situation)
4. Facilities to be provided by the corporate entity, ranging from office space to computers to special test equipment
5. Project management/management information system (MIS) tools and systems
6. Training that might (or might not) be provided in both technical and management disciplines

Figure 2.1 also shows the project management team leader triumvirate (PM, CSE, and PC) at the top of the chart as managing the essential key functions of planning, organizing, directing, and monitoring.

2.3 SYSTEMS ENGINEERING PROCESS AND MANAGEMENT ESSENTIALS

An overview of the essentials of the systems engineering process and its management is provided in Figure 2.2. As with the project management chart of Figure 2.1, the process is initiated by the customer/user with statements of needs, goals, objectives, and requirements. These feed the development of the project plan (as depicted as well in Figure 2.1) and, from a systems engineering perspective, the elements of mission engineering and requirements analysis and allocation. These feed into the element of functional analysis and allocation, which forms the basis for the design/synthesis of the system architecture. Iterations between the synthesis and analysis elements, and the implicit high-level trade-offs that are required, lead to the definition of a preferred system architecture. Confirming the validity of this architecture also requires both a life-cycle costing and an analysis of the risks associated with this architecture. At this point, the systems engineering team, with the interaction of the project manager, has developed what is believed to be a cost-effective architectural design for the system (Box 1).

We note that many of these elements are shown as being supported by other system considerations (Box 4). These considerations are at a top level that is appropriate to architectural, in distinction to subsystem, design. They

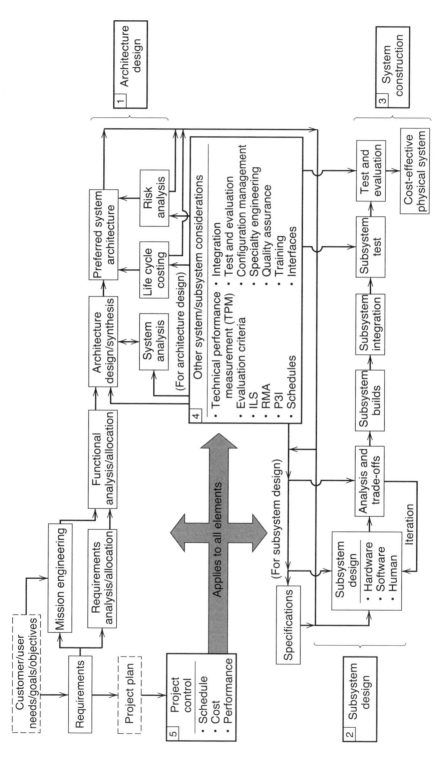

Figure 2.2. Overview of systems engineering process and management.

represent a first-order examination of system factors that include such elements as:

1. Technical performance measurement (TPM)
2. Evaluation criteria
3. Integrated logistics support (ILS)
4. Reliability-maintainability-availability (RMA)
5. Preplanned product improvement (P3I)
6. Schedules
7. Integration
8. Test and evaluation
9. Configuration management
10. Specialty engineering (e.g., safety or security)
11. Quality assurance
12. Training
13. Interface control and compatibility, and others

A more complete articulation of these other system/subsystem considerations is provided in Chapter 8. They are extremely important because they complete the (approximately) thirty elements that comprise systems engineering and its management.

When a cost-effective system architecture has been defined, the next step is to enter the process of subsystem design (Box 2). This set of activities goes into the levels of detail necessary to both design and build the system. Thus, we see a generic process of design defined at two distinct levels, namely, (1) architectural design, and (2) subsystem design.

One might draw an analogy, in this regard, between this process and how an A&E (architect and engineering) firm might design and construct a large building or an airport. The architect part of the firm first develops an overall architecture of the "system" in question. Only after a satisfactory architecture has been designed is it possible to engineer all the details of the architecture and then, in fact, build the system. This two-step approach is mirrored in the systems engineering process shown in Figure 2.2. In addition, the architectural design is usually performed at a "functional" level, whereas the subsystem design is executed at the specific hardware/software/human engineering level. This distinction is an important part of the systems engineering process. In general, functional design describes "what" is to be done in some detail at the top level of a system, with secondary consideration to "how" it is to be done. Subsystem design accepts what is to be done as a given, and fleshes out the "how" details by selecting specific hardware, software, and human engineering components. More will be said about this subject in Chapter 8, with examples that will help clarify these concepts.

The architecture design allows for a more complete definition of the system in the form of a specification. Actually, specifications can be written at several levels, but it is necessary for a specification to be written in order to begin the process of subsystem design. At times, a top-level specification is provided as a derivative of the system requirements and is an input to the mission engineering and requirements analysis and allocation activities that begin the overall systems engineering process.

As suggested before, subsystem design (Box 2) involves the detailed selection of components embodied in hardware, software, and human parts of the system. These design alternatives are analyzed, and traded off, in order to select the best mix of components that will cost-effectively carry out the prescribed system functions. Iteration between analysis and design (synthesis) at the subsystem level parallels the iterative analysis and design/synthesis activities that are an integral part of architectural design. As an example, if "information storage" is a functional element in the top-level architectural design, at the subsystem design one must choose between the various ways in which information storage might be accomplished, to include tape storage versus conventional disk storage versus CD-ROM storage. These different implementations have different cost and performance attributes that have to be balanced while satisfying the requirements and specifications for the system.

When the subsystem design has been completed, the team is ready to begin the formal building of the system. System "builds" is used here as a generic name for such formal construction, although other names such as "configuration items" are used extensively. Each build must be tested to assure that it meets the requirements and specifications, and combinations of builds are tested as an upward process of integration occurs. Progressive integration of components, configuration items (CIs), and builds all require testing in order to verify the performance of the system. As the top levels of the system are constructed, we enter a top-level "test and evaluation" activity that can have a rather formal structure in various implementations of systems engineering. In principle, many cycles of "integration and test" must be carried out as the system is constructed until a "final" set of "test and evaluation" activities confirm that the system satisfies development and operational requirements. Here again, these notions are revisited in Part III of this book. Finally, a cost-effective physical system is constructed (Box 3) that meets all the stated requirements of the system.

A project control function (Box 5) is executed throughout the process in order to assure that overall schedule, cost, and performance profiles are appropriate. This control function is embedded in the project directing and monitoring functions shown in Figure 2.1. The Chief Systems Engineer (CSE) is the key player in the overall systems engineering process described in Figure 2.2. The Project Manager, in conjunction with the CSE and the Project Controller, assures that the project control function is properly executed.

This overview of the systems engineering process and its management shows a sequence of three essential steps:

1. Architecture design
2. Subsystem design
3. System construction

These, in turn, are facilitated and augmented by (1) other system/subsystem considerations, and (2) project control.

It therefore can be observed that the preceding five steps constitute an overview of the systems engineering process and how it is managed. However, certain elements of systems engineering and the life-cycle phases of a system have not as yet been covered. These include system production, installation, operations and maintenance, and system operation and modification in the field. These additional elements are dealt with explicitly in Chapter 8, and are considered to be part of the thirty key elements of engineering and fielding a real-world system. The rationale for these additional elements is further reinforced by the following section, which deals with the nature of the systems acquisition process and some of the issues that are embedded in such a process.

2.4 HISTORICAL OVERVIEW OF ACQUISITION NOTIONS

The matter of project management and systems engineering can also be approached from the perspective of the customer interested in acquiring a system. Such a customer needs to do a considerable amount of planning in order to do so, even if another party is to actually carry out most of the project and the systems engineering that is part of the project. A typical example might be a government agency, such as the Department of Transportation (DOT) or the National Aeronautics and Space Administration (NASA), that wishes to procure systems that will be responsive to its needs and requirements. In the case of the DOT, through its subordinate Federal Aviation Administration (FAA), it may need to acquire a new radar as part of its air traffic control charter. NASA, as an example, may need to procure a new or upgraded satellite that will carry out a portion of its "mission to planet earth" initiative. In both cases, a typical plan calls for following a systems acquisition process set forth by the government, with a major role to be executed by a large systems contractor.

Most government agencies with needs such as those just described have evolved an acquisition process that has been institutionalized. This provides for clear communications and understandings both within the agency as well as between the agency and other groups such as systems engineering contractors. A generic acquisition process usually consists of phases such as those defined in Exhibit 2.1. These phases are normally sequential in time.

Exhibit 2.1: Typical System Acquisition Phases

Phase Name	Phase Activities or Purposes
Prephase 0:	Define mission need
	Validate need
	Assure that new system is required to fulfill need
Phase 0:	Identify alternative concepts
Concept	Evaluate feasibility of alternative concepts
Definition	Determine most promising concepts
Phase 1:	Design alternative feasible systems
Concept	Demonstrate critical processes
Validation	Demonstrate critical technologies
	Build and test early prototypes
Phase 2:	Finalize preferred system design
Engineering	Build system
Development	Test and evaluate system
Phase 3:	Produce system
Production and	Install system
Deployment	Establish system logistics support
Phase 4:	Operate and maintain system in field
Operations and	Monitor performance of system
Support	Modify and improve system as necessary

The Department of Defense (DoD) presented its acquisition phases as the "DoD 5000 Acquisition Model" [2.1]. This model shows four sequential phases, namely:

1. Concept and technology development
2. Systems development and demonstration
3. Production and deployment
4. Operations and support

Technology opportunities and user needs feed into these phases, as necessary and appropriate. Mission needs are articulated concurrently with the concept and technology development phase. An interim operational capability (IOC) is achieved during part of the production and deployment phase. In a parallel manner, a final operational capability (FOC) is present when entering the operations and support phase.

The DoD Defense Acquisition System, as it is called, employs a set of policies and principles that may be examined in the following five categories [2.2]:

1. Achievement of interoperability
2. Rapid and effective transition from science and technology to products

3. Rapid and effective transition from acquisition to deployment and fielding
4. Integrated and effective operational support
5. Effective management

Interoperability, especially between the various services, but also including with our allies, has been a key issue within the DoD for many years. With more rapid deployments and joint forces, these matters are of special and continuing importance. The two transition issues cited above indicate the determination of the DoD to assure technology transfer as well as the compression of schedules that have otherwise been unacceptable. The last above-cited item emphasizes the need for providing support in an integrated fashion. Integration cannot be properly achieved without interoperability. Finally, effective management is critical to satisfying needs in the other four category areas. Managing within an organization as large as the DoD is a major challenge that must be addressed on a continuing basis.

Other topics that are highlighted as parts of the above five areas include:

1. Time-phased requirements and communications with users
2. Use of commercial products, services, and technologies
3. Performance-based acquisition
4. Evolutionary acquisition
5. Integrated test and evaluation
6. Competition
7. Departmental commitment to production
8. Total systems approach
9. Logistics transformation
10. Tailoring
11. Cost and affordability
12. Program stability
13. Simulation-based acquisition
14. Innovation, continuous improvement, and lessons learned
15. Streamlined organizations and a professional workforce

A key paragraph under item (8) above, in terms of its relevance to this text, is [2.2]:

Acquisition programs shall be managed to optimize total system performance and minimize total ownership costs by addressing both the equipment and the human part of the total system equation, through application of systems engineering. Program managers shall give full consideration to all aspects of system support, including logistics planning; manpower, personnel and training; human, environmental, safety, occupational health, accessibility, survivability, and security factors; and spectrum management and the operational electromagnetic environment.

2.4.1 Specific Focus for Systems Acquisition Agent

Given the phases outlined in Exhibit 2.1, the acquisition agent for the customer or user of a system must focus on certain activities in order to assure that the acquisition process is as effective as possible. The following areas of focus are suggested, with emphasis that depends on the particular phase that is being approached:

1. Restatement of needs/goals/objectives (of that phase)
2. Reiteration of requirements
3. Preparation of tasks statements, statements of work (SOWs), and work breakdown structures (WBSs)
4. Key schedule milestones
5. Budget limitations and constraints
6. Project reviews

At each and every phase depicted in Exhibit 2.1, it is necessary for the acquisition agent to restate the needs, goals, and objectives of the customer or user. These may change from phase to phase, but are required to provide new guidance to a project and systems engineering team. The requirements for each phase will almost certainly change throughout the acquisition process, so it is necessary to state explicitly what the requirements are for the forthcoming phase. Such requirements are *work* requirements as contrasted to *system* requirements, which may remain relatively unchanged. As suggested in Chapter 1, requirements become an important point of departure for a systems engineering contractor, for example, to design and develop the system. Similarly, each new phase carries with it a new statement of work and associated WBS that must be conveyed in writing to the system designer. With respect to schedule, the customer usually defines when it is that the system must ultimately be fielded and how that impacts the schedule of the phase under consideration. It therefore falls on the acquisition agent to establish key schedule milestones. Relative to budget matters, monies are allocated to the execution of each phase and these allocations become constraints within which a project team must perform.

The issue of project reviews is often considered as an integral part of the acquisition process. At least four reviews have become more or less standard, namely

1. The system requirements review(s) (SRR)
2. The system design review (SDR)

3. The preliminary design review (PDR)
4. The critical design review (CDR)

with the following general practice:

- Various SRRs are carried out during concept definition and validation.
- The SDR is executed during concept validation.
- Both the PDR and CDR are implemented during the engineering development phase.

It is also standard procedure that a formal review and approval is required to move from one phase to the next phase in the sequence. In addition, it is expected that an interim operational capability (IOC) is achieved at some point in the production phase and a final operational capability (FOC) is confirmed before entering the formal operations phase. Changes to the system in the form of engineering change proposals (ECPs) are considered during this operations phase. Additional guidance relative to these matters is provided in the next section in which certain standards for system acquisition are explored. Further information on system acquisition is presented in Chapter 12.

2.5 SELECTED STANDARDS

Standards that may be applied in the project management or systems engineering arenas provide guidance to the Project Manager or the Chief Systems Engineer. Various domain-specific fields (e.g., computers, communications) provide detailed standards that also suggest to manufacturers of hardware and software how they may be able to assure compatibility with each others' equipment. For example, producers of software must select the operating system with which their software will be compatible. In this section, we deal only with standards that are operative at the level of the overall project management and systems engineering activities because domain-specific standards would fill many volumes of text.

2.5.1 Military-Standard-499A

An engineering standard was produced in 1974 that, at that time, supported and guided the systems engineering process and its management [2.3]. This "engineering management" standard served as a touchstone for systems engineering, especially for systems developed for the military. The standard was "developed to assist government and contractor personnel in defining the system engineering effort in support of defense acquisition programs". Exhibit 2.2 lists the (two-digit) table of contents of this standard.

Exhibit 2.2: Table of Contents of Mil-Std-499A (Two-Digit)

1. SCOPE
   - 1.1 Purpose
   - 1.2 Application
   - 1.3 Implementation
   - 1.4 Tailoring
2. REFERENCED DOCUMENTS
3. DEFINITIONS
   - 3.1 Engineering Management
   - 3.2 Technical Program Planning and Control
   - 3.3 System Engineering Process
   - 3.4 Engineering Specialty Integration
   - 3.5 Technical Performance Measurement
4. GENERAL CRITERIA
5. DETAILED REQUIREMENTS
   - 5.1 System Engineering Management Plan (SEMP)
   - 5.2 Review of Contractor's Engineering Management
6. NOTES
   - 6.1 Relationship of Technical Program Planning to Cost and Schedule Planning
   - 6.2 Relationship of Technical Performance Measurement (TPM) to Cost and Schedule Performance Measurement
   - 6.3 Relationship of Integrated Logistics Support (ILS) to System Engineering
   - 6.4 Minimum Documentation
   - 6.5 Data
10. APPENDIX A
    - 10.1 Technical Program Planning and Control
    - 10.2 System Engineering Process

We note from the preceding that this standard defined systems engineering management in terms of three essential elements

1. Technical program planning and control
2. The system engineering process
3. Engineering specialties and their integration

Although there is much to discuss with respect to the details of this standard and its relationship to project and systems engineering management, we bypass such discussion because Mil-Std-499A was superseded by Mil-Std-499B, which is described in some detail in what follows. The latter has influenced both project and systems engineering management for many years. Further, although it was developed in the context of military systems, it is applicable in terms of its basic structure to the design and development of both civil and commercial systems. Specific emphasis is placed on tailoring the provisions of this standard to the unique requirements of a particular program or project.

2.5.2 Mil-Std-499B (Draft)

This draft standard, entitled "sytems engineering," defined systems engineering as "an interdisciplinary approach to evolve and verify an integrated and optimally balanced set of product and process designs that satisfy user needs and provide information for management decision making" [2.4]. Systems engineering management is cited as "the management, including the planning and control for successful, timely completion of the design, development, and test and evaluation tasks required in the execution of the systems engineering process." The (three-digit) table of contents of this significant standard is provided in Exhibit 2.3 to display its broad and far-reaching scope.

Exhibit 2.3: Table of Contents of Mil-Std-499B (Three-Digit)

1. SCOPE
   1.1 Scope
   1.2 Application Guidance
   1.3 Order of Preference
2. REFERENCED DOCUMENTS
3. DEFINITIONS
   3.1 Acronyms
   3.2 Fundamental Definitions
       3.2.1 System
       3.2.2 Life Cycle
       3.2.3 User
       3.2.4 Primary Functions
       3.2.5 Systems Engineering
       3.2.6 Systems Engineering Process
       3.2.7 Item
       3.2.8 Requirements
       3.2.9 Design
       3.2.10 Output Views
   3.3 Supplementary Definitions
       3.3.1 Baseline
       3.3.2 Configuration Item (CI)
       3.3.3 Environment
       3.3.4 Evolutionary Acquisition
       3.3.5 Exit Criteria
       3.3.6 Function

The essence of the systems engineering process, as far as this standard is concerned, is describable in four parts:

1. Requirements analysis
2. Functional analysis/allocation
3. Synthesis
4. Systems analysis and control

These four elements are also shown in Figure 2.2, but they are placed in a context that includes a variety of other elements. The standard, in effect, says that "needs statements" are formulated as inputs to the system engineering process (the preceding four elements) and a life-cycle-optimized set of products and processes is the resultant output.

The life cycle of a system, as depicted before with respect to the system acquisition process, is described in terms of "primary functions," which are:

1. The development function
2. The production function
3. The test/verification function
4. The deployment/installation function
5. The operations function
6. The support function
7. The training function
8. The disposal function

This standard also defines major reviews that should be carried out for a system. In addition to the four reviews cited earlier (i.e., SRR, SDR, PDR, and CDR), the following additional reviews are called for:

1. Software specification review
2. Functional system audit
3. Functional configuration audit
4. Physical configuration audit

Mil-Std-499B deals with planning, in the main, through the formulation of a system engineering management plan (SEMP). The SEMP is a key responsibility of the Chief Systems Engineer (CSE), whereas the overall project plan lies primarily in the hands of the Project Manager (PM). The two plans must of course be consistent. The elements of the SEMP (para. 5.3 of Exhibit 2.3) include the items listed in Exhibit 2.4.

Exhibit 2.4: The Systems Engineering Management Plan (SEMP) Elements

   5.3.1 Systems Engineering Process

   5.3.2 Systems Analysis and Control

      5.3.2.1 Systems Analysis

      5.3.2.2 Technical Performance Measurement (TPM)

         • Parameters

         • Technical Parameter Planning Data

      5.3.2.3 Technical Reviews

         • Review Success Criteria

   5.3.3 Technology Transition

   5.3.4 Technical Integration Teams

We note the technical orientation of the SEMP, but also must acknowledge the interaction between the SEMP and the project plan in such items as the need for a master schedule, which would be part of both plans.

There are many who view this and other standards, especially those drawn up by the U.S. Department of Defense, as "overkill," arguing that such standards add unnecessary tasks, reports, and formality, thereby increasing cost without much, if any, benefit. This argument is countered, at least in part, in the foreword of Mil-Std-499B, which states explicitly that "this standard must be appropriately tailored to ensure that only cost-effective requirements are cited in defense solicitations and contracts." This is sufficient guidance to provide flexibility in the real-world acceptance and implementation of the details of this important standard. In short, the provisions of this standard, or derivatives thereof, should not be blindly applied; they have to be modified appropriately to suit the particular and peculiar characteristics of the system that is being acquired and the circumstances surrounding the procurement of that system. Such circumstances include schedule, cost, interfaces with other systems, use of commercial-off-the-shelf (COTS) components and nondevelopmental items (NDIs), and other relevant factors.

2.5.3 IEEE P1220

This is a "Standard for Systems Engineering," as developed by the Institute of Electrical and Electronics Engineers (IEEE) [2.5]. Examining this standard in detail shows that it has its roots in Military-Standard-499B, as described in the previous section. The four steps in the systems engineering process have been expanded to five, which are:

1. Requirements analysis
2. Functional analysis
3. Synthesis

4. Systems analysis
5. Verification and validation

The latter two new terms are defined as:

Verification. A process of determining whether or not the products of a given phase of development fulfill the requirements established during the previous phase

Validation. A process of evaluating a configuration item, subsystem, or system, to ensure compliance with system requirements

These two important activities will be revisited in Chapter 7 where the thirty elements of systems engineering are presented.

This standard [2.5] also has a short-form definition of systems engineering as "an interdisciplinary collaborative approach to derive, evolve, and verify a lifecycle balanced system solution which meets customer and public acceptability." This definition may be compared with those provided in the previous chapter.

Another interesting feature of this standard is its definition of a system architecture, namely, the "composite of the functional, physical, and foundation architectures, which form the basis for establishing a system design." In this text, as we formulate concepts and procedures for architecting a system, shown later in Chapter 9, the reader will likely wish to reexamine this standard's notion of a system architecture.

As suggested above, this standard does not represent a major departure in principle from Mil-Std-499B. Further, it illustrates the fact that the IEEE has moved into the systems engineering arena. Over the years, this will certainly help in trying to assure that systems engineering takes its place among other well-accepted engineering disciplines (electrical, mechanical, chemical, etc.).

2.5.4   EIA-632

This standard, with the title "Processes for Engineering a System," has been promulgated by the Electronic Industries Association (EIA) [2.6]. Although an earlier version of this standard (1994) looked quite a lot like Mil-Std-499B, the later version represented a considerable departure.

Several important points can be made about this standard. First, it was basically developed through the combined efforts of the EIA and the International Council on Systems Engineering (INCOSE). Second, it represents a shift from systems engineering to the *processes* that are required in order to engineer any type of system. This may be viewed as related to the notions of business process reengineering, which holds that systems may be enhanced by improving the processes that lead to the design and development of these systems. Third, and most significant, is the overall structure of the standard.

This structure identifies thirteen processes that are critical to the engineering of systems, with these processes cited under the five categories listed below:

A. **Acquisition and Supply**
   1. Supply process
   2. Acquisition process
B. **Technical Management**
   3. Planning process
   4. Assessment process
   5. Control process
C. **System Design**
   6. Requirements definition process
   7. Solution definition process
D. **Product Realization**
   8. Implementation process
   9. Transition to use process
E. **Technical Evaluation**
   10. Systems analysis process
   11. Requirements validation process
   12. System verification process
   13. End products validation process

Some thirty-three requirements are also related to the above thirteen processes. This extremely interesting approach shows, among other things, that there are many ways to look at the issue of the engineering of systems.

2.5.5 EIA/IS-731

This is another standard provided by the Electronic Industries Association (EIA), focusing upon the Systems Engineering Capability Model (SECM) [2.7]. The standard itself was the result of the joint efforts of the EIA, IN-COSE, and the Enterprise Process Improvement Collaboration (EPIC). Previously, the Systems Engineering Capability Maturity Model, developed at the Software Engineering Institute of Carnegie-Mellon University, had produced the first such model, based upon the structure of their software model. This was called the SE-CMM. INCOSE then formulated their version of such a model, namely, the Systems Engineering Capability Model (SECAM). The SECM then became the consequence of harmonizing these two earlier models. The standard itself is divided into two parts. One is the model itself, and the other is the SECM appraisal method.

Just as the original software capability maturity model addressed the matter of how to assess and improve the capability of an organization to develop and utilize all aspects of software, this standard had basically the same intent, except as applied to the field of systems engineering. An important step in developing all of these models is to decide upon a series of focus areas, or process areas, which has been done in all cases.

Additional information regarding capability maturity models will be provided in later chapters, in particular 10 and 12. Further variations on these themes, including integrated models, will also be described.

2.5.6   ISO/IEC 15288

This standard, titled "Systems Engineering—System Life Cycle Processes" [2.8], is international and was issued in 2002 under the ISO (International Organization for Standardization) and the IEC (International Electrotechnical Commission). As the name suggests, a main focus is a set of life-cycle *processes* for systems. There are some twenty-five of these processes under four overview categories dealing with agreements, enterprises, projects, and technical matters. In addition, the standard presents life-cycle *stages* for systems, all of which are in an overview format. Further, a technical report is provided [2.9] that is a guide for the specific application of ISO/IEC 15288. The twenty-five featured processes may be compared with the definition of the thirty elements of systems engineering shown in Chapter Seven.

In looking at the various standards in systems engineering, INCOSE (International Council on Systems Engineering) decided, in version 3 of its *Systems Engineering Handbook* [2.10], to create a "document consistent with the international standard ISO/IEC 15288". This was a significant decision, establishing a quite explicit and stronger connection between this important council and the international systems engineering community. These types of efforts help to unify and integrate our overall knowledge base with respect to systems engineering.

2.5.7   Selected Software Standards

Software development is a critical aspect of building and fielding a system and, as such, is dealt with separately and in detail in Chapter 10. A very well-known standard for software development was the Department of Defense standard 2167A (DoD-Std-2167A) [2.11], which had the title "Defense System Software Standard." The key phases of software development defined in that standard were

1. System requirements analysis/design
2. Software requirements analysis

3. Preliminary design
4. Detailed design
5. Coding and computer software unit (CSU) testing
6. Computer software component (CSC) integration and testing
7. Computer software configuration item (CSCI) testing
8. System integration and testing

From this list, we note the emphasis on requirements, design, integration, and testing. These are indeed extremely critical elements because we know from experience that untested software is a formula for failure. The standard appears to support the industry conventional wisdom for software of "build a little, test a little." Because most of our modern large- and small-scale systems today contain large amounts of software, how it should be developed is a mandatory part of the training of today's Project Manager and Chief Systems Engineer. Of special interest is the manner in which the software development process "its into both project management and systems engineering. Again, this subject is discussed at length in Chapter 10.

The long-standing 2167A standard was replaced by Military Standard 498 in December 1994 [2.12]. The two-digit detailed requirements in this standard were articulated in the following subject areas:

- Project planning and oversight
- Establishing a software development environment
- System requirements analysis
- System design
- Software requirements analysis
- Software design
- Software implementation and unit testing
- Unit integration and testing
- CSCI qualification testing
- CSCI/HWCI (hardware configuration item) integration and testing
- System qualification testing
- Preparing for software use
- Preparing for software transition
- Software configuration management
- Software product evaluation
- Software quality assurance
- Corrective action
- Joint technical and management reviews
- Other activities

Two additional standards that relate specifically to software are:

1. IEEE/EIA 12207—Software Life Cycle Processes
2. IEEE P1471—Recommended Practice for Architectural Description

The second of these is of particular interest as it relates to the architecting of software and systems. The latter is a central theme of this book, with a full Chapter 9 devoted to providing a prescriptive method for the process of architecting a system. Additional information regarding these two standards is provided in Chapter 10.

2.5.8 International Organization for Standardization (ISO)

The International Organization for Standardization (ISO) has been producing standards that apply in the international arena. Companies that wish to do business outside the United States are paying a great deal of attention to such standards, recognizing that compliance is essential in order to compete successfully.

An example is the ISO 9000 Series, which deals with various aspects of product and service quality, as follows:

1. ISO 9000: Provides basic definitions and concepts; explains how to select and use other standards in the series
2. ISO 9001: Deals with external quality assurance (QA) situations; ensures conformance with requirements during design, development, production, installation, and service
3. ISO 9002: Deals with external QA situations; used when production and installation conformance are of concern
4. ISO 9003: Deals with external QA situations; focuses on ensuring conformance in final test and inspection
5. ISO 9004: Deals with internal QA situations; provides guidelines on technical, administrative, and human factors affecting the quality of products and services

Firms interested in providing products and services in Europe and countries outside the United States should pay special attention to what the ISO has established and is working on with respect to international standards in systems and software engineering and their related topics.

2.5.9 Other Standards

Other standards have been promulgated that can be used by the project management team in order to assist in the design and development of systems. Military Standard 499B, discussed earlier, contains a list of such standards, which is reproduced here as Table 2.1. From this, we see a variety of

TABLE 2.1 Standards Cited in Mid-Std-499B

Technical Discipline	Reference	
Configuration management	Mil-Std-480/481/482/483	
Climatic Information	Mil-Std-210	
Computer-aided acquisition and logistics support	Mil-Hdbk-59	
Corrosion prevention and control	Mil-Std-1250	Mil-Std-1568
Environmental analysis	Mil-Std-810	
Electromagnetic compatibility	Mil-Std-1541	Mil-Std-461
	Mil-E-6051	Mil-Hdbk-237
Electrostatic discharge	Mil-Std-1686	
Human factors	Mil-Std-1472	Mil-Std-1794
	Mil-Std-1800	Mil-Hdbk-763
	Mil-H-46855	
Maintainability	Mil-Std-470	Mil-Std-1843
	Mil-Std-2184	Mil-Hdbk-791
Manufacturing	Mil-Std-1528	
Nondestructive inspection	Mil-Hdbk-728	Mil-Hdbk-731
	Mil-I-600	
Parts control	Mil-Std-965	
Producibility	Mil-Hdbk-727	
Quality	Mil-Q-9858	Mil-I-45208
Reliability/durability	Mil-Std-785	Mil-Std-1530
	Mil-Std-1543	Mil-Std-1783
	Mil-Std-1796	Mil-Std-1798
	Mil-Std-2164	
System safety engineering	Mil-Std-882	
Software	DoD-Std-2167	Mil-Std-1803
	Mil-Std-1815	
	Mil-Hdbk-287	
Software quality assurance	DoD-Std-2168	DoD-Hdbk-286
Supportability	Mil-Std-1388	
Survivability	Mil-Std-1799	Mil-Std-2069
	DoD-Std-2169	Mil-Hdbk-336
System security	Mil-Std-1785	
Telecommunications	Mil-Std-188-xxx	
Testability	Mil-Std-2165	
Thermal design /analysis	Mil-Hdbk-251	
Transportability	Mil-Std-1367	Mil-Hdbk-157
Value engineering	Mil-Std-1771	
Technical reviews and audits	Mil-Std-1521	
Work breakdown structure	Mil-Std-881	
Statement of work preparation	Mil-Hdbk-245	
Technical data package	Mil-T-3100	
Specification practices	Mil-Std-490	Mil-S-83490

military standards and handbooks on important subjects such as configuration management, environmental analysis, human factors, manufacturing, quality, reliability, safety, security, and value engineering. For those who may be operating as acquisition agents for systems, standards are available that can provide inputs to such activities as reviews and audits, work breakdown structures (WBSs), statements of work (SOWs), technical data packages, and specifications.

Increasingly, standards are being produced in various domain-specific areas (e.g., telecommunications, information security) under the sponsorship of professional organizations such as the Institute of Electrical and Electronics Engineers (IEEE) and the Electronic Industries Association (EIA). The IEEE has been very active in systems and software engineering and has produced a variety of standards in both subjects. Work in the latter area has been especially significant (see Chapter 10), but systems engineering has not been neglected (see Chapter 12). Other government agencies outside the Department of Defense have also been concerned with the formulation of various kinds of standards. In this category, the National Institute for Standards and Technology (NIST) has a specific charter to develop standards. The responsible Project Manager and Chief Systems Engineer will assure themselves that standards that might be applicable to their project are examined in detail to make sure that, where necessary, the project is in conformance with such standards. Finally, INCOSE has been a major force in assisting with standards that address systems engineering and related disciplines.

QUESTIONS/EXERCISES

2.1 Identify three activities not shown explicitly in Figure 2.1. Where might they fit into this chart?

2.2 Identify three activities not shown explicitly in Figure 2.2. Where might they fit into this chart?

2.3 Investigate the history of a real system and its acquisition. From this history, put the phases shown in Exhibit 2.1 on a time line.

2.4 Identify six items that are typically part of the agenda for a critical design review (CDR).

2.5 Compare Military Standards 499A and 499B with respect to
   a. three similarities
   b. three differences

2.6 Discuss the similarities and differences between a project plan and a systems engineering management plan (SEMP).

2.7 Compare the approaches to defining systems engineering as represented in this chapter with those of the previous chapter. Which do you find most/least satisfying? Explain.

2.8 Based upon the material in this and the previous chapter, write your own three to five sentence definition of:

    a. project management

    b. systems engineering

2.9 Write a two-page summary overview of INCOSE's *Systems Engineering Handbook,* version 3 (see reference 2.10).

2.10 Write a two-page summary overview of IEEE P1471, the software standard dealing with architecture descriptions.

REFERENCES

2.1 Department of Defense (DoD) Website: web2.deskbook.osd.mil

2.2 Department of Defense Directive Number 5000.1 (2000). Washington, DC: Under Secretary of Defense (Acquisition, Technology and Logistics), U.S. Department of Defense, Pentagon, October 23.

2.3 *Engineering Management*, Military Standard 499 A (1974). Washington, DC: U.S. Department of Defense.

2.4 *Systems Engineering*, Military Standard 499B (1971). Washington, DC: U.S. Department of Defense.

2.5 *Standard for Systems Engineering*, IEEE P1200 (1994). New York: Institute of Electrical and Electronics Engineers.

2.6 *Processes for Engineering a System*, EIA Standard 632 (1998). Washington, DC: Electronic Industries Association, Engineering Department; available also through Global Engineering—see http://global.ins.com.

2.7 *Systems Engineering Capability Model (SECM)*, EIA/IS-731 (1998). Washington, DC: Electronic Industries Association, Engineering Department.

2.8 "Systems Engineering—System Life Cycle Processes" (2002). ISO/IEC 15288, 2002-11-01. Geneva, Switzerland: ISO Copyright Office.

2.9 "Systems Engineering—A Guide for the Application of ISO/IEC 15288 (System Life Cycle Processes)." (2003). Technical Report 19760. 2003-11-15. Geneva, Switzerland: ISO Copyright Office.

2.10 *Systems Engineering Handbook*, version 3 (2006). INCOSE—TP 2003-002-03, June.

2.11 *Defense System Software Development*, DoD-Std-2167A (1988). Washington, DC: U.S. Department of Defense.

2.12 *Software Development and Documentation*, Mil-Std-498 (1994). Washington, DC: U.S. Department of Defense.

PART II
PROJECT MANAGEMENT

_____3
THE PROJECT PLAN

3.1 INTRODUCTION

The project plan (PP) is at the core of the planning function for the project team, and is a blueprint for the work to be performed as well as the proposal to the customer, if such a proposal is indeed required. The project plan has seven essential elements:

1. Needs, goals, objectives, and requirements
2. Task statements, statement of work, and work breakdown structure
3. Technical approach
4. Schedule
5. Organization, staffing, and task responsibility matrix
6. Budget
7. Risk analysis

All project plans should contain these essential elements, although it is recommended that a project plan should be as short and concise as possible. If the customer requires great elaboration of the preceding elements, then a "short form," or summary of the project plan, should be prepared that will serve the project team on a day-to-day basis. The project plan should be used in at least the following ways:

1. To allow all project team members, including newly assigned personnel, to understand the essentials of the project

2. To provide corporate management, to whom the project reports, with an understanding of the project

3. To convey to the customer the project essentials, as perceived and formulated by the project team

4. To form the basis for a proposal to the customer, where such a proposal is called for

Updates to the formal project plan should be considered on a quarterly basis. Changes to various portions of the plan, such as the schedule, should be carried out when necessary.

In this chapter, use is made of various items in the literature, and in requests for proposal, to show examples of the type of material that has been used on real projects and programs. In addition, an example of a project with eight tasks is postulated so as to convey how the various elements of the project plan are interrelated.

3.2 NEEDS, GOALS, OBJECTIVES, AND REQUIREMENTS

This first part of a project plan can be divided into two parts, the first consisting of needs, goals, and objectives, and the second constituting the requirements. Requirements are of two types, project and system, with the latter being quite voluminous and usually a part of the formal contract between the customer and the system developer. Statements of needs, goals, and objectives can be rather variable.

The statements of needs, goals, objectives, and requirements come from the customer and acquirer of the system to be developed. The outside system developer, therefore, is a recipient (rather than an original source) of these statements. For at least this reason, they are much condensed or simply reiterated or incorporated by reference when a Project Manager deals with them in a project plan.

3.2.1 Needs

The Department of Defense (DoD) acquisition directive [3.1] states that three key aspects of acquisition management are

1. Translating operational needs into stable affordable programs
2. Acquiring quality products
3. Organizing for efficiency and effectiveness

With respect to the first key, the statement is made [3.1] that "prudent management also dictates that new acquisition programs only be initiated after fully examining alternative ways of satisfying identified military needs."

Mission needs are also "identified as a direct result of continuing assessments of current and projected capabilities in the context of changing military threats and national defense policy." Examples of possible military needs that are identified in the acquisition directive are

1. The need to impede the advance of large armored formations 200 kilometers beyond the front lines, or
2. The need to neutralize advances in submarine quieting made by potential adversaries

Whereas the DoD may have rather formal procedures and processes for identifying and documenting needs, other government agencies and potential commercial clients are likely to be much less structured in their approach to this issue.

The bottom line, with respect to needs, is simply that the system acquirer must make sure that a true need exists for the system in question. If that is not the case, the project may ultimately not be able to be sustained. The reader who wishes to explore this matter in greater depth may obtain the DoD directive cited or examine the commentary of J. Davidson Frame, who spends an entire chapter in his book [3.2] on the matter of "making certain that a project is based on a clear need."

In terms of the program plan, the statement of need can be abstracted from the needs as represented by the acquisition agent. It can be very short and expressed in only a few lines. This is in distinction to the needs analysis and confirmation carried out by the acquirer of the system. As indicated by the earlier DoD directive, such a needs analysis and assessment can be rather formal and substantial, following the guidelines of the DOD.

3.2.2 Goals and Objectives

Goals and objectives are usually short declarative statements, with goals being rather broad and objectives under each goal being somewhat more specific, although some treat goals and objectives in reverse order. They are often established for programs in distinction to projects. For this reason, they may not be a firm requirement as part of a project plan. An illustrative set of goals and objectives is shown in Table 3.1, in relation to an overall defense science and technology strategy.

Another example of a stated objective, as articulated by the U.S. Coast Guard (USCG) in a real-world project procurement [3.3], is

1. To support marine safety and law enforcement activities
2. To record activities and resource usage
3. To analyze mission performance
4. To monitor program effectiveness and resource usage

TABLE 3.1   Illustrative Defense Science and Technology Goals and Objectives

---

Goal A: Deterrence

Objectives

A.I Deploy weapon systems that can locate, identify, track, and target strategically relocatable targets.

A.2 Attain worldwide, all-weather force projection capability to conduct limited warfare operations (including special operations forces and low intensity conflict) without the requirement for main operating bases, including a rapid deployment force that is logistically independent for 30 days.

A.3 Eliminate the threat posed by nuclear ballistic missiles of all ranges, through non-nuclear methods and in compliance with all existing treaties.

Goal B: Superiority

Objectives

B.1 Attain affordable, on-demand launch and orbit transfer capabilities for space deployed assets with robust, survivable command and control links.

B.2 Regain the substantial antisubmarine warfare advantages the United States enjoyed until recent years.

B.3 Achieve worldwide, instantaneous, secure, survivable, and robust command, control, communications, and intelligence (C3I) capabilities within 20 years to include: (a) on-demand surveillance of selected geographical areas; (b) real-time information transfer to command and control authority; and (c) responsive, secure communications from decision makers for operational implementation.

B.4 Field weapon systems and platforms that deny enemy targeting and allow penetration of enemy defenses by taking full advantage of signature management and electronic warfare.

B.5 Deploy enhanced, affordable close combat and air defense systems to overmatch threat systems.

B.6 Field affordable "brilliant weapons" that can autonomously acquire, classify, track, and destroy a broad spectrum of targets (hard fixed, hard mobile, communications nodes, etc.).

Goal C: Affordability

Objectives

C.1 Reduce operations and support resource requirements by 50% without impairing combat capability.

C.2 Reduce manpower requirements for a given military capability by 10% or more by 2010.

C.3 Ensure the affordability, producibility, and availability of future weapons systems.

5. To exchange information between USCG offices, other government agencies at the federal and state level, natural resource trustees, and certain private organizations
6. To fulfill specific statutory requirements

We note that these statements of objectives are concise and to the point and refer specifically to the project work that is being procured by the customer, in this case from the Coast Guard.

3.2.3 Requirements

Requirements, as alluded to in earlier chapters, is a dense and important subject. At the outset, we make a distinction between two types:

1. Requirements to be fulfilled by the project (project requirements)
2. Requirements of the "system" that the project addresses (system requirements)

Project requirements refer to all the work to be performed on the project. System requirements are applicable to the "system" that is being addressed by the project. To illustrate the difference, let us assume that the project is to design, but not build, a new "subway" system for a city. That is, the project is limited to design and does not include the construction of any hardware or software for the system. The project requirements, therefore, are limited to all the work to be accomplished as part of the design process only. This may include estimating the cost of the subway system for each year during its entire life cycle. However, the cost of the project itself, limited as it is to the design phase, is clearly a subset of that total cost, and is likely to be only a minor part of the life-cycle cost of the subway system. The system requirements describe, at increasing levels of detail, the full characteristics of the subway system, from initial design to operations and support.

Thus, the Project Manager (PM) must keep in mind this distinction when constructing the project plan. The latter should be focused on the *project* requirements, with the system requirements carried forth, as defined by the customer, in an ancillary document to be used in engineering the system in question. Further discussions of the ways in which the system requirements are to be handled are found in Chapter 8.

Project requirements are often stated by a customer in the main body of a request for proposal in both broad and specific terms, whereas system requirements can be described in several volumes of reports. An example of the former is shown in Exhibit 3.1, drawn from a requirements statement in a U.S. Coast Guard procurement dealing with mission-oriented information systems engineering (MOISE) support.

Exhibit 3.1: U.S. Coast Guard Statement of Requirements [3.3]

The contractor shall furnish the equipment, software, maintenance, services, and support required under the terms and conditions of the contract. The major requirements are

1. Technical, functional, data, programmatic, and strategic integration of information systems
2. System and software development services within the information system (IS) life-cycle phases defined by the U.S. Coast Guard
3. Management, quality assurance, and requirements metrics
4. Cross-cutting requirements that span more than one life-cycle phase
5. Limited quantities of hardware, services, and software to support system development and to transition the systems to an operating and maintenance provider
6. Establishment, staffing, operation, and management of the System Development Center (SDC) to support the development of ISs by a phased approach using task orders
7. Contractor management and personnel requirements, and security procedures

In a similar vein, the Federal Aviation Administration (FAA), in its procurement of technical assistance contract (TAG) services related to its advanced automation program (AAP) and automation program (ANA), has documented the summary of requirements set forth in Exhibit 3.2.

Exhibit 3.2: FAA Summary of Requirements for TAC Support [3.4]

a. The contractor shall support FAA's Advanced Automation Program (AAP) and Automation Program (ANA) in an engineering and technical assistance capacity. The contractor's efforts shall complement, support, and extend those of the AAP/ANA engineering staff, which are responsible for the overall technical and contractual direction of the AAP and ANA programs.
b. The contractor shall provide assistance to the FAA in critical technical areas at key periods throughout the contract. A broad range of talents are required to address complex technical problems, for varying periods of time and under stringent response constraints. The contractor's skills must have particular depth, breadth, and quality as follows: [as listed].
c. The contractor's experience and performance must be at a level to provide the highest quality of analytic expertise and, if required, to support the provision of expert testimony.

The two exhibits focus on project requirements. The Project Manager will normally accept these customer statements of requirements for integration into the project plan. Also note that they are relatively brief and differ from system requirements when a large-scale system is being procured. If provided, the system requirements can be cited by reference in the program plan rather than being an integral part of the plan itself.

Other project requirements may also be stated in a procurement that may be regarded as subordinate requirements. These may be described as minimum position qualification requirements, or estimated staffing requirements, or even special contract requirements. For example, Exhibit 3.3 shows the estimated staffing requirements associated with a NASA procurement [3.5].

Exhibit 3.3: Illustrative Estimated Staffing Requirements

Labor Category	Estimated Hours Required
1. Program Manager	100
2. Senior Instructor	300
3. Instructor	120
4. Senior Facilitator	100
5. Facilitator	80
6. Course Development Specialist	240
7. Administrative/Clerical	120
Total	1060

These subordinate requirements impact the staffing of the project and, of course, the estimated cost of the project, as described in later parts of the project plan. A list of other special contract requirements from this same NASA procurement follows:

1. Printing and duplicating
2. Task ordering procedure
3. Key personnel and facilities
4. Observance of legal holidays
5. Protection of information
6. Level of effort (cost)
7. Property administration
8. Contractor's program manager
9. Special provisions regarding travel costs
10. Minimum position qualifications
11. Identification of contractor employees
12. Advance understanding: nonfee-bearing costs
13. General-purpose equipment
14. Contractor-acquired property: submission of vouchers

These subordinate requirements should only be cited by reference and not be included in detail in the requirements portion of the project plan. An alternative is to list them as an appendix to the plan.

3.3   TASK STATEMENTS, STATEMENT OF WORK (SOW), AND WORK BREAKDOWN STRUCTURE (WBS)

Task statements are usually part of the statement of work (SOW) provided by the customer. Thus, these are used interchangeably in this text. These statements are normally accepted and reiterated by the system developer in the program plan. Changing the customer's SOW is not a recommended action because it may turn into a point of contention later in the process. Work breakdown structures (WBSs) may or may not be part of the customer's definition of the work to be performed. When it is provided by the customer, it almost always should be accepted and used by the system developer.

An example of task statements (sometimes also called task areas) is shown in Exhibit 3.4, drawn from a real-world request for proposal from the FAA to industry. Under each of the task areas listed in the Exhibit, there is a short description of the tasks to be performed. Thus, task statements are embedded in the defined task areas.

Exhibit 3.4: Example of Task Statements and Task Areas from the FAA [3.4]

Task Area 1
  Requirements Analysis and Documentation
  Computer-Human Interface (CHI)
  Specification Development
  Reliability, Maintainability, and Availability (RMA) and Fault Tolerance (FT)
  System Modeling and Performance Analysis
  Local Communication Network (LCN)
  Algorithm Evaluation
  Interfaces
  Standards
  Technical Planning and Risk Analysis
  Software Engineering
Task Area 2
  Advanced Automation System (AAS) Contractor Requirements Traceability and Compliance Tracking
  Technical Support
  Engineering Change Proposals (ECPs)
  Testing and Evaluation
  Implementation

Division/Branch Segment
Integrated Logistics Management
Task Area 3
Terminal Automation Program Management
En Route/Traffic Management Systems Automation Program Management
Maintenance Automation Program Management
Task Area 4
Planned Product Improvement Design Evaluation
Planned Product Improvement Operational Test and Evaluation (OT&E)
Operational Test and Evaluation
Site Implementation
Manufacturing and Production
Factory Testing
Task Area 5
Configuration Management
Financial Management
Data Management
Office Automation
Program Control

A similar example for work to be performed under a solicitation from the Defense Systems Management College (DSMC) is provided in Exhibit 3.5. Here, again, task areas are defined and elaborated to constitute task statements.

Exhibit 3.5: Example of Task Areas from the Defense Systems Management College [3.6]

Task Area 1: Curriculum Design and Development
Task Area 2: Program Management Course Curriculum Material Update
Task Area 3: Development of Automated DSMC Management and Teaching Tools
Task Area 4: Statistical Analysis Support
Task Area 5: Executive Education Curriculum Development
Task Area 6: Quality of Instruction
Task Area 7: Define Teaching Quality
Task Area 8: Establishment of Baselines
Task Area 9: Reports

A third illustration of how a statement of work may be formulated is found in Exhibit 3.6. This exhibit lists seven "detailed items of work" with a breakdown of more specific work under the first item.

Exhibit 3.6: Example of Items of Work from the Department of Transportation's Volpe Center [3.7]

1. Project Planning and Scheduling Support
    1. User and System Requirements Analysis
    2. Technical Meeting Support
    3. Technology Assessment
    4. Project Management Support
    5. Transition Planning
    6. Environmental Impact Assessments
    7. Cost Benefit Analyses
2. System Design and Development Support
3. Software Development Support
4. Systems Integration Support
5. Testing and Evaluation Support
6. Training Support
7. Documentation and Configuration Management Support

The work breakdown structure (WBS) is also a formal exposition of work to be performed and is illustrated in Figure 3.1 for a NASA program known as Earth Observing System Data and Information System (EOSDIS) Phase B [3.8]. The most convenient form of a WBS is one in which there is a one-to-one correspondence between the tasks and the WBS. In such a case, the WBS elements can be considered a further breakdown of the major tasks into subtasks and each WBS element is identical to a subtask. If a WBS has not been defined by the customer, it is a recommended procedure to have the WBS correspond directly to the tasks and subtasks. However, if the customer provides a WBS, it may be necessary to develop a "cross-walk" between the task statements and the WBS. Such a cross-walk is shown symbolically in Table 3.2. This usually creates a layer of complexity that is not really desirable but may be necessary in order to satisfy the instructions of the customer.

3.4 TECHNICAL APPROACH

The technical approach is a task-by-task description of how the project team intends to execute the tasks and subtasks in the SOW, from a technical perspective. The technical approach is usually formulated, in detail, in response to a request for proposal (REP) if such an REP is a precursor to the project. In such a situation, this technical approach can be abstracted and used in this section of the project plan.

The technical approach changes, of course, for each project and the domain of the project. The approach to building a management information system

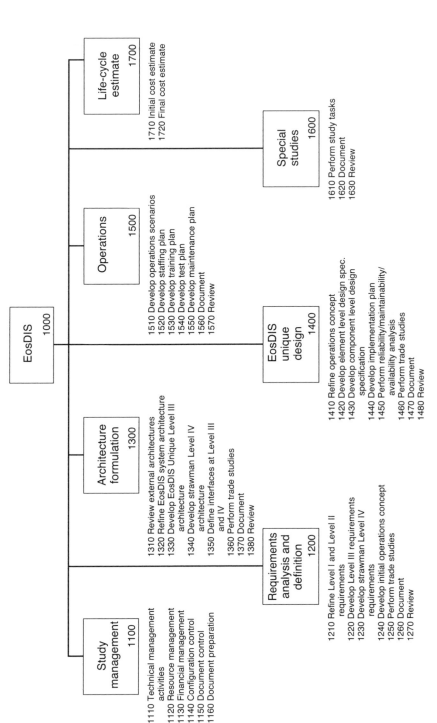

Figure 3.1. EOSDIS Phase B work breakdown structure.

TABLE 3.2 Cross-Walk Between SOW and WBS

		Work Breakdown Structure				
		WBS 1.0	WBS 2.0	WBS 3.0	WBS 4.0	WBS 5.0
	Task 1	X	X		X	
Statement	Task 2		X	X		X
of Work	Task 3		X		X	
(SOW)	Task 4	X		X	X	
	Task 5	X	X			X
	Task 6			X		X

(MIS) is different from the approach to constructing a subway or transit system. However, some questions that might be common to all projects, with respect to the technical approach, are cited in Exhibit 3.7. These questions assist the project team in addressing a wide variety of technical issues.

Exhibit 3.7: Two Dozen Selected Questions for Technical Approach

1. How do we plan to execute this task/subtask?
2. How will we employ a systems approach and systems engineering process?
3. What is special or unique about our approach?
4. What technology do we plan to utilize or transition?
5. Is this technology at or pushing the state of the art?
6. How can we be most productive and efficient?
7. What computer tools will we be using?
8. How can we demonstrate that we will, as a minimum, satisfy all customer requirements?
9. Do we plan to exceed requirements in certain areas?
10. Are certain requirements vague, incorrect, or inconsistent?
11. What special facilities will we need?
12. What aspects of our previous work can be brought to bear?
13. Do we plan to use any special models or simulations in order to assess system performance?
14. How will we execute a coherent technical performance measurement program?
15. What is our approach to system and subsystem testing?
16. Do we have a unique knowledge base to support our approach?
17. Can our independent research and development (IR&D) program results be utilized for this project?
18. What specialty engineering capabilities will we be using?

19. How does our technical approach correlate with our Systems Engineering Management Plan (SEMP)?
20. Do we plan to use special processes such as concurrent engineering, business process reengineering, or Total Quality Management (TQM)?
21. How can we approach software development with the most up-to-date methods?
22. What types of technical support will be needed from the rest of the company?
23. How will we find the most cost-effective solution?
24. How will we prove that we have the most cost-effective solution?

Question 19 of Exhibit 3.7 raises the issue of the relationship between the Systems Engineering Management Plan and the technical approach. These two documents are not the same, but certain elements of the SEMP, as a minimum, should be addressed in the technical approach. These include the systems engineering process, technical performance measurement, methods of systems analysis, technology transitioning, and technical integration.

3.5 SCHEDULE

A schedule is an expression of the tasks and activities to be performed along a time line. Two main methods of describing a schedule are in use today, namely, (1) a Gantt Chart and (2) a program evaluation and review technique (PERT) Chart. Figures 3.2 and 3.3 show examples of these types of schedules. Both figures are constructed for a hypothetical project of eight principal tasks that involve selecting commercial-off-the-shelf (COTS) software for use by a project team. Such software might be, for example, a project management package or some other package (e.g., geographic information system, executive information system) that would be needed by a project team.

Both figures map the tasks against a time line for the eight weeks of the total project. The Gantt chart is quite straightforward and easy to read. Each bar represents a single task. The PERT chart is somewhat more complex but is also simple to grasp. Each circle represents an event—a specific point in time at which a measurable activity is either started or completed. The lines between the events are activities during which resources are expended to achieve the designated end event. Further details regarding how a PERT chart is developed are provided in Chapter 4. However, one major point is that the PERT chart places in evidence the longest path through the network, which is known as the *critical path*. By definition, all other paths are at most as long

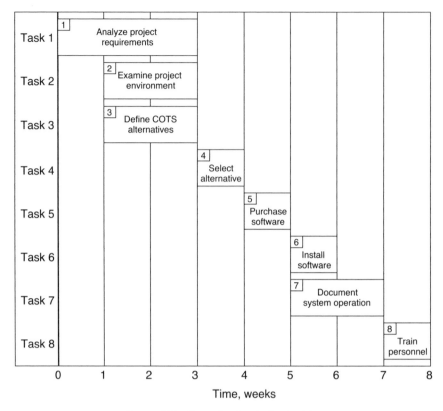

Figure 3.2. Example of project schedule.

as the critical path. Those parallel paths that are shorter have "slack" in them. Slack represents an opportunity for moving some tasks or subtasks forward or backward to utilize resources in more efficient ways.

The schedule for a complex project can literally take up the space of an entire wall. The schedule in the program plan should be an overview schedule, emphasizing major tasks and milestones. Too much detail is not warranted in an overview project plan. A full computer-generated schedule might be included only as an appendix to the plan.

The schedule must be ultimately consistent with the customer's delivery requirements. This applies to interim dates as well as the final project completion date. To the extent that the schedule drawn up by the project team does not do this, it has to be continually reworked until all customer requirements with respect to the schedule are met. If this is not possible, then the plan is not viable and there is an impasse that must be negotiated and resolved before work can begin.

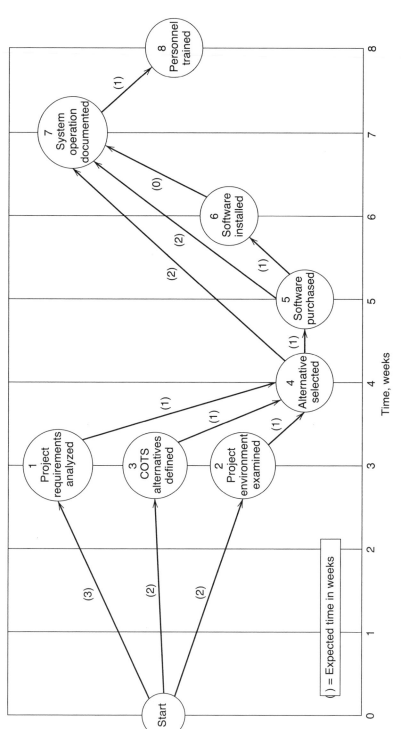

Figure 3.3. Example of a project PERT chart.

TABLE 3.3   Task Responsibility Matrix (TRM)

	Personnel (Staff) Categories								
	A Senior Software Engineer		B Software Engineer		C Documentation Specialist		D Training Specialist		Total
	Number of People	Person-Weeks	Number of People	Person-Weeks	Number of People	Person-Weeks	Number of People	Person-Weeks	Person-Weeks
Task 1	1	3							3
Task 2			1	2					2
Task 3	1	2	1	2					4
Task 4	1	1	1	1					2
Task 5			1	1					1
Task 6	1	1	1	1					2
Task 7	1	2	1	2	1	3	1	1	8
Task 8	1	1	1	1			1	1	3
Totals:		10		10		3		2	25

3.6   ORGANIZATION, STAFFING, AND TASK RESPONSIBILITY MATRIX (TRM)

The project organization can be simply the organization chart (see Figure 1.2 in Chapter 1), supplemented by a short description of key roles and responsibilities. Staffing refers to the next step of actually assigning categories of personnel to the various project tasks. This results in a *task responsibility matrix* (TRM), as illustrated in Table 3.3, based on the schedule in Figure 3.2. We note that person-week totals by task and by category of personnel are easily derived from the assignment of personnel types to the tasks. As indicated previously, if a WBS is part of the project, and the relationship between the tasks and the WBS has been developed, then it is possible to also develop a profile of which personnel are expected to execute the elements of the WBS.

It is considered optional for the project plan to contain the specific names of the people that represent the various personnel categories. Clearly, this step must ultimately be taken, but it does not necessarily have to be part of the program plan document. In many ways, it is preferable to treat the assignments by category rather than by the names of specific individuals.

3.7   BUDGET

From the previously developed information, and some additional cost data, it is then possible to formulate a budget for the project. Such a project budget

TABLE 3.4 Project Budget

Direct Labor	Rate/Week	Person-Weeks	Cost
Senior software engineer	$1,000	10	$10,000
Software engineer	700	10	7,000
Documentation specialist	500	3	1,500
Training specialist	600	2	1,200
Fringe benefits @ 30%		25	$19,700
			$ 5,910
Overhead @ 70%		Subtotal 1	$25,610
			$17,927
Other direct costs		Subtotal 2	$43,537
1. Software			
2. Training materials			$8,000
			1,000
General & administrative @ 15%		Subtotal 3	$52,537
			7,881
		Total cost	$60,418
		Fee (profit) @ 10%	6,042
		Cost and fee (Price)	$66,460

is illustrated in Table 3.4, utilizing the same data from the task responsibility matrix in Table 3.3. Cost items may also be broken down by week, as illustrated in Table 3.5. This facilitates cost tracking, as discussed in the next chapter.

From Table 3.4, a project budget is prepared by first examining the direct labor costs. These costs are incurred as a result of project personnel working on the various tasks of the project. As shown in the figure, and using the person-week data provided in Table 3.3, we list the four categories of personnel together with their labor rates per week and the person-weeks that each has been assigned. This yields the direct labor cost by category. The summation of these costs ($19,700) is the total direct labor cost for this project. This is augmented by adding the fringe benefits, in this case 30% of the direct labor costs. The resultant sum, shown as subtotal 1, is $25,610. The next item of cost to consider is the overhead cost. This example shows the overhead rate to be 70%, or a total of $17,927. Some companies embed fringe benefits into the overhead percentage and therefore there is no reason to consider fringe and overhead separately. In this model, they are constructed as separate rates.

Subtotal 2 shows the direct costs with fringe and overhead costs added, yielding an amount of $43,537. At this point in the process, other direct costs (ODCs) are considered. These may be a variety of cost items, such as materials, consultants, subcontractors, and services provided from outside the company. In Table 3.4, two such costs are listed: the cost of purchasing software, estimated at $8,000, and the cost of training materials, shown as $1,000. Adding these costs to subtotal 2 leads to subtotal 3, which is $52,537.

TABLE 3.5 Cost Budget by Week

Labor Category	Week 1		Week 2		Week 3		Week 4		Week 5		Week 6		Week 7		Week 8	
	P-wk	Cost	P-wk	Cost	P-wk	Cost	P-wk	Cost	P-wk	Cost	P-wk	Cost	P-wk	Cost	P-wk	Cost
Senior software eng'r	1	1,000	2	2,000	2	2,000	1	1,000	1	1,000	1	1,000	1	1,000	1	1,000
Software eng'r	—	—	2	1,400	2	1,400	1	700	2	1,400	1	700	1	700	1	700
Document specialist	—	—	—	—	—	—	—	—	1	500	—	—	1	500	—	—
Training specialist	—	—	—	—	—	—	—	—	—	—	1/3	500	1/4	600	1/3	600
	1	1,000	4	3,400	4	3,400	2	1,700	4	2,900	3	2,200	4	2,800	3	2,300
Fringe @ 30%		300		1,020		1,020		510		870		660		840		690
Subtotal 1		1,300		4,420		4,420		2,210		3,770		2,860		3,640		2,990
Overhead @ 70%		910		3,094		3,094		1,547		2,639		2,002		2,548		2,093
Subtotal 2		2,210		7,514		7,514		3,757		6,409		4,862		6,188		5,083
ODCs: Software train, mat'ls		—		—		—		—		8,000 / 1,000		—		—		—
Subtotal 3		2,210		7,514		7,514		3,757		15,409		4,862		6,188		5,083
G&A @ 15%		331		1,127		1,127		564		2,311		730		928		763
Total cost		2,541		8,641		8,641		4,321		17,720		5,592		7,116		5,846
									Total cost $60,418							
Cumulative cost		2,541		11,182		19,823		24,144		41,864		47,456		54,572		60,418

Next, general and administrative (G&A) costs are added. These are represented as a percentage, in this case 15%, or a total of $7,881. Summing this cost with subtotal 3 yields the total estimated project cost of $60,418. This, of course, is a critical number. If all went according to plan, the overall project, for this example, would cost $60,418. Finally, a fee is added, in this case 10%. This is the profit that the company wishes to make by engaging in this effort. Adding the fee to the cost results in the overall estimated price of $66,460.

In order to calculate the bottom-line cost without ODCs, and do it quickly, we introduce the notion of a "cost factor." This cost factor (CF) is a multiplier on the direct labor costs that results in the total project cost, without any ODCs. The cost factor is

$$CF = (1 + FR)(1 + OH)(1 + GA)$$

where FR = fringe rate (expressed as a decimal)
              OH = overhead rate
              GA = general and administrative rate

Thus, in the example shown in Table 3.4, the cost factor is

$$CF = (1 + 0.3)(1 + 0.7)(1 + 0.15) = (1.3)(1.7)(1.15) = 2.54$$

This means that every dollar of cost at the direct labor line translates into $2.54 in bottom-line cost, exclusive of ODCs.

In a similar vein, the "price factor" (PF) translates direct costs into bottom-line prices and is

$$PF = (CF)(1 + PR)$$

where PR is the profit (fee), expressed as a decimal. By using numbers from Table 3.4, the price factor is

$$PF = (2.54)(1 + 0.1) = (2.54)(1.1) = 2.79$$

Again, both the cost and price factor are rapid ways to estimate bottom-line costs and profits, but without other direct costs.

We note that some elements of cost are estimated by the Project Manager (more likely in concert with the Project Controller and the Chief Systems Engineer). These include the original person-week estimates, and the software and training materials costs (ODCs). All other costs are derivable from the fringe, overhead, and G&A costs. These three costs are characteristic of the corporate structure and generally are not under the control of the Project Manager. In a similar vein, company management will likely select the fee (profit) that is to be bid, obtaining an input from the Project Manager.

The PM must control the project to the overall cost number and not the total bid price. The fee is not to be spent by the PM unless some agreement has been reached within the company's decision-making apparatus to do so for this project. There are even times when the PM is not aware of the fee (profit) that has been proposed.

The customer may also wish to see the estimated project costs broken down by time period (as in Table 3.5) or by project task. The task breakdown of cost would be derived from using the task person-week data shown in Table 3.3.

An inviolate notion in project as well as corporate management is that cost and price are not the same. If it turns out that the project is successful, then the enterprise would receive the bid price from the customer, which then would be booked as company sales or revenues. The difference between project revenue and project cost is, of course, the profit that is made by the project. If that number is positive and equal to or greater than the bid profit, then all is well. If it is negative, then the PM may be in some difficulty, with a lot of explaining to do.

3.8   RISK ANALYSIS

In order to avoid future difficulty, the project triumvirate, as a minimum, should carry out a risk analysis, the results of which become part of the project plan. This analysis attempts to focus on trouble spots before the fact, developing risk-mitigation strategies prior to actual work on project tasks.

In general, it can be said that there are four kinds of risk that the Project Manager would be concerned with. These are

1. Technical performance risk
2. Schedule risk
3. Cost risk
4. Administrative risk

Technical performance risk (TPR) flows from items of design, development, and construction of the system that result in not meeting the technical requirements set forth by the customer. These might include situations where one is pushing the state of the art, not meeting system response times, experiencing harsh environments that degrade system performance and numerous others. TPR is by far the most complex area and must be examined by domain-knowledge personnel to identify what tasks appear to be most difficult from the perspective of system performance. Background information regarding the details of how to carry out a technical performance risk analysis can be found in the literature [3.9, 3.10, 3.11, 3.12]. It is also considered again in Chapter 8. Technical performance risk is also a primary factor in creating schedule and cost risks.

Schedule risk, of course, involves not meeting project milestones. If internal milestones are not met, then the PM may be able to get back on schedule by the deployment of additional resources and other means. Customer delivery and review dates are viewed as much more serious, especially the delivery of parts of the system and special reports to the customer. If penalty clauses for late delivery are operative, schedule risks in this regard are considered critical. Analysis of the project PERT chart is usually a good place to start in assessing schedule risk. Activities and events on the critical path should be examined in considerable detail because this path is the controlling factor in the overall project schedule. Long lead-time items that come from other vendors should be subject to scrutiny and it may be necessary to have backup plans for such items. Cases in which the customer has to provide an input or pieces of hardware and software should also be examined in detail, the question being: What happens if the customer fails to provide these inputs when required? Schedule risks, although numerous, are subject to analysis and can, in a great many cases, be accurately predicted.

Cost risk is often experienced when not enough effort has gone into the early cost-estimation processes. "Guesses" are accepted as hard data and not discovered as incorrect until the situation is investigated. For example, in the sample costing shown in Table 3.4, software costs are estimated to be $8,000, representing about 13% of the total project cost. If these costs turn out to be $12,000 instead of the original $8,000 estimate, an increase in cost of $4,600 would be experienced (the $4,000 increase plus the G&A of $600). There is very little room in this illustrative project to make up a cost increase of this amount.

Risk to the project may be present when the overhead and G&A rates in the enterprise are not stable. If these rates increase in the middle of the project, they will impact total cost even though the PM may be doing everything correctly. In such a case, the PM is somewhat "off the hook," but may still be asked to try to make up for these cost increases. This type of risk may be called an administrative risk, but its effect results in a cost risk. Another type of administrative cost risk involves the failure of the company to hire on time for the project, or the unexpected loss of a key person to another firm halfway through the project.

As suggested by this discussion, there are often many risks to the success of the project and it behooves the project triumvirate to attempt to anticipate these risks and establish risk-mitigation strategies. Therefore, the risk-analysis portion of the project plan would consider the previous four types of risks in terms of the following questions:

1. What specific risks are present for this project?
2. What are the likelihoods of experiencing these risks?
3. What are the likely consequences if indeed the risks occur?
4. Based on 2 and 3, how can we prioritize the risks that have been identified?
5. What can we do to minimize the likelihood of occurrence as well as the consequences of high-priority risks?

The answers to the last question represent the risk-mitigation strategies.

3.9   THE PROPOSAL

The Project Manager is often faced with the matter of writing a proposal in order to have the opportunity to be awarded the project contract. In such a case, it is recommended that the previous project plan be constructed, in rough form, as a precursor to the actual proposal-writing process. Thus, the project plan becomes a critical input to proposal preparation because it deals with most of the crucial issues.

The format of the proposal is very often structured as follows:

1. A technical proposal
2. A cost proposal
3. A management proposal

It is very important that the proposal manager, often but not always the proposed Project Manager, follow the request-for-proposal (RFP) instructions to the letter in order to score as high as possible. The bases for evaluating proposals are usually described in a portion of the RFP citing the "evaluation criteria."

Writing a high-quality and winning proposal is a complex matter that is likely to be a part of the careers of the Project Manager, the Chief Systems Engineer, and the Project Controller. Some of the rules and vagaries of developing high-quality proposals are examined in greater detail in Chapter 6.

3.10   SEMP AND SEP

The project plan, discussed in some detail at the beginning of this chapter, is a generic document that describes the seven elements of an overview plan at the project level. Some attention is paid to technical matters of the project, but the systems engineering approach is not the main focus of the plan at the project level. One might say, however, that the Chief Systems Engineer (CSE) must take on the challenge of producing, with the approval of the Project Manager (PM), a technical systems engineering plan. In the remainder of this chapter we look at two related plans, one called the SEMP (systems engineering management plan) and the other is the SEP (systems engineering plan).

3.10.1   SEMP

In Exhibit 2.4 of Chapter 2 we cited the key elements of the SEMP, from the perspective of the Department of Defense (DoD). In Exhibit 3.8, we list a DoD and a NASA view of the SEMP. In this format we are able to look briefly at the similarities and differences between these two views.

Exhibit 3.8 DoD and NASA Views of a SEMP

a. DoD View (see Exhibit 2.4)
  1. Systems Engineering Process
  2. Systems Analysis and Control
    2.1 Systems Analysis
    2.2 Technical Performance Measurement (TPM)
    2.3 Technical Reviews
  3. Technology Transition
  4. Technical Integration Teams
b. NASA View [3.13]
  1. Purpose and Scope
  2. Applicable Documents
  3. Technical Summary
  4. Technical Effort Integration
  5. Common Technical Processes Implementation
  6. Technology Insertion
  7. Additional SE (systems engineering) Functions and Activities
  8. Integration with the Project Plan Resource Allocation
  9. Waivers
  10. Appendices

The DoD and the NASA views are by no means the same, nor are they entirely different. Both emphasize *process* and, to that extent, are in line with many project and systems engineering notions. For example, the DoD leads off with an overview of the systems engineering process, and NASA looks at the (seventeen-element) set of common technical processes and their implementation. In Chapter 2 we cited the ISO/IEC 15288 standard that also emphasized processes (twenty-five of them). Getting the process correct is now an important ingredient in achieving success in our program endeavors.

Both the DoD and NASA also deal with technology issues in an explicit way in their SEMPs. The DoD calls it "Technology Transition" whereas NASA refers to it as "Technology Insertion." As we will see in Chapter 12, our approach to system acquisition emphasizes technology and its role in building better systems. This is especially important as new technologies seem to be appearing at faster rates and are crucial in order to be competitive in both military and commercial worlds.

NASA also emphasizes technical effort integration, recognizing that this is a critical part of how to approach systems engineering and project management. It also leaves room for additional systems engineering functions and activities. On the DoD side, special attention is paid to technical integration teams. The full flavor of the DoD and the NASA approaches can be gained only by reading, in some detail, the full documentation provided by these two agencies with respect to a SEMP.

3.10.2 SEP

The DoD, in 2004, continued down the road of explaining its needs in terms of a SEP. In the policy statement for systems engineering, the office of the Under Secretary of Defense declared that programs will have an SEP such that each plan [3.14]:

> [S]hall describe the program's overall technical approach, including processes, resources, metrics, and applicable performance incentives. It shall also detail the timing, conduct and success criteria of technical reviews.

In a SEP preparation guide [3.15], eight points were articulated:

1. The SEP is the blueprint for the conduct, management, and control of the technical aspects of an acquisition program.
2. The SEP defines the methods by which system requirements, technical staffing, and technical management are to be implemented.
3. A sound systems engineering strategy needs to be defined.
4. The SEP shall be updated continuously.
5. Linkages are to be established between other technical and programmatic efforts (e.g., test and evaluation, risk management, etc.).
6. The SEP should be tailored to the specific needs of the individual program or project.
7. Technical questions need to be set forth and answered, such as: What are the key technical issues, and how are these issues to be solved and managed?
8. The SEP shall be submitted for approval at each major program milestone.

There is a preferred format (which can be tailored) for the SEP. That format is reproduced here [3.15] as Exhibit 3.9.

Exhibit 3.9 Preferred Format for DoD's SEP

1. Title and Coordination Pages
2. Table of Contents
3. Introduction
   3.1 Program Description and Applicable Documents
   3.2 Program Status as of Date of this SEP
   3.3 Approach for SEP Updates

4. Systems Engineering Application to Life Cycle Phases
    4.1 System Capabilities, Requirements, and Design Considerations
        • Capabilities to Be Achieved
        • Key Performance Parameters
        • Certification Requirements
        • Design Considerations
    4.2 SE Organizational Integration
        • Organization of IPTs (Integrated Product Teams)
        • Organizational Responsibilities
        • Integration of SE into Program IPTs
        • Technical Staffing and Hiring Plans
    4.3 Systems Engineering Process
        • Process Selection
        • Process Improvement
        • Tools and Resources
        • Approach for Trades
    4.4 Technical Management and Control
        • Technical Baseline Management and Control (Strategy and Approach)
        • Technical Review Plan (Strategy and Approach)
    4.5 Integration with Other Program Management Control Efforts
        • Acquisition Strategy
        • Risk Management
        • Integrated Master Plan
        • Earned Value Management
        • Contract Management

By comparing Exhibits 3.8 and 3.9, we can see how DoD thinking may have changed in proceeding from the SEMP to the SEP. Such a comparison is posed as question 3.10 below.

QUESTIONS/EXERCISES

3.1 Identify additional elements you might add to a project plan. Explain why.

3.2 Compare the work breakdown structure (WBS) in Figure 3.1 with the project structure in Figure 1.2. Should the latter be more or less the same as the former? Why?

3.3 Cite three features of the PERT approach that are not present for Gantt charting. When and why are these features significant?

3.4 Calculate the fully loaded hourly rate for each of the four labor categories in Table 3.4. What is the average hourly rate for the entire project?

3.5 From Table 3.5, enter the person-week and cost expenditures for each week into a spreadsheet and print out graphs showing:
  a. expenditures each week
  b. cumulative expenditures by week

3.6 Identify and discuss three areas of significant cost risk. What might be done to mitigate these risks?

3.7 Identify and discuss three areas of significant for schedule risk. What might be done to mitigate these risks?

3.8 Cite and discuss three suggestions for writing an outstanding proposal.

3.9 Define a numerical approach to evaluating a proposal. Show by example how a given proposal might receive a set of evaluation scores. See if you can compare your approach to that taken by a government agency on a real procurement.

3.10 In this chapter we see how the Department of Defense approached both a systems engineering management plan and a system engineering plan. Write a two-page discussion of similarities and differences between these two plans.

REFERENCES

3.1 *Defense Acquisition*, DoD Directive 5000.1 (1991). Washington, DC: U.S. Department of Defense.

3.2 Frame, J. D. (1987). *Managing Projects in Organizations*. San Francisco: Jossey-Bass.

3.3 *Mission Oriented Information System Engineering (MOISE) Solicitation* (1992). Washington, DC: U.S. Coast Guard, Department of Transportation.

3.4 *Technical Assistance (TAG) Request for Information (RFI)* (1993). Washington, DC: Federal Aviation Administration, Department of Transportation.

3.5 *Management Training and Organizational Development Support Services Solicitation* (1993). Washington, DC: National Aeronautics and Space Administration, Headquarters Acquisition Division.

3.6 *Curriculum Support for the Defense Systems Management College (Solicitation)*. Fort Belvoir, VA: DSMC: Directorate of Contracting, Contracting Division.

3.7 *Specialized Quick Response Support*, Solicitation DTRS-57-93-R-00047 (1993). Washington, DC: Department of Transportation.

3.8 *Earth Observing System Data and Information System (EOSDIS), Phase B Study Statement of Work {SOW)* (1988). Greenbelt, MD: NASA, Goddard Space Flight Center.

3.9 Eisner, H. (1988). *Computer-Aided Systems Engineering*. Englewood, NJ: Prentice Hall, Chapter 13.

3.10 Marciniak, J. J., and D. J. Reifer (1990). *Software Acquisition Management*, New York: John Wiley, Chapter 6.

3.11 Charette, R. N. (1989). *Software Engineering Risk Analysis and Management*. New York: McGraw-Hill.

3.12 Henley, E. J., and H. Kumamoto (1981). *Reliability Engineering and Risk Assessment*. Englewood, NJ: Prentice Hall.

3.13 *NASA Systems Engineering Handbook* (1995). SP-610S. Washington, DC: NASA, June.

3.14 *Policy for Systems Engineering in DoD* (2004). Washington, DC: U.S. Department of Defense, February 20.

3.15 *Systems Engineering Plan (SEP) Preparation Guide*, Version 0.95 (2004). Washington, DC: U.S. Department of Defense, December.

4

SCHEDULE, COST, AND SITUATION ANALYSIS

4.1 INTRODUCTION

This chapter deals mainly with the monitoring function of the Project Manager (PM), the Chief Systems Engineer (CSE), and the Project Controller (PC), as previously depicted in Figure 2.1 in Chapter 2. A key element of this function is to continuously review and analyze the status of the project with respect to schedule, cost, and performance. In this chapter, we focus on schedule and cost, leaving technical performance to the systems engineering chapters. As shown in Figure 2.1, once a problem area is discovered, feedback is provided to the project team, which, together with the project triumvirate, will analyze the issue in greater depth in order to define and implement corrective action.

Monitoring of a project is often used synonymously with the word "control." Indeed, the PM does in fact attempt to control the project, steering it through times of difficulty by continuous measurement and corrective action. Thus, in several other texts on the subject of project management [4.1, 4.2, 4.3], the reader will find many of the topics of this chapter listed under the heading of project control.

The last subject of this chapter deals with the topic of "situation analysis." This is a procedure, defined herein, to be used both in real-world situations as well as in teaching the subject of project management. In effect, it is a minicase study of a situation that might arise on any given project. An example of how this procedure may be applied completes this chapter.

4.2 SCHEDULE ANALYSIS AND MONITORING

Matters relating to the project schedule were presented briefly in the last chapter in discussing the project plan. Every project plan must have a schedule, although such a schedule is generally an overview of the project, with a detailed schedule included perhaps as an appendix. In this chapter, we examine some of the "nuts and bolts" of project scheduling, exploring particularly the characteristics of the PERT (program evaluation and review technique) approach.

PERT is the preferred scheduling procedure for a large-scale system in which there are large numbers of events and activities that must be identified and tracked. This technique, in distinction to Gantt charting, deals explicitly with dependencies between the various tasks and activities. A PERT network is normally devised by starting with known "end" events and milestones and asking the question: What activities need to be accomplished before this event or milestone can be achieved? By working backward in this fashion, eventually an entire network is developed.

The PERT procedure leads to a network of serial and parallel paths of events and activities. A simple example of such a network was shown in Figure 3.3 of the previous chapter. We use this example here to examine the network itself as well as some of the data that are required to formulate the network.

We first work backward from the end event (number 8) and redraw the network so as to identify the critical path, which is the longest path through the network. This path, by definition, has no slack in it, and slippage along this path leads to slippage in the project end date unless corrections are made. The redrawn network, based on Figure 3.3, is shown here as Figure 4.1. We note that the critical path now consists of the following sequence of events:

$$\text{Critical Path} = 0\text{--}1\text{--}4\text{--}5\text{--}7\text{--}8$$

In this example, this path is 8 weeks long. We note that slack exists in various subpaths:

Subpath 0–3–4 has slack of 1 week
Subpath 0–2–1 has slack of 1 week
Subpath 4–7 has slack of 1 week
Subpath 5–6–7 has slack of 1 week

Using the convention that, where slack exists, all activities start as late as possible, we have the network in Figure 4.1. If we used the convention that, where slack exists, all activities will start as early as possible, the network would change by events 2, 3, and 6, all moved to 1 week earlier. If we compare Figures 3.3 and 4.1, we see that in the latter, event 6 has been moved 1 week

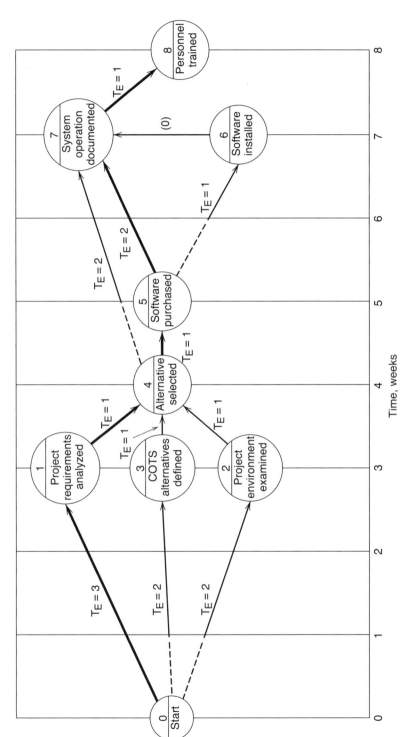

Figure 4.1. Illustrative project PERT chart.

to the right to reflect the convention adopted here of starting an activity as late as possible.

The basis for determining the critical path is the set of time estimates for the various activities in that path. These time estimates are designated as "expected times" (T_E) for these activities. In the original PERT procedure [4.4], expected times were derived from three time estimates for each activity:

- An optimistic time (T_O)
- A most likely time (T_L)
- A pessimistic time (T_P)

Under an assumption of a "beta" distribution for the activity times, the expected time for each activity is calculated as

$$T_E = \frac{T_O + 4T_L + T_P}{6}$$

Some projects still use this three-time estimate procedure in order to develop expected value estimates, but many do not. In the latter case, the expected times are estimated directly and associated with the various project activities, as in Figure 4.1. It is also possible, given the three time estimates, to calculate the activity time variance utilizing the following relationship:

$$\sigma^2 = \left[\frac{T_P - T_O}{6} \right]^2$$

The basic purpose of this calculation is to sum the variances along the critical path and thereby calculate the variance associated with the project end date. This estimate (or its square root, the standard deviation) yields some measure of the "uncertainty" of this end date. For those who wish to explore uncertainty in this fashion, the preceding relationships are available. For the remainder of this discussion, we limit our scope to the use of expected values of activity times.

Analysis of a schedule, given the basic network and the time estimates, starts with the determination of the critical path. If we are able to shorten the critical path, then it may be possible to reduce the overall project schedule. Thus, each activity along the critical path should be reexamined to see if reductions in time are feasible. This is especially important if, in a project that has missed earlier milestones on the critical path, the Project Manager is looking for ways to get back on schedule. If the reduction of activity times along the critical path leads to another path becoming critical, then the same procedure would apply to the new critical path.

The next aspect of analyzing a network schedule is the examination of slack along noncritical subpaths so as to establish where the slack is to be placed. The Project Manager has that prerogative, and the decisions in this regard are deceptively complex. The PM can use the conventions discussed before or can attempt to base a decision on personnel availability. If personnel are available, the PM may elect to start as soon as possible; if they are not available, the PM will likely start an activity when they become available as long as the critical path is not increased. Thus, there is an interaction between the schedule network and the availability of personnel. In general, the PM tries to "level" the resources across a project as long as such leveling does not force an increase in the overall critical path.

Another aspect of the interaction between personnel assignments and schedule is the obvious fact that, within limits, it may be possible to reduce activity times by adding resources. Thus, we have the PERT "anomaly" that allows one to estimate activity times without explicit consideration of the resources that will be applied to these activities. Working back and forth between activity times and personnel assignments can be a complex analysis task. It may be facilitated by using project management software, as discussed in Chapter 12. In general, considerable effort may be necessary to "optimize" a complex project schedule, that is, to derive a schedule that meets internal and customer requirements and also makes the best possible use of personnel and other project resources (e.g., facilities). Best use includes minimizing personnel down time (maximizing personnel utilization).

A final project schedule should be based upon the foregoing considerations and the Project Controller is often the person who attempts to optimize this schedule with respect to both time and utilization of resources. Unfortunately, this process is normally repeated numerous times as the inevitable changes and slippages occur. Most projects require juggling and rescheduling as the differences between reality and plan begin to surface. For this reason, a good PC who is able to reschedule quickly and efficiently can have a major influence on the success, or lack of it, of a complicated project.

We pause here to comment upon a subject that does not generally receive enough attention in the world of schedule and cost analysis and tracking. That subject is the manner in which input time and dollar estimates are provided. The GIGO (garbage-in, garbage-out) principle clearly suggests that we can be in serious difficulty if we are not careful about how we obtain input data.

Input estimates, of course, are provided as the first order of business as we prepare schedules and budgets. Once a project is up and running, an information system provides automated reports on current status, but we still require new inputs, at least every month, on time and cost to complete. Treating these inputs in a cursory manner can be a disaster waiting to happen.

Following one simple rule can help us solve a host of problems with respect to poor input estimates. That rule is that we require *multiple independent* estimates in all areas in which we are likely to run into problems. A more conservative approach is simply to apply that rule for all schedule and cost estimates. Following the rule means that we obtain inputs from more than one person and, further, that the folks who provide the inputs are required to not discuss the matter with the other estimators before they make their estimates. A simple "rule of 2 or 3" is that two people are asked for inputs for most situations, and three folks provide independent inputs for very important or critical situations.

For example, if we are approaching the last 3 months of a one-year project being carried out on a firm fixed price (FFP) contract, we might consider invoking the "rule of 3" for time and cost to complete estimating. If we get similar results from the 3 estimators, we can have considerable confidence in these inputs. If we get disparate inputs, it's time to call a meeting to resolve the disparity. The objective, of course, is to zero in on the best input data we can produce as we move into the final stretch of an important project. Another example is that of estimating the cost and schedule of a software development project. Chapter 10 will define the COCOMO (Constructive Cost Model) method for providing these cost and schedule estimates, both of which are seriously dependent upon estimates of delivered source instructions (DSI). Therefore, if the DSI estimates are incorrect at the beginning, much trouble is in store for the project. Applying the "rule of 3" for the initial (as well as updated) estimates is likely to pay great dividends during the life of the project.

4.3   COST ANALYSIS AND MONITORING

As suggested in the previous chapter dealing with the project plan, direct labor costs are developed from assigning people to the various activities (tasks) of the schedule. This is often the key element of cost for a project and is achieved by first constructing a task responsibility matrix (Table 3.3) and then a project budget (Table 3.4). In order to track costs as a function of time, a cost budget by week is also developed (Table 3.5). At times, and certainly if required by the customer, the costs for each time period are also broken down by task and subtask. Thus, it is possible for the periodic cost reports for a large project to become quite complex and voluminous.

4.3.1   Cost Monitoring

Bottom-line monitoring of project costs can be carried out by the sample cost report shown in Table 4.1. This cost report is produced at the end of the fourth week in the project. The table shown is based on the bottom-line budgeted costs derived in Table 3.5. Lines 1 and 2 show these budgeted costs for each of the 8 weeks of the hypothetical project. Line 1 is the cost budgeted by week and line 2 carries forth the weekly cumulative budget. The items from the project plan, when accepted, become the budgeted cost numbers against which actual costs are compared. Lines 3 and 4 in Table 4.1 show the actual costs, by week and cumulative. In general, of course, these may be less

TABLE 4.1 Bottom-Line Cost Monitoring: Week 4 of 8

	End of Week								Line
	1	2	3	4	5	6	7	8	
Budget, by Week*	2,541	8,641	8,641	4,321	17,720	5,592	7,116	5,846	1
Budget, Cumulative*	2,541	11,182	19,823	24,144	41,864	47,456	54,572	60,418	2
Actuals, by Week	2,600	8,800	9,000	5,000	—	—	—	—	3
Actuals, Cumulative	2,600	11,400	20,400	25,400	—	—	—	—	4
Budget Less Actuals, by Week	(59)	(159)	(359)	(679)	—	—	—	—	5
Budget Less Actuals, Cumulative	(59)	(218)	(577)	(1,256)	—	—	—	—	6
Percent Deviation, by Week	(3.5)	(1.8)	(4.2)	(15.7)	—	—	—	—	7
Percent Deviation, Cumulative	(3.5)	(1.9)	(2.9)	(5.2)	—	—	—	—	8

*See Table 3.5.

or greater than the budgeted costs. By subtracting the actual costs from the budgeted costs, we are able to produce lines 5 and 6 of the cost report. These data show that actuals have been greater than budget numbers for each and every week of the four weeks of the project. By the end of the fourth week, a total overexpenditure of $1,256 has occurred. The convention adopted here is that overexpenditures are shown with parentheses around them. Finally, lines 7 and 8 convert the deviations to percentages on the basis of budgeted numbers. The sample cost report numbers show a jump during the fourth week to 15.7% and a cumulative percent deviation of 5.2%. This degree of overexpenditure is normally considered significant.

The sample bottom-line cost report signals a problem in that

1. There has been an overexpenditure every week.
2. In the last week (week 4), the percent overexpenditure has jumped to 15.7%.
3. The overall project is now 5.2% overspent.

These observations would normally lead to the following two questions:

1. Where have the overexpenditures occurred?
2. Why have the overexpenditures occurred?

The answer to the former question is usually found by looking at more detailed cost reports, the first of which would be a report similar to Table 3.5, but with the addition of actual costs. By examination of such a report, it is possible to track the cost deviations back to the potential sources, namely:

1. Direct labor costs
2. Fringe costs
3. Overhead costs
4. Other direct costs (e.g., software, training materials, etc.)
5. General and administrative (G&A) costs

Most likely, but not always, deviations occur in the application of direct labor. It is then possible to investigate why more hours were required in relation to the original plan and budget.

Another type of cost report may be provided that indicates costs by task or by work breakdown structure (WBS) element. For such cases, these reports allow tracking back to cost deviations at the task and WBS levels. Ultimately, reasons for why these deviations occurred have to be explored with project personnel closest to the task and work elements.

In at least some cases, the customer designates the form of cost status information that is required. As an example, Table 4.2, drawn from a real-world procurement, shows a cost status report format that tabulates both hours and cost information for the period in question, cumulative sums for that period, cumulative sums since contract inception, authorized hour and cost data, and remaining hours and cost listings. In such a situation, the organization under contract is obliged to provide this type of information to the customer on a periodic basis. Thus, the issue of whether internal hour and cost reports provide such information becomes important for the PM. Ideally, the accounting/finance department can make appropriate adjustments to make such information available in an automated fashion. If this is not possible, then it becomes necessary for the Project Manager, usually in conjunction with the Project Controller, to determine how to provide the needed customer report, augmenting the "standard" internally generated cost reports. Translation of internal reports to necessary external reports thus becomes a problem for the PM. Such specialized customer reports usually have to be signed off as well by other company personnel before they are sent to the customer.

The essence of cost tracking lies in the periodic (weekly, monthly) examination of the foregoing types of cost reports. Budget and actual cost numbers are arrayed as a function of time, element of cost, task and subtask, and WBS element. Analysis and further questioning as to deviations reveal both where the problems are occurring and also why there appear to be cost problems. Minor deviations are often noted but not investigated. Larger deviations (reaching levels of 5% or more) are usually triggers for a detailed examination of sources and reasons.

For many projects, it is standard operating procedure, at the end of each reporting period, to reestimate time and cost to complete. With this new input of data, overall project times and costs are projected and compared to budgets. Where unacceptable end results are projected, a corrective action is triggered. This includes revisions of the original plan for both schedule and cost, resulting in new budgeted values. In short, where there is a schedule or cost problem, future periods have to be examined to see where time or dollars can be reduced to get back on schedule and perform within overall project cost budgets. There is a procedure that assesses the current schedule and cost situation and projects project end results as a function of performance to date. This procedure, called earned value analysis (EVA), is described in what follows.

4.3.2 Earned Value Analysis (EVA)

Basic Relationships.   Earned value analysis (EVA) is a formal procedure for estimating cost and schedule variances during a project and extrapolating these variances to the end of the project. The word *variance* is interpreted as a deviation or difference in distinction to a mean square error as in the field of statistics.

TABLE 4.2 Cost Status Report

Task/Subtask Areas (Col. 1)	Actual Expended		Cumulative for Base/Option Period		Cumulative Since Start of the Contract		Authorized Hours (Col. 8)	Authorized Cost (Col. 9)	Remaining Hours (Col. 10)	Remaining Cost (Col. 11)
	From	To								
	Hours (Col. 2)	Dollars (Col. 3)	Hours (Col. 4)	Dollars (Col. 5)	Hours (Col. 6)	Dollars (Col. 7)				

Total Contract

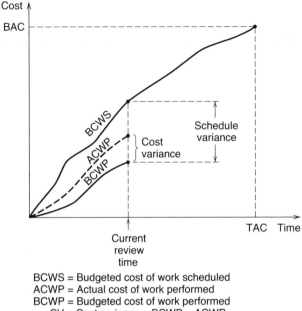

BCWS = Budgeted cost of work scheduled
ACWP = Actual cost of work performed
BCWP = Budgeted cost of work performed
CV = Cost variance = BCWP − ACWP
SV = Schedule variance = BCWP − BCWS

Figure 4.2. Earned value analysis (EVA) terminology.

An overview of the EVA concept can be gleaned from Figure 4.2. Three cumulative cost curves are depicted, each flowing from the project initiation time to the current reporting time. These cost curves are

1. Budgeted cost of work scheduled (BCWS)
2. Budgeted cost for work performed (BCWP)
3. Actual cost of work performed (ACWP)

The EVA concept specifically accounts for the degree to which work that has been scheduled has also been accomplished. In that sense, it does more than simply compare budgeted versus actual costs without regard for the extent to which work has been executed. For example, a project can be at month nine of a 10-month period and also have spent 90% of the budget. In that simple sense, both time and cost are tracking until one realizes that perhaps only 50% of the work may have been accomplished. Many a naive Project Manager has been caught in this trap by not considering the work progress in relation to schedule and budget.

By comparing the actual versus budgeted cost of work performed (ACWP vs. BCWP), at each point in time, we have a true measure of the "cost variance":

$$\text{Cost variance (CV)} = \text{BCWP} - \text{ACWP}$$

In this context, both budgeted and actual costs are computed on the same basis, namely, the work that has been performed. Therefore, the PM and PC ask the question: How much work has been performed (i.e., which tasks or WBS elements have we actually accomplished) at this point in time? The budgeted cost for these tasks/WBS elements is then calculated (BCWP) and compared against actual expenditures for these same tasks/WBS elements (ACWP). If the BCWP is greater than the ACWP, we have underspent; if the BCWP is less than the ACWP, we have a cost overrun. The definition of cost variance, as shown before, will yield positive numbers if we are underspent and negative values if we have overspent. Thus, a negative cost variance indicates a problem. Figure 4.2 shows a negative value for the cost variance and therefore reveals an issue that must be further investigated.

Perhaps a more difficult concept is the meaning of the discrepancy between budgeted cost of work performed (BCWP) and the budgeted cost for work scheduled (BCWS), both of which are shown in Figure 4.2. The difference between these two is defined as the "schedule variance":

$$\text{Schedule variance (SV)} = \text{BCWP} - \text{BCWS}$$

Here it is recognized that the work scheduled and the work performed, at each point in time, may be different. As an example, halfway through the project in time, we may have scheduled to finish fifty WBS elements but actually have completed only forty WBS elements. We budgeted \$100,000 for the fifty elements and \$80,000 for the forty elements. Therefore, the schedule variance is

$$\text{SV} = \$80,000 - \$100,000 = -\$20,000$$

We note that the schedule variance, for the EVA construct, is measured in *dollars*, not time. Clearly, at this point, by completing only forty of the fifty planned WBS elements, we are behind schedule. Our actual and planned rate of completing work elements is the same, namely, \$2,000 per work element. For some reason, possibly because we did not staff the project as quickly as our plan called for, we are some ten work elements behind. In principle, we are not overrun in cost, but lag in the rate at which we have been able to get the work done. This is basically a schedule issue. Further, this lag in time shows up in the negative value for the schedule variance.

The estimates of BCWP, BCWS, and ACWP also allow us to carry out a linear extrapolation as to the estimated cost at completion (ECAC) and the estimated time at completion (ETAC). This can be found through the following relationships:

$$\text{ECAC} = \frac{\text{ACWP}}{\text{BCWP}} \times \text{BAC}$$

$$\text{ETAC} = \frac{\text{BCWS}}{\text{BCWP}} \times \text{TAC}$$

where BAC is the original budget at completion, and TAG is the original time to completion. The BAC is either increased or decreased as it is multiplied by ACWP/BCWP. If ACWP is greater than BCWP, then the budget at completion (BAC) is augmented, representing a linear extrapolation of the current cost overrun to the end of the project. It must be recognized that this is only an extrapolation and is not based on a detailed analysis of the reasons for the current overrun condition.

Similarly, the estimated time at completion (ETAC) is determined by multiplying the original time at completion (TAG) by BCWS/BCWP. If BCWS > BCWP, then ETAC will be greater than TAG. This, too, is a linear extrapolation, but in this case in the time dimension. The PM and the PC are urged to look more deeply into schedule and work performance issues and problems before accepting the new ETAC as a fully accurate representation of the project schedule status.

Illustrative Example of an EVA.   An example of the results of an EVA can be posed by the following situation:

> As a PM, you are at the 18-month point of a 24-month project, with a $400,000 budget. Your original project plan and a review of work performed reveal that BCWS = $300,000, ACWP = $310,000, and BCWP = $280,000. What are your current estimates of the cost variance (CV), schedule variance (SV), cost at completion (ECAC), and time at completion (ETAC)?

This example is depicted in Figure 4.3. From the given data, we calculate the cost and schedule variances as

$$CV = BCWP - ACWP = \$280,000 - \$310,000 = -\$30,000$$
$$SV = BCWP - BCWS = \$280,000 - \$300,000 = -\$20,000$$

The estimated cost and time at completion are

$$ECAC = \frac{ACWP}{BCWP} \times BAC = \frac{310,000}{280,000} \times 400,000 = \$442,857$$

$$ETAC = \frac{BCWS}{BCWP} \times TAC = \frac{300,000}{280,000} \times 24 = 25.7 \, \text{months}$$

Thus, the original budget of $400,000 is now reestimated to be $442,857 and the new time at completion is estimated to be 25.7 months instead of the original 24 months. The example indicates overruns in both cost and schedule and suggests a more definitive analysis to determine status and what can be done to meet the original budget and schedule.

We note here the earlier comment in this chapter regarding the procedure to reestimate time and cost to complete. The EVA process leads automatically to

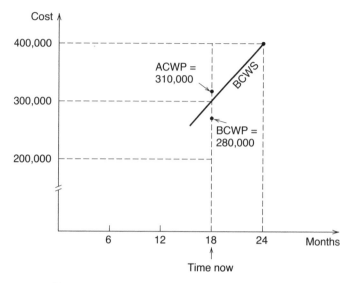

Figure 4.3. Example of earned value analysis (EVA).

such estimates, although they are linear extrapolations of the current situation. In the preceding EVA example, these estimates are

Estimated cost to complete = \$442,857 − \$310,000 = \$132,857
Estimated time to complete = 25.7 months − 18 months = 7.7 months

If the EVA indicates the existence of a problem, as does the preceding example, it is suggested that further detailed and project-specific estimates of cost and time to complete be made, leading, we hope, to necessary corrective actions.

4.3.3   Other Cost Considerations

Monitoring Other Direct Costs (ODCs).   The overall project budget shown in Table 3.4 shows other direct costs (ODCs) as a separate category of costs, which may include such items as travel, computer services, equipment, consultants, subcontractors, mailing, reproduction, telephone, materials, software, and other types of costs. Many of these costs come in late and therefore lag the normal reporting cycle times. The PM must be aware of these costs and commitments and make sure that they are not lost in the reckoning of the project cost picture. Many PMs have been surprised by late inputs of these types of costs simply because they were forgotten or lost in company processing. The PM should assign this tracking responsibility to the PC so that these costs do not appear as a late and not very welcome surprise.

A particular type of ODC requiring special attention is subcontracting. If a PM is in a position where subcontracting is a major part of the project, or critical path events depend on the delivery of a subcontract product or service, then unique steps may need to be taken to assure there is no victimization by the subcontractor. Some very large projects have dozens of subcontractors, thus increasing manyfold the likelihood of a significant problem. Some actions for a PM under these circumstances include:

1. Placing project personnel at the subcontractor's facility to monitor status and progress
2. Establishing interface control and documentation as a more prominent aspect of the systems engineering effort
3. Holding more frequent status review sessions for subcontractors
4. Meeting with the management of the subcontractors to obtain commitment to cost, schedule, and performance requirements
5. Providing parallel developments and backup sources as insurance, if they can be afforded
6. Using incentive award contracts for on-time, high-quality deliveries

Monitoring for Different Contract Types.   The way in which information is aggregated and reported is also related to the type of contract under which the project is being carried out. We discuss some of the vagaries of monitoring for three generic contract types:

1. Cost contracts
2. Fixed-price contracts
3. Time and materials (T&M) contracts

Cost contracts include cost-plus-fixed-fee (CPFF), cost-plus-incentive-fee (CPIF), cost-plus-award-fee (CPAF) contracts and variations on this basic theme. All such contracts mean that the customer pays the basic costs of the contract and the fee can be fixed or variable. Such contracts are prevalent in the world of government contracting and are almost never used in the commercial arena. Under an arrangement where under most conditions all costs are covered and guaranteed, there is sometimes not a strong incentive for a company to control costs to the budgeted numbers. However, it is strongly recommended that the PM adopt a point of view that such control is mandatory. It is generally not a good idea to lose the discipline of cost control, even when there is not a strong penalty for overrunning a contract. Cost reports for a cost contract are precisely those that have been shown in this chapter. Each element of cost is monitored and tracked, and corrective action is taken whenever actual costs begin to exceed budgeted values.

Incentive- and award-fee cost contracts are recommended in order for the system acquisition agent (customer) to make sure the contractor focuses on meeting the cost, schedule, and performance requirements. Incentive- and award-fee parameters can be defined so as to reward contractors for emphasizing the items most important to the customer. Experience has shown that these types of contracts are quite effective in motivating contractors. Explicit evaluations and scoring by the customer also provide periodic feedback to the contractor so that the positions and issues of both parties are known as the contract proceeds. A PM who is not getting good evaluation scores is likely not to be achieving fee (profit) goals. This gets the immediate attention of both the PM and management.

Fixed-price contracts basically mean that the contractor works on the contract until all requirements and specifications are satisfied. Costs in excess of the original budget are borne by the contractor. Thus, if budgeted costs are exceeded, profit dollars are jeopardized. Such contract forms are utilized almost exclusively in the commercial world, and increasingly in the government arena.

A PM working under a fixed-price contract should be aware that every dollar "saved" is one that could be added to the profit made under that contract. There is a tendency, therefore, for such contracts to be monitored extremely carefully, always looking for a better solution that will satisfy requirements within cost and schedule constraints. It should also be noted that, at times, such contracts have penalty clauses. A typical clause of this type penalizes the company for late delivery of the product. In this fashion, the customer makes it clear that meeting schedule is a very important issue, and failure to do so may force both the contractor and the customer to experience increased costs.

Cost reports for fixed-price contracts can take the same form illustrated in this chapter, as each element of cost is tracked on a periodic basis. A PM may wish to shorten the periodicity of such reports, such as getting a weekly reading of costs instead of the more usual monthly report. This can place a strain on the company accounting system, which may not be geared to such rapid reporting. In such cases, PMs have been known to generate their own interim cost reports in order to satisfy their needs. This may be done by capturing weekly time charges on Friday afternoon and feeding them in to a spreadsheet developed by the Project Controller, so that by the Monday following the week in question, a weekly cost report is available. This type of special reporting is recommended for all contracts as they near their completion times when it may be necessary to exercise more stringent controls.

Both cost and fixed-price contracts may also include the requirement of a minimum delivery of hours. At times, a "window" of ±10% of the bid number of hours is established and placed into the contract document itself. Thus, if more experienced personnel than originally proposed are used on the work, it may turn out that the "−10%" requirement is not satisfied. This normally results in some type of fee penalty. Where there is a requirement for delivery of hours, the PM must also be monitoring hours expended, by category, to make sure that the hours are actually delivered as per the contract. In this regard, we note the "hours" columns in Table 4.2. "Hours expended" reports for such a situation must become part of the normal process of tracking projects.

A time and materials (T&M) contract is usually set up on the basis of a customer requirement that a certain number of hours be purchased, at a fixed rate by personnel category. The customer is thus buying expertise, by the hour, for various categories of personnel. The essence of the contract lies in the delivery of such expertise, and the customer paying a fixed amount for each hour delivered. Of course, the expertise must be sufficient to satisfy the customer's requirements. If it is not, the customer will insist upon a change of personnel. Materials called for under the contract (e.g., computers and COTS software) are delivered separately, with an agreed-on markup or at cost. The overall contract, of course, has a ceiling price that the customer has agreed to.

Monitoring a T&M contract normally requires cost reports different from those shown previously in this chapter. Not only does overall cost need to be tracked, but hours by category and associated costs must be monitored in detail. The basic reason is that the actual people assigned to work may be at the "high end" of their categories. Because bid costs by category are often constructed as an average of the people in a given category, high-end personnel cost more than the bid and contracted rate. This leads to losing money (spending nominal profits) for such categories. The more hours that are worked by such personnel, the more money is lost. A PM who is not aware of this possibility will be in for a significant shock when management declares that the contract may not have overrun total budget but has spent corporate profits.

A cost report that might be used to monitor a T&M contract lists the specific people, by name, together with their actual costs, and compares these costs, by hour, with the rates in the contract. When the contract rates are higher, a profit is achieved. When the contract rates are lower, the contract is losing money.

For example, assume that three senior engineers are working full time at hourly rates of $61.18, $67.35 and $64.82, with the contract rate for senior engineers being $65.50. This means that the project makes money each hour for two people (in the amounts of $4.32 and $0.68) and loses money each hour for the third person ($1.85). On a weekly basis, the reader can confirm that the total profit for the three engineers will be $126, but only for all three engineers working 40 hours each week.

It can be seen that considerable up-front planning is required in order to staff a T&M contract. It is not necessarily a terrible circumstance to be losing money on an individual, as long as, in the aggregate, the contract is making its nominal profit across all assigned persons. The mixture of people and rates, and the often dynamic nature of assignments over time, however, make it critical that all new assignments be considered on the basis of cost as well as capability to do the job. If such a contract is long-term, perhaps several years, it is standard practice for a company to escalate the rates from year to year in its bid so as to cover salary increases and thereby avoid the problems cited. Within certain boundaries, customers will accept year-to-year rate increases as a normal way of doing business. In all cases, the PM must be aware of the form of contract and the specific provisions of the contract that must be satisfied. For this reason, as a project plan is about to be prepared, it is recommended that the PM sit down with a "contracts" person and make sure that all the key contract requirements are understood.

Corporate Rate Changes. Many a Project Manager has been surprised when corporate rates have changed at midstream in a project, causing unexpected cost increases. Although these are usually not the responsibility of the PM, increases in corporate fringe, overhead, and G&A rates can force the PM to "make up" for the problem by finding ways to reduce future costs. For at least this reason, the PM should track these rates in cost reports and also try to keep abreast of overall company problems that might cause rate increases. This type of unexpected change also reinforces the prudent action of keeping reserves whenever possible. The cost elements that impact fringe, overhead, and G&A rates are shown in Exhibit 4.1.

Exhibit 4.1: Fringe, Overhead, and G&A Cost Elements

A. Fringe
   1. Sick Leave
   2. Holiday
   3. Vacation
   4. Severance
   5. Compensation Insurance
   6. Unemployment Insurance
   7. PICA Tax
   8. Group Insurance
   9. Travel Expenses
   10. Recruiting
   11. Training
   12. Employee Pension
B. Overhead and G&A
   1. Salaries and Wages
      1.1 Indirect Labor
      1.2 Other Compensation
      1.3 Overtime Premium

    1.4 Sick Leave
    1.5 Holiday
    1.6 Vacation
    1.7 Severance
2. Personnel Expenses (see "Fringe Items")
3. Supplies and Services
    3.1 General Operating
    3.2 Office and Printing
    3.3 Utilities
4. Fixed Charges
    4.1 Depreciation
    4.1 Equipment Rentals
5. Office Space Facility Rental/Mortgage

Reserves. It is strongly recommended that a PM keep a cost reserve, especially on fixed-price contracts. Suggested amounts are (1) 2 to 5% on cost contracts and (2) 8 to 12% on firm fixed-price contracts. The basic idea, of course, is to plan to perform the contract for an amount equal to the base contract value less the reserved amount. Selecting a precise reserve percentage, given the preceding ranges, should be based on the risk analysis and particular circumstances surrounding the project. However, on a cost-reimbursable contract, reserves can be given up as the project approaches completion unless there is a specific incentive fee for underspending the original cost budget.

It is also prudent for a PM to ascertain whether upper management has also taken reserves on the project. This will normally affect how much reserve the PM should set aside.

Cost Item Limitations and Trades. Some contracts contain explicit limits on certain cost items. For example, limits may be placed on travel, consultant services, and various categories of other direct costs (ODCs). Here again, the PM must be aware of these limitations and control these costs so as not to exceed the limits. This also raises the question as to whether the PM may "trade" one type of cost for another. To illustrate the point, suppose it turns out that only $6,000 is needed for consultant services when the original plan and budget estimated a requirement for $8,000. Unless the contract prohibits such action, the PM is normally free to move the $2,000 difference to, for example, the direct labor budget in order to obtain additional hourly work, if necessary. A good rule for the PM to follow, where actual expenditures differ from budgeted expenditures (and they usually do), is to check the contract provisions to assure that all changes are made in consonance with the terms and conditions (Ts and Cs) of the contract document.

Cost Graphs.   It is also recommended that the PM convert project cost report data to graphs of monthly and cumulative expenditures, compared to budget, for various categories of cost. This may be achieved by taking data received from accounting/finance and feeding them into a spreadsheet that has a graphing capability, which most of them do. The purpose is to be able more readily to ascertain cost slopes and trends. This is usually difficult to see with a large set of numbers in a tabular format.

Cost Reporting to Customers.   The customer usually requires some type of cost reporting on a periodic (e.g., monthly) basis. Exceptions might include fixed-price contracts. Such reports are usually prepared by accounting/finance, go to the contracts department, and from there are sent to the customer. Often, the PM and PC are out of the loop for this type of reporting. This is *not* a recommended practice. The PM and PC, as a minimum, should be given copies of all reports sent to the customer in satisfaction of contractual obligations. A more desirable practice is to assure that the PM and PC concur with the cost information that is being sent to the customer.

It is emphasized that all contractual cost reporting requirements should be meticulously satisfied and the PM should attempt to assure that this is achieved, even when company procedures do not keep him or her "in the loop." At the same time, the contracts department should list and check all contractual deliverables and contact the PM and PC to determine and confirm status. For example, some cost contracts require that an anticipated overrun in cost be flagged for the customer when expenditures reach 75 or 85% of the contract value. The PM and PC should be best qualified to determine whether or not such an overrun is expected. If they miss the flagging notification, the customer may not pay the overrun costs. A good contracts department tracks the situation and alerts the PM and PC as to the flagging requirement. Good communication between PM, contracts, accounting/finance, and the customer is thus seen as critical in order to achieve project success.

4.3.4   Special Focus on Firm Fixed-Price Bids and Contracts

We add here to the previous discussion of fixed-price contracts our "top dozen" list of actions to consider in order to reduce the risk associated with bidding and accepting such contracts; the possible actions include walking away from a particularly high risk project. This list is provided below as Exhibit 4.2. Further explanation of the items follows the list.

Exhibit 4.2: "Top Dozen" List for Firm Fixed-Price Bids/Contracts

1. Assure high performance team (HPT) with proven track record.
2. Construct a contingency plan.
3. Set aside reserves.

4. Perform no out-of-scope work.
5. Assure "chunking" of work with associated payments.
6. Establish clear, measurable acceptance criteria.
7. Limit software warranties.
8. Have limits on penalty clauses.
9. Incorporate the proposal by reference.
10. Establish special rewards for the project team.
11. Set up a cost information system.
12. Know when to walk away.

High Performance Team (HPT).   Since you are dealing with a firm fixed-price (FFP) effort, it is critical to use the very best team that you have. This normally means that the core team has worked together for the past three to five years and has proven that it consistently achieves high-level results. Adding new senior people to a proven HPT should be done carefully to make sure the new folks are team players. If not, act quickly to remove them before they're able to cause damage to the team's spirit and productivity. Never micromanage an HPT. Doing so invariably leads to negative consequences. If the team has a proven track record, as you should insist be the case, give them the job to do and manage largely by facilitating.

Contingency Plan.   Even with an HPT starting the project, you still need a contingency plan if the work is being done on a FFP basis. A contingency plan flows from the project risk analysis (see elements of a project plan in Chapter 3) in that it is focused upon the actions that might be taken if various risks actually occur. Although risks to a project can be quite numerous, one of the most serious is the loss of key personnel, for whatever reason. We used to ask the generic question—"What if your lead engineer gets hit by a truck?" That worst case scenario required the PM to think through how the project would be completed without experiencing a serious setback.

A contingency plan does not have to be a long document. It can be a one-pager listing the risks on one side and the plans to mitigate those risks on the other.

Reserves.   As referred to under Section 4.3.3, reserves are especially important in FFP contracts, with recommended amounts in the 8–12% range. The lower end of reserve is used when the project is one of many similar ones that have been carried out in the past. The 12% may be expanded to 15% when there are a few new challenges presented on the project. Such challenges should not be, for example, pushing the state-of-the-art. They should be well within what the project team believes is workable. Project reserves for FFP contracts should be under the control of the PM and signed off by the PM's boss. Other "pots of money" might be made available as reserves above the project level, for example, by the cognizant vice president. Knowing how to deal with reserves in a corporate enterprise context is an important element of overall risk management.

Out-of-Scope Work.   Out-of-Scope (OOS) work is all activity that is carried out in support of project goals and/or objectives that are not in the terms of reference of the contract. It is also any "gold-plating" that was not part of the original bid that was signed off by the customer. The bottom line with respect to OOS work is simply that it should not be permitted since it normally increases risk and spends potential profit dollars. Whether suggested by the internal team or by the customer, OOS work can be kept on a list for consideration in a future contractual effort (not the current one). If there is a question as to whether or not a piece of work is in- or out-of-scope, then a most serious meeting needs to take place with the most senior people of the project in attendance. If you and your customer disagree on this type of issue, an extensive and immediate meeting is necessary to resolve this conflict. Failing to do so usually leads to great difficulties down the road.

"Chunking" of Work.   Work on a project is normally defined in "chunks," whether they are elements of a work breakdown structure (WBS) or task statements that are part of a statement of work (SOW). The added ingredient here, however, is to connect the completion of these work elements to contractual requirements to get partial payments on an "as-you-go" basis. So, for example, if there are ten major and approximately equal effort tasks to be accomplished, it is highly desirable that you get paid one-tenth of the contract value as you complete each of these tasks. Under these circumstances, if there is a controversy at the end of the contract, you will have received 90% of the payments due under the total contract. This, of course, reduces the overall financial risk. It may be compared to the case in which there are no payments made until all ten tasks are completed. In this case, a controversy at the end has quite different dynamics as well as psychology. This overall strategy might well have the subtitle "progress or partial payments."

Acceptance Criteria.   Many FFP contracts give to the customer the exclusive right to determine whether or not a contractual deliverable is acceptable. In this situation, the customer may look at a deliverable and say "This isn't acceptable; go back and work on it until it is." Of course, this would normally lead to a request for specifics as to why it isn't acceptable. All of this hinges upon the existence of written acceptance criteria that form the basis for judging acceptability. If such criteria do not exist, then a fuzzy set of conditions is likely to prevail, and you may be asked to continue to work the problem indefinitely. This argues for establishing firm acceptance criteria for all deliverables, one at a time, or as a group. Limits on review times as well as number of deliverable iterations can often serve adequately when it is difficult to be definitive about acceptance criteria.

Software Warranties. Some customers will request software warranties. One form of such warranty is that the software be free from latent and patent defects. This should be viewed as a red flag since it may lead to (a) downstream rejection of the software product, even after it has initially been accepted, and (b) the invocation of penalty clauses as a result of defective software. A more useful construction is simply to agree to fix software (at no additional cost) found to be defective during the six month period after it has been accepted and paid for. This normally limits the liability in ways that should be acceptable. The potential cost of this type of warranty should be built into the price of the software in the initial bid, usually on an expected value basis. If a "worst case" cost scenario is built into the bid, it may be that the probability of winning such bids will sink to unacceptably low levels.

Penalty Clauses. Penalty clauses require the developer to step up to various types of penalties if the delivered product is not satisfactory. These types of clauses vary in their difficulty, and at times can be viewed as a double red flag. Penalties can be sought when the product is simply not delivered on time. Typically, there would be a charge for each day of delay in the delivery schedule. This can be quite onerous and should be avoided in most situations. A second type of penalty is associated with the discovery of product defects (e.g., bugs in the software). Here again, great care should be exercised in accepting this type of penalty clause. If the product is proven and mature, one can be less nervous about these clauses. Another alternative is to request extra (bonus) payments for early delivery, for example. This can provide offsets and incentives that work for both developer and customer.

Incorporation of Proposal by Reference. Some customers like to incorporate the proposal into the contract between the two parties. This means that the developer must live up to any and all promises made in the proposal. When these are overstated so as to increase the chances of winning, major problems can result when the proposal becomes part of the contract. The two most obvious countermeasures are (a) to not accept incorporation of the proposal into the contract, and (b) to try to go back and "scrub" the proposal so that all questionable promises are deleted. Here's where the company lawyers need to earn their keep by focusing on these promises and working as part of the team, together with the PM, CSE, and PC for the project.

Special Rewards. Since FFP contracts represent special potential risks for the company, it should be worth real dollars to mitigate such risks. One way to do this is to offer bonuses to key project personnel if the project is executed within schedule, budget, and performance specifications. Further, such bonuses should be commensurate with the effort required as well as the benefit to the company. In other words, they should be real motivators for the team to do an outstanding job. At times, possible promotions are connected to high levels of achievement on a particularly important project. Business as usual rewards typically result in business as usual performance. Companies should not be worried about treating some people in a special way when the circumstances call for special treatment.

Cost Information System.   The project cost information system needs to be finely tuned in at least three dimensions for FFP contracts. The first dimension is the cycle time for cost reporting. Monthly cost reporting should be changed to every two weeks for FFP efforts, for example. If the standard cost reporting system cannot respond quickly enough, investing in a special system is likely to be worth it. Second, cost to complete estimates need to be provided on the same two-week cycle time, requesting such estimates from more than one person if necessary. Third, the PM and CSE should ask the PC to take personal responsibility for presenting a project status overview at a meeting every two weeks. This includes looking very carefully at potential schedule slippages that, in turn, might lead to cost increases. The PC should also be in a position to look for biased cost and time to complete inputs that might present an overly optimistic view of project status. In FFP contracts, one needs to search for the truth, and as early as possible.

Walking Away.   If you are unable to successfully negotiate an appropriate number of the above items, especially on a FFP software product delivery that needs to be developed from scratch, it may be time to simply walk away. One form of doing so is to add large reserves to account for all risk areas such that you drive your price into a noncompetitive range. This strategy may backfire if your intent was to truly walk away in that your competitors may have done the same thing, leaving you the winner. Walking away from what many people in your company believe is a real opportunity may be quite difficult. But signing up to a losing contract can be a disaster in that one not only loses money, but also experiences significant opportunity costs. Having your best people working ten to twelve hours a day leaves them unavailable to tackle other ongoing contracts as well as new bid opportunities. Here's a case in which discretion may truly be the better part of valor.

4.4   SITUATION ANALYSIS (SA)

4.4.1   Overall Situation Analysis Process

Maintaining sharp and continuous monitoring of project schedule, cost, and performance is, of course, an essential element of effective project management. Indeed, the "nuts and bolts" of project management involves the continuous positing of these questions:

1. Are we on schedule?
2. Are we within budget?
3. Are we satisfying all performance requirements?

However, if we get a "no" answer to any of these questions, or if other issues surface that could adversely affect them, the question then becomes: What else should the PM be doing? The answer lies in situation analysis (SA), which is fundamentally a problem-solving process at the project level. It is an adaptation of the "case study" approach utilized in some business schools. It may also be viewed as follows: Case studies are to an overall enterprise (or key portion of an enterprise, such as a division) as situation analysis is to a project. Situation analysis is a sequence of steps to be undertaken by the project triumvirate (PM, PC, and CSE) once one or more problems or potential problems have surfaced.

The general situation analysis process is depicted in Figure 4.4. The first step (Box 1) in such a process is, as might be expected, to gather up, or restate, the facts that are known in the given situation. Such facts are usually in the domains of schedule, cost, or performance, but might be other facts not as immediate or obvious. Examples of the latter type of facts might be

1. A strike at the plant of a major supplier or subcontractor
2. Serious conflict between members of the project team
3. Resignation of a key member of the project

After such facts have been identified, two paths are suggested. One leads to a set of evident problems (Box 2) and the other to potential or inferred problems (Box 3). The former represent clear and irrefutable problems, normally of a high priority, that must be dealt with. Examples include:

1. Schedule slippage on the critical path
2. Expenditures greater than budgeted amounts
3. Missed contractual delivery dates
4. System testing failures

All are evident problems, almost by definition. How significant these problems are remains to be investigated in detail.

In the category of potential or inferred problems, we normally find occurrences that may or may not lead to significant problems. In this regard, one finds secondary events that might eventually do serious damage to the project.

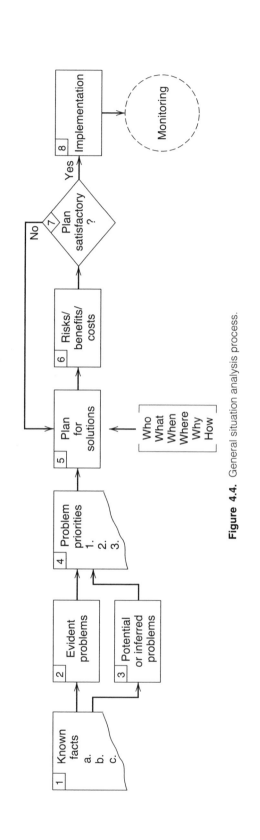

Figure 4.4. General situation analysis process.

Such events might be:

1. Project staff perturbations or conflicts
2. A change in the PM's immediate supervisor
3. Company reorganizations
4. Loss of key people, not on the project team, but in support organizations such as accounting/finance, contracts, and human resources
5. Changes in subcontractor/supplier organizations

The point of separating evident (obvious) problems from potential or inferred problems is to assist in the eventual step of sorting these problems in order of priority (Box 4). A priority list is intended to force a discipline that assures that key problems cannot be ignored or placed on the back burner. Without this discipline, a PM might be otherwise inclined to tackle more tractable issues that are of little or no real importance and avoid handling critical problems that might be difficult to confront. Such behavior may be difficult to understand, but it is part of human nature to not want to face unpleasant and stressful tasks.

Given the problems in priority order (Box 4 of Figure 4.4), the next step is to develop plans for solutions (Box 5). Plans at the top of the list *must* be addressed; plans at the bottom of the list might be deferred until further data are obtained. This is a judgment call that should be decided by the project triumvirate. The usual journalistic questions of who?–what?–when?– where?–why?–how? should be considered, noting that a plan for situation analysis is not the same as a project plan. Plans must be evaluated in terms of risks (assuming that the plan is implemented), benefits, and costs (Box 6). Alternatives are recommended so that all reasonable solutions are at least placed in evidence. Leaping to premature or incorrect "solutions" can be more damaging than the original problem.

An important footnote to the formulation of the plan for solutions (Box 5) is the question of who it is that devises such solutions. Overall responsibility rests with the PM, PC, and CSE, but it is suggested that a team approach to problem solving be undertaken. In other words, information and proposals for solutions should be consciously elicited from members of the project team. Full or partial team meetings are a good way to kick off such a process. In this manner, participative management can be demonstrated in addition to being expounded. More importantly, it usually leads to clearer definitions of problems and more effective solutions. More is discussed in this regard in Chapter 6.

Another implicit question is: When does the PM give an alert to the boss when there is a problem? The recommended answer is, for most situations, after the PM has developed an appropriate plan for solving the problem. In general, do not "hide" problems from bosses. At the same time, it is prudent to come to the boss with a complete plan for solution. This shows the boss that the PM is on top of the problem. It also gives the boss a last opportunity to provide input into the plan, or to modify the plan if necessary. Implementing a solution without consulting the boss carries some risks with it, especially if the problem is severe.

Thus, the situation analysis process shown in Figure 4.4 involves two additional and very important considerations:

1. When and how to involve the project team
2. When and how to involve the PM's immediate supervisor

Interim plans that are not considered satisfactory (Box 7) have to go back around the loop for improvement and consideration of alternatives. Once the plan is approved, implementation starts (Box 8). After that, the normal monitoring function is resumed.

4.4.2 Example of situation analysis

We pose a "situation" facing the PM as follows:

> It is Wednesday afternoon and Jack, the Project Manager, receives a call from the Project Controller who claims that the latest cost report shows the project to be overspent, compared with budget, by 11%. Jack, meanwhile, had been thinking about his lead hardware and software engineers, who incessantly complain to him about each other. Jack now also begins to think about the project review session with his customer that is scheduled for 2 P.M. next Monday. What should Jack do, and in what sequence?

A response to this situation, that is, a situation analysis, follows.

Step 1: This step calls for assembling the known facts, which, at this point, appear to be
  a. The project is 11% overspent.
  b. The lead hardware and software engineers are complaining about each other.
  c. There is a project review session planned with the customer in approximately five days (three working days).
Jack next picks up the phone and asks the Project Controller (PC) and Chief Systems Engineer (CSE) to come to his office immediately. The PC is asked to bring all cost and schedule data that are relevant. The project triumvirate then reviews the facts and overall situation from top to bottom.
Step 2: The cost overexpenditures are identified as an evident problem.
Step 3: The hardware and software engineers issue and planned meeting with the customer are placed in the "potential or inferred" problem category.

Step 4: The project triumvirate identifies the problem priorities as

1 Cost overrun
2 Scheduled customer meeting
3 Hardware/software engineer issue

They decide that the scope of the plan for solution will *not* include the hardware/software engineer issue. They also analyze, to the extent that they are able to, the data they have involving:

1 The cost elements that have been overspent
2 Why these cost elements are overspent
3 Potential effects on schedule
4 Potential effects on technical performance

This activity takes most of the rest of Wednesday afternoon. The project triumvirate agrees that more data are needed. Jack calls for a project team X meeting at 8:30 the next morning (Thursday). He does not reveal the precise purpose of the meeting. Project team X is a subset of the overall project staff and is handpicked for its ability to solve a problem of this type. Jack asks that the PC and CSE think about the situation but not convey it to anyone else until the meeting the next morning.

Step 5: The project team X meets on Thursday morning to discuss the two top-priority problems, in the following sequence:

1 Cost overrun
2 Meeting scheduled with the customer

Reasons for the cost condition are ascertained at this meeting. A basic plan (Plan A) for how to fix the overrun situation is set forth.

Step 6: At the same meeting, which is proceeding through the entire morning, Plan A is reconsidered with respect to risks, benefits, and costs.

Step 7: Based upon the preceding scrutiny, team X does not believe the plan is good enough. It is also 11:30 A.M. on Thursday morning. Jack asks all members of team X to reconvene at 2 P.M., coming to the table with new and hopefully better ideas. Team members are encouraged to talk to other project personnel if it is considered helpful. All project personnel are reminded that the situation is to be kept within the project staff for the time being.

Reiteration of Steps 5, 6, and 7: Team X meets at 2 P.M. and a new plan is devised that is considered satisfactory and, indeed, the best the team can formulate. This is a plan for correction of the cost overrun. Jack now focuses the team on the matter of the project review session with the customer, scheduled for next Monday. The team agrees that:

1 The customer is very likely to accept the plan.
2 There is no good reason to alert the customer to the problem before Monday.
3 They should confirm the Monday meeting with the customer,

Jack suggests to the team that now is the time to alert his supervisor as to the set of problems as well as the plan for solution. The team agrees that this is an important step prior to the implementation of the plan.

Step 8: As a precursor to implementation, Jack, the PC, and the CSE meet with Jack's boss. The boss appreciates the steps taken and being kept informed. He also agrees with the plan, but insists on being present at the meeting with the customer on Monday. Jack and his boss agree on how to make the presentation to the customer, who else should attend, and what the roles of all participants should be.

This example illustrates the SA process as well as some of the vagaries of that process. The basic issue for SA is not how well one can analyze schedule charts and cost reports. The issue, rather, is how to mobilize a team effort to prioritize and find solutions to problems that invariably arise during the course of a project. The reader is invited to practice situation analysis for the situations that are included in the questions and exercises that follow.

QUESTIONS/EXERCISES

4.1 For the activity data related to a small project, as shown, draw the PERT chart and find

    a. the critical path and its expected time

    b. the slack in all other paths

    c. the standard deviation associated with the project end date

Activity	Three Time Estimates (weeks)
A–B	1–3–5
B–D	1–2–3
A–C	1–2–3
C–D	2–4–6
C–F	4–6–8
C–E	1–4–7
D–F	1–2–3
E–F	1–3–5

4.2 Design a cost monitoring report that expands the data provided in Table 4.1.

4.3 You are at the 18-month point of a 24-month project with a $400,000 budget. The schedule variance has been estimated as $30,000 and the cost variance as $20,000.

    The BCWS is $300,000.

    a. ACWP

    b. BCWP

    c. ECAC

    d. ETAC

    Compare these results with those in the EVA example in the text. Why are they different?

4.4 In general, for a project:

    a. If BCWP > BCWS, is the project early or late? Explain.

    b. If ACWP > BCWP, is the project over or under cost? Explain.

    c. If ACWP < BCWS, what conclusion, if any, can you draw? Explain.

4.5 What role does the Project Manager play in controlling fringe, overhead, and G&A costs and rates? How does the PM deal with increases in these rates that might result in a cost overrun?

4.6 Enter the data from Table 4.1 into a spreadsheet. Print out graphs of the results. Do the graphs provide insights into cost status that might not be revealed in columns of numbers? Explain.

4.7 The chapter provides twelve suggestions for dealing with firm fixed price bids or contracts. Identify three additional actions that might be employed in this regard in order to reduce risk. Explain your rationale.

4.8 Carry out a situation analysis for the following situations:

    a. As a PM, you discover, at month 8 in a ten-month project, that you are 8% over cost. Your Chief Systems Engineer tells you that you can meet budget if the entire project team works a 48-hour week. Your Project Controller estimates only a 50% chance of success with that strategy, but claims that 5% of the work done has been "out of scope." You have a project status review planned with the customer in two days. What should you do, and in what sequence?

    b. As a PM, it is Friday afternoon at 4 P.M. and you receive a call from your customer complaining about the quality of your company's last report and a bad attitude on the part of your on-site lead engineer. Your customer wants to see you in his office at 9 A.M. next Monday. What should you do and in what sequence?

4.9 Locate a document produced by the Department of Defense that provides guidance to a Project Manager regarding how to perform "schedule and cost" analysis of a project. Write a two-page discussion of the key points in this document.

4.10 Locate a document produced by NASA that provides guidance to a Project Manager regarding how to perform "schedule and cost" analysis of a project. Write a two-page discussion of the key points in this document.

REFERENCES

4.1 Frame, J. D. (1991). *Managing Projects in Organizations*. San Francisco: Jossey-Bass.

4.2 Kezsbom, D. S., D. L. Schilling, and K. A. Edward (1989). *Dynamic Project Management*. New York: John Wiley.

4.3 Kerzner, H. (2000). *Project Management: A Systems Approach to Planning, Scheduling and Controlling*, 7th edition. New York: John Wiley.

4.4 Malcolm, D. G., J. H. Roseboom, C. E. Clark, and W. Fazar (1959). "Application of a Technique for Research and Development Program Evaluation," *Operations Research* **7**: 646–669.

___5

THE PROJECT MANAGER AND LEADERSHIP

5.1 INTRODUCTION

This chapter focuses more sharply on the specific attributes of a Project Manager (PM) that are needed in order to be both a manager and a leader. Much has been written about the differences between a manager and a leader, making the point that even a good manager is not necessarily a leader. Students in my classes regale me with stories of their bosses who do not seem to have even the barest minimum of skills and awareness to deal with the human aspect of managing other people. At times, it appears as if this is the rule rather than the exception, leading to a great deal of frustration in the work environment. Bad managers may know the "nuts and bolts" of project management, but if the execution is devoid of an understanding of how to relate to people, the project will usually get into trouble. People just do not put forth their best efforts for a bad boss.

This chapter, then, attempts to shed some light on three basic questions:

1. What are the personal skills that the project manager must have or develop, especially with respect to human interactions?
2. How can the project manager deal more effectively with a poor or bad boss?
3. What are some of the key issues in dealing with the customer?

Going beyond these questions, the chapter ends with a discussion of transcending the basics of management into the domain of leadership.

5.2   PROJECT MANAGER ATTRIBUTES

The first place to look in terms of exploring the desired attributes of a PM is the set of tasks that a PM must be able to carry out. These tasks, reiterated from Chapter 1, are:

1. Planning
2. Organizing
3. Directing
4. Monitoring

Planning and organizing are important tasks of the PM and tend to be done well by a person who enjoys process. Although a good plan is more often than not the results of a team effort, a PM who likes to plan may have a tendency to take on the entire plan. The PM who enjoys planning will see organizing as just another part of a good plan, which indeed it is. Such a person may tend toward introversion and requires considerable order and discipline. When we move on to the tasks of directing and monitoring, however, we see a requirement for another type of perspective. These tasks involve interactions with people. Directing requires that people be given assignments and that they be guided through these assignments through monitoring and feedback. These "people" interactions are often best accomplished by an extroverted type of person who likes to discuss situations with people and may not enjoy the paperwork associated with planning and reporting. Through this simple discussion, we note that the PM is called upon to do many things that require a balanced personality. The PM is a well-integrated person who can shift gears as well as pay attention to and keep in balance the many issues that inevitably come across the desk.

Although it is possible to further analyze the attributes of a good project manager in terms of the required skills associated with planning, organizing, directing, and monitoring, we prefer to list, in Exhibit 5.1, twenty critical aspects of a PM. A brief discussion of each follows. These same twenty attributes may also be interpreted as the characteristics of a good boss.

Exhibit 5.1: Twenty Attributes of a Project Manager

1. Communicates well and shares information
2. Delegates appropriately
3. Is well-organized
4. Supports and motivates people
5. Is a good listener
6. Is open-minded and flexible
7. Gives constructive criticism
8. Has a positive attitude
9. Is technically competent

10. Is disciplined
11. Is a team builder and player
12. Is able to evaluate and select people
13. Is dedicated to accomplishing goals
14. Has the courage and skill to resolve conflicts
15. Is balanced
16. Is a problem solver
17. Takes initiative
18. Is creative
19. Is an integrator
20. Makes decisions

5.2.1 Communication and Sharing Information

One of the major complaints of people working on a project is that they are not kept informed. The Project Manager must pay special attention to letting people know what is happening on the project, in all of its dimensions. The PM must also try to assure that the cross-communication between members of the project team is effective and that all personnel share important information.

5.2.2 Delegation

The effective PM must be careful not to take on all the key tasks. If this is done, the project team members will quickly learn that they are not trusted to do anything important, which will soon lead to nonteam playing and disaster. Effective delegation is a critical task of the PM, and it is coupled with resisting any temptation to "micromanage." Once a task is delegated, the PM should stay in touch and guide rather than hover over, criticize, and redo.

5.2.3 Organization

In this context, being well-organized is to know where everything is (project status), know where it is going, assure that all members of the project have what they need to do their jobs, and to be prepared to solve problems. Some project managers take on too many internal project tasks and cannot pay attention to the project as a whole. From this position, they become disorganized and struggle to keep up with the overall project needs. Being well-organized will keep the overall project moving forward and will help assure that other people's efforts are efficiently performed.

5.2.4 Support and Motivation

Project managers sometimes underestimate the power that they have in the eyes of the project team. Team members usually look to the PM for attention and support. Without this, some personnel begin to feel that nobody cares what they do or don't do. In some cases, the PM may have to use special motivations to assure performance, recognizing that all people are different and respond in different ways to the pressures of a project. The PM must also behave impartially so that all members feel they are being fairly treated.

5.2.5   Listening

Effective listening is so important that it is singled out as a separate attribute. Listening, or not listening, can be very subtly executed. A good listener sustains eye contact and responds at the right times with a nod of the head or appropriate words. Good listening conveys the message that what is being said is important, and also that the person is important enough to be listened to. This helps to build trust and encourages further communication and sharing of information. Not listening leads to frustration and gives the message that you and what you are trying to say are not significant. This can be devastating, especially to a young project member, and often leads to poor performance.

5.2.6   Open-Mindedness and Flexibility

Open-mindedness relates to listening but is the next step in the process. It implies that new information is received appropriately and used to make adjustments. Studies by psychologists have shown that some managers work on mental models based only on past experiences and are not able to absorb new information. Complaints about such managers are that their "minds are already made up." In this context, the PM must be aware of prior prejudices and be open to the input of new data. Such a person will then be able to respond to each situation on its own merits and will be able to behave in a much more flexible manner.

5.2.7   Constructive Criticism

Giving constructive criticism is a crucial job of the PM and it is surprisingly difficult to do. If a project task is not being adequately performed, it is often up to the PM to critique the work in such a way as to encourage change without destroying the ego and motivation of the person working on the task. This is usually best achieved one on one, but at times may be accomplished through peer pressure in a group setting. All of this requires careful consideration of the recipient of the criticism and the most effective way of reaching him or her. Words must be chosen with considerable skill so as to get the message across in a supportive and encouraging manner.

5.2.8 Positive Attitude

The PM must be a positive person who reflects a "can-do" attitude to the project team, the boss, and the customer. This helps to propagate such a perspective to all parties and generally leads to an overall sense of accomplishment and moving forward. There are many hurdles that the project team has to deal with and a positive PM makes the job of getting over these hurdles a learning experience. Conversely, a negative attitude on the part of the PM likewise tends to spread like wildfire and the team members find themselves either preferring to be victimized or intensely frustrated.

5.2.9 Technical Competence

This attribute refers to the nuts and bolts of the project demands, from the domain knowledge of the project to the skills required to read and understand schedules and cost reports. Without these capabilities, the PM soon drowns and loses the confidence of the entire project team. A person without these basic skills should not take on the difficult job of Project Manager.

5.2.10 Discipline

Many projects go astray simply because the PM has not assured a disciplined approach on the part of all team members. Each and every task must be viewed as critical in terms of adherence to schedule, cost, and technical performance requirements as well as the impact that they have on other tasks. Company and project standards, methods, and procedures should be followed unless there is an excellent reason not to. Experience has shown that projects with a large component of software development have to be particularly well-disciplined in order to be successful. If there ever was a "silver bullet" for the process of developing software, discipline is its name.

5.2.11 Team Builder and Team Player

We explore the matter of team building in the next chapter because it plays such a central role in the success of a project. Given that perspective, a critical skill of the PM is to build a team as an effective and organic element of a project that succeeds. At the same time, the PM must assume the position of being part of a larger team, the enterprise within which the project is being executed. Project Managers who take the view that the project is "we" and the rest of the enterprise are "they," and that "we" do everything right and "they" are incompetent are very likely to fail. The project exists and works within the context of the corporate enterprise and, ideally, the two entities should be mutually supportive.

5.2.12   Evaluation and Selection of People

A key activity of the project is to staff it with people who function as a team and who are competent in the various disciplines required by the project. The PM should have a special eye for this activity, knowing whom to select and where to have them assigned. A weak team member tends to pull the entire project down as people spend their valuable time fixing problems that have been created by such a person. Work elements or tasks that have to be continually redone create resentment as well as schedule and cost problems. In addition, when a weak person cannot be brought up to minimum standards of performance, the PM also has to recognize that action has to be taken to fix the problem.

5.2.13   Dedication to Accomplishing Goals

This attribute might also be called determination. Because it is the rule rather than the exception that problems will arise on essentially every project, the PM must be dedicated to getting through each and every problem in order to accomplish the stated project goals. This means not allowing oneself to be victimized by the "system," management, or the customer. After a frustrating day, the PM should have the energy and determination to come back the next day with a renewed sense of how to relieve the frustration and find effective work-arounds, if necessary.

5.2.14   Courage and Skill to Resolve Conflicts

Because projects are staffed by people, and people have conflicts, then it is expected that projects will experience and be affected by conflicts. These can be internal conflicts, as with two members of the project team not getting along, or external conflicts, as between a member of the team and a support organization such as accounting/finance, contracts, human resources, and so forth. The management of conflict, which is examined in some detail in the next chapter, requires both courage and skill. Courage is necessary because it can be quite difficult to confront the conflict, especially if the PM is part of the conflict situation. Skill is needed because conflict is delicate and wrong moves can exacerbate the conflict rather than calm it down and resolve it.

5.2.15   Balanced

The PM must be balanced, as a minimum, in terms of handling people with a basic sense of fairness and equity, and also in terms of balancing the effort that goes into satisfying cost, schedule, and performance requirements. In the former area, for example, prima donna behavior should not be accepted while other personnel are working with dedication for the common good of the project. In the latter domain, balance involves understanding how to respond with equilibrium to the forces that would tend to push the project off course. The PM is "steering the ship," and demands come from many directions. A balanced approach to these often conflicting demands will be respected and supported and will increase the likelihood of success.

5.2.16 Problem Solver

The PM must be a problem solver. This means going beyond the discussion of a problem and its symptoms and causes. It also requires driving toward a solution and then implementing that solution. Too many PMs behave as if the mere examination of a problem is the same as taking the actions necessary to solve the problem. Also, too many PMs develop a series of solutions but then procrastinate in the implementation because the solution involves doing some difficult things such as confronting a supervisor or approaching the customer. A problem solver is action-oriented and is not fearful of making a mistake.

5.2.17 Initiative

The PM is active rather than passive and is always aware of the need to take action when it is appropriate to do so. This can apply to interactions with the customer as well as with the project team. The PM is always asking the question: What can be done to improve the project and the situation in which the project is being carried out? Opportunities do not pass by the PM with initiative. Such a PM is also on the lookout for ways and means to achieve continuous improvement.

5.2.18 Creative

The best PMs are creative people who look for new ways of solving problems or new approaches to the project tasks. This type of creativity is disciplined and does not resort to new methods unless they show promise of bringing higher efficiency and productivity to the project. This type of PM also does not impose his or her creativity on the project team. Rather, there is an understanding that creativity can and should be expressed by any member of the project, and all good ideas are solicited and welcomed. By being a creative person, creativity in others is recognized and valued. This attitude pervades the project and the excitement of creative solutions is experienced by the entire team.

5.2.19 An Integrator

The PM must be able to integrate in many dimensions. One such dimension is the synthesis of technical inputs, finding ways to blend such inputs to construct an overall solution. Another dimension is to see where and how people can be utilized so that they are challenged rather than bored. A third dimension is to see the project as a whole, seeking balance in terms of cost, schedule, and performance. Yet another dimension is to integrate the human side of managing with the nuts and bolts of planning, organizing, directing, and monitoring. This type of person is able to perceive relationships between different parts of a project.

5.2.20 Decision Maker

Finally, but not by any means last in importance, is that the capable PM makes decisions when it is necessary to do so. The PM knows when it is time to stop analyzing the problem at hand and bring any given situation to closure. This type of PM has a sense of when further examination of an issue brings diminishing returns. The PM is also keenly aware of the level of urgency associated with all situations and behaves accordingly. Project personnel react particularly well to a PM who makes decisions and moves forward. If a decision was made that later is shown to be incorrect, the effective decision maker is not afraid to admit to a mistake and backtrack, as necessary, to rectify a bad decision.

Whereas the preceding attributes have been associated with the Project Manager, it should be recognized that the Chief Systems Engineer (CSE) is also a manager, taking responsibility for the system design team and the overall engineering effort. Therefore, most of the attributes discussed here apply as well to the CSE. The difference lies mostly in the scope, orientation, and focus of the work to be performed. Above all, the PM and the CSE must be able to work harmoniously together and both be dedicated to the success of the project.

5.2.21 Learning from the Negative

The above text described and briefly discussed twenty positive attributes of a Project Manager. A study of the leadership characteristics of American Project Managers [5.1] explored significant aspects of effective PMs, but also looked at factors that contribute to making a PM ineffective. These factors give us some insight into what the current or prospective PM needs to avoid where and whenever possible. At times, we learn more from the negative than we do from a long list of positives. Thus, the top five negative factors for the PM are, with the top-listed item the worst:

- Sets a bad example for the team
- Is not self-assured
- Does not have sufficient technical expertise
- Is a poor communicator
- Is a less than acceptable motivator

Thus, if these problems have been related to issues that you have had to struggle with as a PM, you probably would do well to commit yourself to making improvements in these areas. If you do not do so, you may well be on the road to failure as a PM.

In addition to the above personal attributes that a PM might have, the cited study [5.1] also explored organizational factors that had a negative effect upon the effectiveness of the PM. These factors are, with the most negative listed first:

- Lack of the commitment and support of top management
- Overall resistance to change
- A reward system that is inconsistent
- Reactive behavior instead of planning in advance
- Insufficient resources

The implication of the above listing is that if you, as a PM, find yourself in an organization that exhibits these types of behavior patterns, you have an increased likelihood of getting into trouble. It also may be that efforts you put forward toward solving these types of organizational problems may serve you and others in good stead. However, for a PM to be the force behind the solution of rather large organizational problems is a rather daunting task.

Finally, and in relation to the same study cited above, we can look at reasons why projects may tend to experience problems in terms of completion within budget and schedule. The top five reasons identified, with the most cited at the top of the list, are:

- Tools to manage the project in a systematic manner are not employed.
- The PM is a poor leader.
- The customer/client is slow to respond.
- Decisions and corrective actions are not taken in a timely manner.
- Interorganizational communication is poor.

Here again, the above items provide a "view of the negative" that might be helpful to the project triumvirate in terms of trying to increase the chances of success.

5.3 SELF-EVALUATION

The effective PM knows himself or herself. Such a person has a level of awareness of tendencies toward certain types of behavior and is sensitive to the behavior patterns of others. To some people, it is not difficult to achieve this type of sensitivity and understanding. To others, "knowing oneself" is an alien concept. However, it is possible to gain a better understanding through processes of self-evaluation and taking the time to examine one's behavior in a variety of situations. This yields a keener sense of self-awareness, which is always helpful to a PM. By doing so, it becomes easier and easier to deal with problem people and problem situations, both of which improve the likelihood of success. In this section, we explore a few formal procedures that the PM or the prospective PM (or CSE) can employ in order to carry out a rudimentary self-evaluation.

5.3.1 Scoring Yourself

Given the attributes of a successful PM, as discussed earlier in this chapter, it is a simple matter to do a first-order self-evaluation by scoring oneself against this set of attributes. Table 5.1 places the twenty attributes in a scoring context, and the reader is asked to take a moment to evaluate himself or herself directly on the scoring sheet of the figure. A score of "5" should be given if the reader almost always behaves according to the stated attribute, and so forth as listed in the table. Take some time now to score yourself.

If your aggregate score is in the range 80-100, you are likely to be an excellent project manager. Essentially, no fine-tuning is necessary and you

TABLE 5.1 Evaluation versus Attributes

	Attributes	Scores*
1.	Communicates/shares information	[ ]
2.	Delegates appropriately	[ ]
3.	Well-organized	[ ]
4.	Supports and motivates people	[ ]
5.	Good Listener	[ ]
6.	Open-minded and flexible	[ ]
7.	Gives constructive criticism	[ ]
8.	Positive attitude	[ ]
9.	Technically competent	[ ]
10.	Disciplined	[ ]
11.	Team builder and player	[ ]
12.	Able to evaluate and select people	[ ]
13.	Dedicated to accomplishing goals	[ ]
14.	Courage and skill to resolve conflicts	[ ]
15.	Balanced	[ ]
16.	Problem solver	[ ]
17.	Takes initiative	[ ]
18.	Creative	[ ]
19.	Integrator	[ ]
20.	Makes decisions	[ ]
	Total:	

*Scoring: 5: almost always; 4: most of the time; 3: often; 2: sometimes; 1: rarely; and 0: never.

should be pleased that you have all the necessary skills to be successful in just about any management role. If you scored between 60 and 79, you are doing well but probably have a few areas that need improvement. Those are likely to be represented by the attributes that you scored yourself as a "2" or lower. If your score was in the range 40–59, you still may be a good candidate for a PM or manager position but need to work in a disciplined way to improve your skills. This may involve more substantial training to develop these skills as well as a deeper sense of self-awareness of your own behavior and the way that it might be affecting others. If your score was between 20 and 40, some type of continuous training program is recommended, depending on whether the score was closer to 40 or in the 20 range. If your score was less than 20, you have a long way to go to become an effective PM or manager. This does not mean that you cannot get to a PM or manager position, but it is likely to take a lot of hard work over an extended period of time.

Developing PM and manager skills is rarely achieved by reading a few books on the subject. Reading is only one component of the process. Broadly speaking, there are two other critical elements. One has to do with the afore-mentioned self-awareness of your own behavior as well as the behavior of others. Without this consciousness, one is not absorbing and assessing data in the real world. The other critical component is experiential learning. This may be achieved through workshops in which you are asked to carry out exercises that simulate situations in the real world. By actually *experiencing* the pro-cesses that evolve from this type of training, learning occurs relatively quickly and it is possible to improve skills and awareness rather rapidly. Low-scoring readers who aspire to PM or management positions are urged to consider some type of experiential training program over a long period of time. Many such programs also utilize "personality tests," such as those briefly described in what follows.

5.3.2 The Myers-Briggs Type Indicator (MBTI)

The Myers-Briggs Type Indicator (MBTI) is a very well-known personality construct [5.2, 5.3] that is based upon the following four polarities:

Extrovert (E)	versus	Introvert (I)
Sensing (S)	versus	Intuitive (N)
Thinking (T)	versus	Feeling (F)
Judging (J)	versus	Perceptive (P)

By filling out a questionnaire, the user obtains a scored self-profile in each of the eight dimensions. Scores can be very close to the center of a given polarity or they can show a very distinct preference for one dimension over another. Each of these aspects of the MBTI is briefly discussed in what follows. The reader is urged to see if he or she can identify with the various types.

Extrovert (E).   The extrovert is what you might expect from a simple dictionary definition. This person is sociable and enjoys a multiplicity of relationships. He or she is usually gregarious, outgoing, energetic, and conveys a breadth of interests. According to data collected over a long period of time with respect to the MBTI, about 75% of the population falls into this category [5.3].

Introvert (I).   This type of person is more closed, turned inward, territorial and conservative of expending energy. It may be difficult to find out what such a person is thinking because there is a tendency toward not speaking, especially in a group situation. Such a person may be concentrated, watchful, and have limited interactions and relationships with only a few good friends or colleagues. About 25% of the population at large exhibit this type of behavior.

Sensing (S).   In this category, one finds people who focus on facts, figures and real-world data and experience in order to grasp and relate to what is going on around them. Such a person is very practical, down to earth, sensible, realistic, and prone to adopt the perspective that if it cannot be seen or measured, it is not likely to exist or be true. Past experience is very important and has a strong influence on views of the current or future situations. Some 75% of the population exhibit this tendency.

Intuition (N).   The intuitive person likes to speculate about the future and is often imaginative, inspirational, and ingenious. This type of individual may be prone to fantasizing and searching for new ways of doing things. "Gut" reactions may be much more important than facts and figures, which may be discounted in considering what to do in a project situation. A high "N," with respect to this point, may have a serious clash with a high "S" because the latter will not understand how gut reactions play a role in evaluation and remediation of problems. About 25% of the population have an "N" score.

Thinking (T).   This type of person takes pride in using analytical skills in puzzling through problems. Some might refer to such a person as "left-brained," relying on abilities to be objective and impersonal to analyze and resolve situations. Such a person attempts to use standards, policies, and laws to create order and a sense of equity and fairness. Subjective evaluations might make such a person uncomfortable because objective measurements are distinctly preferred. The literature suggests that 50% of the population would tend to qualify as a "T" in the MBTI.

Feeling (F).   Representing the opposite polarity from "thinking," the "feeling" person relies on visceral reactions and human connections. Often, the "F" person looks behind and beyond the words at such things as reactions, facial expressions, and body language to try to understand what is actually happening in a given situation. Such a person is comfortable with subjectivity and emotions in others and himself or herself. As a manager, he or she tends to empathize with the situation of subordinates and shows a great deal of patience and understanding. This type of person likes harmony and spends time to try to persuade other people on a preferred position. About 50% of the population score in this category.

Judging (J).   The judging person likes to make decisions and move on to the next problem. This person insists on closure and has an internal sense of urgency about almost all matters. He or she responds very well to deadlines and works very hard to assure that all milestones on a project are met. Such a person likes to plan and then proceed with measuring against the plan. The "J" person is not very patient and likes to converge to core issues as quickly as possible. About 50% of the population have a "J" profile.

Perceiving (P).   As an opposite from the "J," the perceiving individual is happier with open-ended assignments and situations that allow more flexible responses. Such a person might tend to want to collect more data about a given problem, with little consciousness regarding the time that it takes to do so. The "go with the flow" position of the high "P" often drives a high "J" to anxiety and anger. Whereas the "P" person may adopt a "wait and see" attitude, the "J" wants to "get the show on the road" [5.3]. About 50% of the population have this type of characteristic.

One of the conclusions that might be drawn from the MBTI is that people with extreme and opposite scores for a given polarity may have a tendency to clash with one another. For example, if a PM is a high "S" and a project staff member is a high "N," these two people, with their different views of the world and behavior tendencies, may frustrate each other. Although this is not a hard-and-fast rule, it is a point that may explain certain personal antipathies. This may be generalized to a main application of the MBTI as well as other such tests, namely, that it may be used to try to understand why people do not get along on a project and what each may do to try to better understand why and what might be done to bridge such a gap of understanding. Further, if any two people have absolutely opposite MBTI profiles, as for example an ENTJ versus an ISFP, the gap may be even broader and deeper. Having an awareness of these natural differences helps to explain various types of conflict. It may also provide a basis for an appreciation of differences, which could serve to strengthen a project team and its overall performance.

Another obvious question regarding the MBTI: Is there a preferred MBTI profile for a project manager? Some investigators of the field of management believe that this is true. For example, J. Davidson Frame [5.4] appears to select the ESTJ profile as the preferred type for a PM. However, he also points out that, for research projects, "ESTJ project managers who are unaware of the differences in psychological type are likely to be exasperated by their workers," based on the differences in how they deal with and see the world. The key word, from this author's point of view, is "awareness" and that people who have a strong awareness of both similarities and differences can make both a strength in a project situation. Research-oriented people can be doing the research tasks and extroverts can be doing the marketing and project presentations. Facts-and-figures people can be happily devoted to the project control activities and high Js can help to bring focus and closure to interminable meetings. In short, we are all different and we perform best when we are working in areas of strength rather than weakness. The aware Project Manager knows this and spends the time necessary to understand differences and assure that individual strengths and tendencies are fully utilized and that the effects of weaknesses are minimized. Thus, even if a PM is an INFP, in distinction to Frame's ESTJ, excellent results can be achieved on the project if such a PM has the awareness, skill, and discipline to use a team approach that fully utilizes the complementary capabilities of that team.

5.3.3 Other Personality Considerations

The MBTI is not the only "test" that might be employed to carry out some type of self-evaluation. There are literally dozens of others that will be helpful in a lifelong process of trying to better understand oneself. Another such approach is cited as a Communication Self-Assessment Exercise [5.5] and it is based on numerical scoring for the following "styles":

- Action
- People
- Process
- Idea

The maximum score for any one style is 20 and the total score for all the styles adds to 40. Thus, a completely balanced score would be 10–10–10–10. A summary of what the various people with these styles like to talk about as well as how they tend to behave is provided in Table 5.2 [5.5]. We see from these descriptions that there is a great potential for conflict between people who have widely differing scores. For example, the action-oriented person may have little patience with the process-oriented person, who may be perceived to be too slow and interested only in form rather than substance. Similarly, the latter may regard the people-oriented person as too emotional and subjective and not understand how such attitudes fit into the disciplined world of project management. Potential trouble spots on a project may be predicted from the examination of the interaction of these different styles, which is explored in Section 5.4.

TABLE 5.2 Communication Styles [5.4]

Styles/ Features	Content		Process
Action (A)	*They talk about:* • Results • Objectives • Performance • Productivity • Efficiency • Moving ahead • Responsibility	• Feedback • Experience • Challenges • Achievements • Change • Decisions	*They are:* • Pragmatic (down to earth) • Direct (to the point) • Impatient • Decisive • Quick (jump from one idea to another) • Energetic (challenge others)
Process (PR)	*They talk about:* • Facts • Procedures • Planning • Organizing • Controlling • Testing	• Trying out • Analysis • Observations • Proof • Details	*They are:* • Systematic (step by step) • Logical (cause and effect) • Factual • Verbose • Unemotional • Cautious • Patient
People (PE)	*They talk about:* • People • Needs • Motivations • Teamwork • Communications • Feelings • Team spirit • Understanding	• Self-development • Sensitivity • Awareness • Cooperation • Beliefs • Values • Expectations • Relations	*They are:* • Spontaneous • Empathetic • Warm • Subjective • Emotional • Perceptive • Sensitive
Idea (I)	*They talk about:* • Concepts • Innovation • Creativity • Opportunities • Possibilities • Grand designs • Issues	• What's new in the field • Interdependence • New ways • New methods • Improving • Problems • Potential • Alternatives	*They are:* • Imaginative • Charismatic • Difficult to understand • Ego-centered • Unrealistic • Creative • Full of ideas • Provocative

(Reprinted by permission-Reference 5.4)

5.3.4   Psychological Decision Theory

The personality types represented by the above Myers–Briggs test, as well as the action–people–process–idea notion, indicate that we have different tendencies in the way that we look at the world and the problems with which we are faced. These usually result in different approaches to our jobs, and in the case of this text, different ways of dealing with the many issues that arise in managing a project. However, various researchers have studied human behavior patterns and have found that there are certain tendencies that are more-or-less common to large numbers of us. These can be thought of as fitting within the general category of psychological decision theory, championed by D. Kahneman, P. Slovic, and A. Tversky [5.6], among others. These behavior patterns are relevant to our subject because they may affect how we might behave as project managers or as members of a project team. Three aspects of the results provided by these researchers are:

- Regression to the mean
- Representativeness and availability
- Loss avoidance

Regression to the mean refers to a general tendency to let down after a stellar performance and to improve after a poor performance. Its application to the world of project management might suggest that after a high-performing core team of software engineers has been working overtime for months in order to complete a software system, it is likely that for the next assignment they might well "regress to the mean." So what might be done about this? Depending upon the situation, one might consider (a) giving the team some time to decompress, and/or (b) starting the next difficult assignment off with a different core team, if possible.

Representativeness and availability both refer to setting up a mental model that is based upon prior experience rather than current facts and likelihoods. Its relevance to project management has to do with what people might do when trying to solve a particularly knotty problem. A thought pattern, for example, might be expressed as: "When we saw a problem like this before, we cut back the staff and that solved the problem." The person with this particular thought and suggested solution might be viewed by other members of the team as (a) not seeing how the current problem differs from the previous problem, (b) stubborn in not being receptive to other solutions, and (c) headed down the road of doing some serious damage to the project by removing some people from the team. We all learn valuable lessons from prior experience, but we must also be open to new solutions that are based upon the data and information of the situation we are in today.

Loss avoidance is the possible tendency for people to avoid virtually sure losses in favor of cases where expected value losses may be much greater. This might well account for why people are reluctant to sell stock holdings at a loss, hoping (against hope) that all will be better in the future. In terms of project management, it may be relatable to not being willing to take a poor performer off the job, or moving on to a different product or customer even though you are not being successful with the ones at hand. Losses are hard to accept in many situations, but the project triumvirate needs to be able, at certain times, to take a loss and then move on. Perhaps you can think of project management situations you have been involved with in which such an approach would have been the correct one.

5.4 INTERACTIONS WITH YOUR SUPERVISOR

Poor interactions between boss and subordinate often appear to be the rule rather than the exception in projects as well as all domains within the broad field of management. Indeed, this is the most dominant and important problem yet to be solved in the management arena. Almost everyone seems to have a "bad" boss and the worlds of industry, government, and academia are rife with ferocious complaints about the boss.

As a means of exploring this issue, the reader is asked to page back to Table 5.1 and now score your boss with respect to the cited attributes and scoring system. Compare your own score with the score that you gave your supervisor. Are you one of the very large population that indeed has a bad boss?

5.4.1 The Bad Boss

In the context of a project, the "bad" or difficult boss shows up in two relationships: between the PM and his or her supervisor and between the PM and the project personnel. The PM may have a bad boss who interferes with and fails to support the activities of the PM. The project staff, similarly, may have a bad boss (the PM or CSE) who has few of the attributes discussed earlier in this chapter. Beyond the realm of these attributes, the boss may be a "pathologic" type who sabotages his or her people at every turn.

Although there are numerous studies and descriptions of bad bosses, a particularly interesting treatise is presented by Robert Bramson [5.7], who identifies seven bad boss types, namely:

1. Hostile-aggressives
2. Complainers
3. Silent and unresponsives
4. Super-agreeables
5. Know-it-all experts
6. Negativists
7. Indecisives

Bramson describes the hostile-aggressives as Sherman tanks, snipers, or exploders. Such bosses are not likely to be sympathetic to the many problems that invariably arise on a project. Instead, they are prone to being actively abusive and to run over all people who are perceived to have caused the problem or who are not able to provide an instantaneous solution. Interactions with such a boss are enervating and leave one either exhausted or extremely angry. Hostile-aggressives are very toxic people.

The complainers tend to find everything incomplete and inadequate and adopt the position that all would be well if only he or she had competent people on the job. The complainer can also be less negative about immediate subordinates, but focused instead on his or her boss and other people in the enterprise that are not cooperating and doing their jobs. This type of person prefers to be a victim and contaminates all who would listen with incessant complaining. In its most virulent form, the targets of the complaints are the subordinates.

The silent and unresponsive bosses appear to soak up inputs and requests for help but provide no feedback or assistance. In distinction to the hostile-aggressives, these people may be passive aggressives or they may be simply unable to keep up with the numerous issues and problems of project management. Their own inadequacy may be reflected in their unresponsiveness because they may be fearful of appearing to be stupid or uninformed.

Super-agreeables are pleasant to a fault and avoid ruffling feathers and confronting difficult situations and people. They therefore refuse to deal with controversial issues for fear of making someone else angry or, indeed, coping with their own submerged anger. Because many project-oriented problems require straight talk and confrontation of problems, such bosses are likely to be of no help whatsoever. At best, they may be empathetic but will not engage in even a minor battle to move a project forward. As PMs, super-agreeables find it extremely difficult to carry out a complex negotiation with a customer or with superiors.

The know-it-all expert tends to undermine the work of all subordinates. Either as PM or CSE, this type of boss frustrates subordinates by always having the "best" answer to a problem, whether it be administrative or technical in nature. This behavior pattern often leads to a "clamming up" by subordinates because they perceive the boss as someone who is not able to listen to and elicit a variety of opinions and solutions. "Because the boss knows all the answers," they reason, "let's withhold our views and any constructive thinking about the problem." This, of course, can lead to disaster in terms of putting best efforts forward, which, in turn, leads to sabotaging a project.

The negativists cannot find something good in anything that is done on a project. They embody this negative attitude that causes subordinates to avoid interactions with them. They reflect the opposite of a can-do viewpoint and therefore can be deadly in dealing with customers as well as subordinates. This type of behavior, of course, takes its toll on a project staff and inevitably leads to loss of interest, productivity, and performance. Subordinates want to transfer to a different project as soon as possible.

Indecisive bosses are invariably frustrating because they cannot bring themselves to a point of closure. By trying to keep all options open all the time, they fail to commit themselves and therefore fail to make progress. Such bosses are often fearful of making mistakes, which paralyzes them as well as the overall project. Projects run by such bosses tend to bog down and overrun schedule. They often also want to "study a problem to death," leading to serious diminishing returns and missing key milestones. Subordinates soon learn that they should make decisions themselves and ask for forgiveness rather than permission, if they have the wherewithall to do so.

If you have a boss that scores very low with respect to the attribute evaluation of Table 5.1, or fits one of the seven types just discussed, you have a serious problem. If you are a PM, the success of your project is in some jeopardy. If you are a worker on a project, you are likely to be frustrated and be engaged in a project that almost certainly will fail. The question that presents itself then becomes: What can you do to more effectively "manage your boss" so as to minimize your frustration and anger and maximize the chances of your own personal success and the success of your project?

5.4.2 Managing Your Boss

Effective management of your boss requires an awareness of both how you tend to behave and an understanding of the patterns of your boss's behavior. Further, rather than insisting that your boss change behavior, you should change your behavior, which will have the effect of forcing your boss to deal with something new. This is likely to change the dynamic of what might be going on between both people. As an example, if you have been passive as a response to the wilting onslaught of a hostile-aggressive boss, you might try a more aggressive response. This change of behavior is likely to create a new dynamic that may lead to some changes in how your boss deals with you. Remember, the only way to deal with a bully is to fight back.

A well-considered and rational analysis of the situation is a good first step in trying to manage your boss by changes in your own behavior. An example of how one might do that is shown in Table 5.3 in relation to the action–people–process–action paradigm previously discussed. This table shows various combinations of boss and subordinate types and calls for information about how both types might complain about each other. Try filling in the blanks. This will help in trying to understand how you might view the situation both as a boss and as a subordinate.

Another perspective regarding the management of a boss can be found by a careful reading of Bramson's book [5.7], particularly in relation to what to do about the seven bad bosses that he describes. His prescriptions for these extreme cases provide some valuable insight into new ways of behaving. They

TABLE 5.3   Predicting Boss-Subordinate Trouble Spots

		List two complaints from:	
Boss*	Subordinates	Boss	Subordinate
A, PE	PR, I	1.	1.
		2.	2.
PR, I	A, PE	1.	1.
		2.	2.
PR, PE	A, I	1.	1.
		2.	2.
A, PR	PE, I	1.	1.
		2.	2.
A, I	PR, PE	1.	1.
		2.	2.
PE, I	PR, A	1.	1.
		2.	2.

*A: action; PE: people; PR: process; and I: idea.

also implicitly reinforce the point that changes in your own behavior are the most effective ways of coping with a bad boss.

Finally, we list some more moderate actions that might be taken with a not very good but less than pathologic boss:

1. Keep all interactions on a formal basis.
2. Provide short but regular status reports on your activities.
3. Develop lists of items you think are important to accomplish and present these to your boss for agreement.
4. Demonstrate your capabilities with respect to
   - Your judgment
   - Your creativity and competence
   - Your responsiveness and responsibility
5. Look for opportunities to build trust.
6. Do not confront in public situations.
7. Do not allow yourself to be victimized.
8. Speak to a trusted colleague who knows your boss to try to get another point of view.
9. Take your boss to lunch to explore better ways of interacting.
10. If these do not work, speak to your boss's supervisor or the human resources people in your organization.

In all cases, try not to resign yourself to living with a seriously bad boss. Life is too short to not try to fix the problem.

5.5 CUSTOMER INTERACTION

In recent years, it has finally been recognized that the "customer is king." That is, the customer may not always be right, but he or she is paying the bills and desires to have the results of a project provide appropriate value for the money that is being spent.

In broad terms, from the perspective of a PM or CSE, there are several types of customers, namely:

1. The outside direct customer
2. The outside surrogate customer
3. The internal customer

The outside direct customer is the entity outside the organization for which the project is being executed, with such a customer directly using the results of the project. As an example, if the project involves the building by a systems contractor of an on-line transaction-processing (OLTP) system for a bank, the bank would be the outside direct customer. The user needs and requirements are defined directly by the bank, and the bank must be satisfied that the system ultimately meets these needs and requirements.

In the case of the outside surrogate customer, an agency or group is serving as a surrogate customer, representing the end user needs and requirements. This situation is typified by many government agency customers whereby one group is the acquisition agent and another group is the ultimate user. One may see this situation, for example, in the U.S. Navy, where one agency or center serves as the acquisition agent for the end user, which is the Navy's fleet. Thus, the needs and requirements of the fleet are represented and translated by the surrogate customer who is, in some sense, a "middleman" in the overall acquisition process.

The internal customer case may be described by a project being carried out totally *within* the confines of an organization or company. For example, some enterprises have a Management Information Systems (MIS) or Information Resource Management (IRM) Department that produces information systems for the rest of the organization. The project is thereby responsible to a user inside the company. At the same time, the PM is likely to have a supervisor who reports upward to a group that is parallel (e.g., the corporate information officer, CIO) to the customer group organization.

The point is that whereas each of the preceding customer situations presents somewhat different problems to the PM, all must be treated with due respect to the customer, and due regard for the customer's needs and requirements. To do otherwise is to create problems rather than solve them.

A summary of a dozen points that the PM and the CSE should keep in mind in dealing with a customer are listed in Exhibit 5.2.

Exhibit 5.2: Guidelines for Dealing with Your Customer

1. Your customer has a MBTI profile; try to figure it out and behave accordingly.
2. Focus on the needs and requirements as stated by your customer.
3. Imagine yourself in your customer's position.
4. Listen intently to what your customer is saying.
5. "Sell" your approach and end product or service to your customer.
6. Speak to your customer at least once a week.
7. Be thoroughly professional in all interactions.
8. Live up to all commitments.
9. If your customer is headed in the wrong direction, gently suggest alternative directions and actions.
10. Demonstrate your technical and management skills.
11. Maintain customer contact and interaction in parallel channels above the level of the PM (e.g., vice president to vice president).
12. Treat your customer with honesty and respect.

Following these "rules" helps to establish a long-term trusting relationship with your customer. This does not guarantee success, but provides an overall environment and relationship that helps to foster success. Perhaps the most important rule is the last one dealing with honesty and respect. This is the most critical aspect of dealing with a customer.

5.6 LEADERSHIP

The headline of an article in a newspaper on computers suggested that the purchase by Loral of IBM's federal operations raised "leadership questions." This headline echoed what has become a critical issue for many of our organizations and enterprises—that of leadership. Whereas not too many years ago, corporate executives were under close scrutiny for their management skills, or lack thereof, today the key word is leadership. There is little question, therefore, that a great deal of attention has been focused on the top management, in both government and industry, having the requisite leadership attributes. Some of that focus, indeed, is seeping downward into the domain of the Project Manager and the Chief Systems Engineer. Companies are looking for people who are not only outstanding managers, but are leaders as well. The conventional wisdom is that leaders are a small "subset" of good managers; leaders have extra qualities that transcend the skills of even the best manager.

From experience, leaders are not necessarily born to such a capability, but can be taught and can grow into leaders. For the PM or CSE who has achieved a high level of competence in this type of position, and who aspires to become a leader at the project and ultimately at higher levels in an organization, we add this short perspective on leadership.

5.6.1 Situational Leadership

One of the well-accepted models of leadership is the so-called "situational" leadership paradigm [5.8]. The premise of this model is fundamentally that leaders choose or select a leadership "style" that depends upon the situation in which they find themselves. The situations are characterized by two basic behavior dimensions, namely, (1) task or directive behavior, and (2) relationship or supportive behavior. In both cases, the leader correctly perceives the situation and modifies behavior to suit the circumstances.

If we form a scale from low to high for both task and relationship behavior, we can visualize the following four situations that describe the fundamentals of the situational leadership model:

Situation 1 (SI): High task, low relationship
Situation 2 (S2): High task, high relationship
Situation 3 (S3): High relationship, low task
Situation 4 (S4): Low relationship, low task

For SI, there is a high need to direct the behavior of subordinates, who generally are characterized by a low level of maturity. At the same time, the situation does not call necessarily for a close or supportive relationship during the execution of the work. In this type of situation, the leader is "telling" the followers what has to be done and is closely supervising the work as it is being performed. The argument is that the leader is selecting this mode of behavior because that is what is called for in this type of situation.

In situation 2 (S2), the task behavior is also high (follower maturity is low), but there is a high need for relationship and supportive behavior. In such a case, also recognized by the leader, he or she is "selling" by making sure that decisions are understood and that all questions are appropriately answered. The followers need to be "sold," so to speak, partly because they are not mature and partly because they require close contact with the leader.

In the third situation (S3), relationship and supporting behavior remains high, but the task behavior is low (maturity of follower is high). Here the leader is "participating" with the subordinates by sharing ideas and encouraging inputs and ideas to facilitate the decision-making process. The leader and followers are more in a collaborative type of relationship, with each making distinct progress through such an interaction.

Finally, in the last situation (S4), both the relationship (supporting) and task (directive) behaviors are low and the follower level of maturity is high. Here the leader is "delegating" a great deal of responsibility to the subordinates, feeling confident that they are capable of carrying out the various required tasks without much supervision. The leader is more of an observer and monitor, and the followers have the skills and perspectives for almost independent progress.

This situational model, then, is characterized by a conscious change of behavior on the part of the leader, adapting a leadership style that is tuned to the situation at hand. If the followers are not homogeneous in their capabilities and needs, the leader treats certain of them in one way (e.g., telling) and others in another way (e.g., delegating). In summary, the leader assumes the following roles for the four situations:

- Telling (SI)
- Selling (S2)
- Participating (S3)
- Delegating (S4)

We note that there is little emphasis, in this model, on the specific *attributes* of the leader. The qualities or traits of a leader are basically not addressed, other than that he or she is able to perceive situations and modify behavior in response to these situations. The following section explores the matter of the characteristics of a leader.

5.6.2 The Attributes of a Leader

There have been numerous investigations of the attributes of a leader. Indeed, this has been the main thrust of recent analyses of the leader and how he or she behaves. In broad terms, these explorations have taken two interrelated tacks. One has been to examine known leaders and to see how they tend to behave and what their personal characteristics are or have been. The other has looked at the demands placed on the leader, functioning in all domains (i.e., industry, government, academia).

In a survey of more than two dozen sources [5.9], some of the documented leadership investigations that have defined requisite leadership attributes have been summarized. The results are listed in Exhibit 5.3, with the order going from most to least important.

The attributes listed in the top seven all had different scores with a natural breakpoint between numbers 4 and 5. The most critical attribute, from the exhibit, was outer-directed and dealt with empowering, supporting, motivating, and trusting others. The issue of having a vision, so dominant in the news, scored number 2. The third most critical attribute was cooperating, sharing, team building, and team playing. This is distinctly opposite to some of the competitive behavior that we see in enterprises today, much of which is destructive and leads to burnout. Also, we note that such a leader is not only able to build a team, but can function easily as part of someone else's team

Exhibit 5.3: Results of Survey of Leadership Attributes

Critical Attributes
 1. Empowering, supporting, motivating, trusting
 2. Having a vision, long-term viewpoint
 3. Cooperating, sharing, team playing, and team building
 4. Renewing, learning, growing, educating

Extremely Important Attributes
 5. Being communicative
 6. Having culture and values, serving as a role model
 7. Being productive, efficient, determined

Significant Attributes
 8. Demonstrating time management, prioritization
 9. Being action-oriented
10. Making a contribution, commitment, legacy
11. Being innovative, imaginative
12. Having integrity, morality, humanity
13. Demonstrating skill, knowledge, substance

(if that someone else is also a leader). Rounding out the critical attributes is that of renewing, learning, growing, and educating. Such a leader is dedicated to individual as well as corporate growth, believing that without a constant renewal process, the organization will stagnate and ultimately fail.

In the second category of extremely important attributes, the list is led by the communicative leader. We stressed the importance of this characteristic and have more to say about it as well in the next chapter. Inculcating a culture and value system is next on the list. Many organizations take on the mantle of the culture supported by a strong leader (e.g., Tom Watson at IBM, Henry Ford at Ford Motor Company, Bill Gates at Microsoft). The culture is usually reflected in the personal behavior of the leader serving as a role model. Finally, and completing this category, the leader is productive, efficient, and determined. Many leaders, through their constant doggedness and determination, are able to achieve their desired results for themselves as well as their organizations. They do not allow themselves to be stopped by obstacles and initial setbacks.

There are six attributes in the significant category. Having the ability to prioritize and manage their time heads this list. Next, the leader is action-oriented, preferring to move ahead even when it may be an errant direction. Such a leader is able to make mistakes, learn from them, and retrace steps, if necessary. Next, the leader has a sense of the contribution that all are making to the overall well-being of the organization. Such a leader is committed to the enterprise and wishes to leave a legacy and mark on the organization. The leader is also innovative and imaginative and is able to try new modes of behavior, even if he or she is not the originator of the new idea. Number

12 on the list is having a definitive and positive sense of integrity, morality, and humanity. Many despotic "leaders," especially those who have led their countries down destructive paths for themselves and others, would fail this test of leadership. Finally, and curiously last on the list of thirteen, the leader has the skills, knowledge, and substance in the domain of the enterprise, whether it be business, engineering, politics, or some other arena.

Another more recent exploration by this author [5.10] of the attributes of leaders examined the writings of twenty-four investigators of this topic. The overall conclusion was the set of five attributes listed below as the most significant.

Exhibit 5.4 Additional Selected Attributes of Leaders [5.10]

- Practical visionary
- Inclusive communicator
- Positive doer
- Renewing facilitator
- Principled integrator

The practical visionary is able to focus upon distant goals, but does not have his or her head in the clouds. The inclusive communicator has the critical skill of being able to communicate, and makes sure that everyone is and feels part of the team. The positive doer maintains a positive attitude in the face of all kinds of obstacles, and keeps moving forward, accomplishing real things in the real world. The renewing facilitator helps other members of the team reach their goals, crossing bridges as necessary from the old to the new. Finally, the principled integrator is able to synthesize important pieces to construct the whole, maintaining an ethical perspective from beginning to end.

The reader with a further interest in the attributes of a leader can refer especially to the two sources cited in this section [5.9, 5.10] as well as to other significant sources [5.11, 5.12, 5.13, 5.14, 5.15].

5.6.3 The Project Manager as Leader

One might argue that requiring the PM and the CSE to be leaders, in the previous context, is somewhat of a stretch. However, the PM and the CSE *are* de facto leaders of their respective project teams. They may function well or they may function poorly in these capacities. The objective is to try to grow so that they can become leaders in the full sense of the word. This requires an understanding of what leadership means and a clear and determined receptivity to leadership ways of being and behaving. An abstraction of leadership attributes in terms of the Project Manager and Chief

Systems Engineer perspectives is cited Exhibit 5.5, using the notion and mnemonic of remaining receptive to what it might take in terms of changes in behavior and ways of interacting that require close examination.

Exhibit 5.5: A Leadership Mnemonic for the PM and CSE

R: Results and process-oriented
E: Encourages positive change
C: Communicates
E: Empathizes and trusts
P: People developer
T: Team builder
I: Integrates and synthesizes
V: Visionary
E: Exhibits a can-do attitude

Keeping the attributes of a true leader in focus at all times gives the PM and the CSE opportunities to grow from simply being a manager into becoming a leader. This goal is within the grasp of the receptive person who can embrace and deal with internal and external change and growth processes.

QUESTIONS/EXERCISES

5.1 Develop a score for a Project Manager you have worked for using the format of Table 5.1.

5.2 Develop a score for yourself as a Project Manager using the format of Table 5.1.

5.3 Make an educated guess at the Myers–Briggs profile for your boss. Do the same for yourself. What might this suggest in terms of your relationship with your boss?

5.4 Repeat the preceding exercise for the action-people-process-idea assessment.

5.5 Fill in the blanks in Table 5.3. Discuss the results.

5.6 Evaluate your boss in terms of the cited critical leadership attributes in Exhibit 5.3. Do the same for yourself. Then evaluate your boss in terms of the five leadership attributes listed in Exhibit 5.4. Do the same for yourself.

5.7 Construct your own "top five" list of leadership attributes that you believe are the most important for success as a Project Manager. Explain your choices for this list.

5.8 Examine the three aspects of psychological decision theory shown in the text of this chapter and for each of them cite an example of how it might apply to managing a project.

5.9 Develop a score for your best "subordinate" using the format of Table 5.2.

5.10 Identify six of your own attributes that you believe might help you continue to assume leadership positions in your company or enterprise. Note "why" for each of the six.

REFERENCES

5.1 Zimmerer, T., and M. Yasin (1998). " A Leadership Profile of American Project Managers," *IEEE Engineering Management Review* (Winter).

5.2 Myers, I. Briggs, with P. B. Myers (1980). *Gifts Differing*. Palo Alto, CA: Consulting Psychologists Press.

5.3 Keirsey, D., and M. Bates (1978). *Please Understand Me*. Del Mar, CA: Prometheus Nemesis Books.

5.4 Frame, J. D. (1987). *Managing Projects in Organizations*. San Francisco: Jossey-Bass.

5.5 Casse, P. (1981). *Communication: A Self-Assessment Exercise.* Washington, DC: International Bank for Reconstruction and Development, the Economic Development Institute, and the World Bank.

5.6 Kahneman, D., P. Slovic, and A. Tversky, eds. (1982). *Judgment Under Unceratinty: Heuristics and Biases*. Cambridge, MA: Cambridge University Press.

5.7 Bramson, R. M. (1981). *Coping with Difficult People*. New York: Ballantine Books.

5.8 Mersey, P., and K. Blanchard (1977). *Management of Organizational Behavior Utilizing Group Resources*, 3rd edition. Englewood Cliffs, NJ: Prentice Hall.

5.9 Eisner, H. (1993). *Leadership Imperatives: Do They Support Creativity and Innovation?* Paper read at the American Society for Engineering Management (ASEM) 1993 Annual Conference, Dallas, October 22–25.

5.10 Eisner, H. (2000). *Reengineering Yourself and Your Company: From Engineer to Manager to Leader*. Norwood, MA: Artech House.

5.11 Maccoby, M. (1981). *The Leader*. New York: Simon and Schuster.

5.12 Bennis, W., and B. Nanus (1985). *Leaders*. New York: Harper & Row.

5.13 Leavitt, H. J. (1986). *Corporate Pathfinders*. New York: Viking.

5.14 Zaleznik, A. (1989). *The Managerial Mystique: Restoring Leadership in Business*. New York: Harper & Row.

5.15 Gardner, J. (1990). *On Leadership*. New York: The Free Press.

_____6
TEAM BUILDING AND TEAM INTERACTIONS

6.1 INTRODUCTION

This chapter deals principally with issues surrounding team building and interactions between team members. The most natural area of focus is the project team as a whole and the Project Manager (PM) as the head of that team. In point of fact, however, there are actually several teams within the context of a project team. The Chief Systems Engineer (CSE) has the engineering group as his or her team. That group is usually broken down into subgroups (see Figure 1.2), so that there may also be a hardware engineering team and a software engineering team, and so on. If the project is large enough, the Project Controller (PC) is likely to have a team of people working on project measurement and control matters.

A variety of other activities that are often carried out within a team context are considered in this chapter, as depicted in Figure 6.1. Included are the activities of holding meetings and making presentations. These are forums for both internal and external communications and doing them correctly enhances overall project operation. In addition, and possibly inevitably, team members will come into conflict with either one another or the team leader. Thus, we also deal with the matter of conflict resolution and management in this chapter. In today's world, writing proposals to potential customers is a critical part of an enterprise and is almost always carried out by teams. Some ground rules for this important activity are presented and explored. Finally, the chapter is completed with a brief discussion of some practical ways to motivate and create incentives for members of a project team.

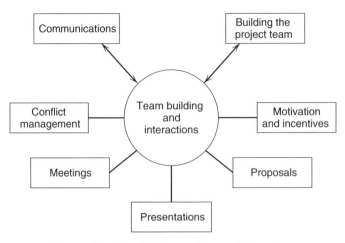

Figure 6.1. Team building and areas of interaction.

6.2 COMMUNICATIONS

We have stressed the importance of strong and effective communication both within and external to a project. We have also seen that communication skills are essential to the success of a manager and a leader. Poor communicators are likely to fail at the challenging job of running a project or a systems engineering team. It is an axiom of management that there can never be too much effective communication. It is almost always true that there is too little positive and honest communication.

Strong and well-considered communication is at the heart of building a productive team. Thus, we pause at this point to highlight some of the critical aspects of being an effective communicator. These are listed in Exhibit 6.1 and briefly discussed in what follows.

Exhibit 6.1: Essentials of an Effective Communicator

1. Listen.
2. Adopt a management by walking around (MBWA) way of being.
3. Assure participation by all team members.
4. Synthesize and integrate.
5. Meet with all key project personnel every week.
6. Insist on information "flow-down."
7. Hold short "information" meetings.
8. Communicate with boss and other project support people.
9. Talk to customer at least once a week.

10. Maintain a positive and supportive attitude.
11. Offer training for poor communicators.
12. Assure that communications is part of personnel evaluation.

As indicated in the previous chapter, listening is a crucial part of communicating. It gives respect to the person who is talking to you and conveys the message that he or she has something to say that is of value. Regarding management by walking around (MBWA), this is an informal way for the PM and CSE to obtain and convey information in an easy and nonintrusive manner. Coming to the workplace of a subordinate also suggests that the manager is comfortable with and wishes to be in contact with the "innards" of the project. The PM should also make sure that the more reticent of the project team are invited to participate. Otherwise, the dominant members of the team may monopolize the discussion, both formally and informally, and the more laid-back people will not put forth their ideas. Many people want to be asked what they think. A smart and sensitive PM understands that and does the asking. True communication also involves listening, absorbing what was said, integrating it with other information, and providing the results to those one is communicating with.

As another ground rule, the PM and CSE should be in touch with all key project people, not necessarily all personnel, at least once a week. MBWA is but one way to accomplish that. A short telephone call, or a short meeting, also assures that contact is continuous and productive. The flow-down of information is sometimes assumed, but often not carried out. Many PMs are surprised to find that what they convey to lead engineers stops there. Information flow-down must be assured to avoid isolation ("No one tells me anything around here") and let people know what they need to know to be part of the team. As suggested before, not all meetings are "decision" meetings. A short "information" meeting lets team members know that you are specifically interested in keeping them informed.

Open channels of communication are also crucial with your boss and with support department personnel that have a role to play on the project. As suggested in the last chapter, it is a good idea to contact your customer every week, if only to assure that everything is on track. Communication that is negative and nonsupportive is worse than no communication at all. The PM must adopt a positive and supportive position, except under the most radical of situations. Here again, we see that type of position, for example, for sports coaches that are successful. For those with lead positions on the team that have difficulty in communicating, the PM should establish some type of training program to build skills in this most important area. Finally, make sure, especially with these same lead people, that communications is part of your (at least annual) evaluation of them. This brings the point home in ways that most people understand and value.

6.3 BUILDING THE PROJECT TEAM

Team building and being part of a team are critical issues in project and systems engineering management. Teams, however, do not spontaneously appear; they must be built. As suggested before, a prerequisite to team building is to follow the communication "rules" identified in Exhibit 6.1. These are necessary but not sufficient conditions for building an effective team.

We see, in various parts of our lives, numerous real-world examples of team building and lack thereof. Perhaps three examples stand out and are visible, at least in part, to the population at large. One has to do with the President of this country. We can track, through impressions obtained in the newspapers, how the President has built a team and the extent to which bridges have, or have not, been built to the Congress. This is a massive team-building undertaking and context, and the more successful Presidents have been broad and inclusive in their interpretation of the team that must be built. Another example is that of a coach of a football team. Some coaches appear, year after year, to get the best out of the talents present in the members of the team. Indeed, one measure of the success of a team is precisely whether all team members are doing the best they can do. When this is achieved, even if the team does not win every game, there is a strong and positive sense of team effort and achievement along with strong ties and camaraderie between team members as well as the coach. Thus, team building and coaching are very similar. The effective team builder must be a good coach. Finally, most of us are part of some type of team in our work environments. We thus can observe team interactions in that context, whether we are teachers, engineers, administrators, middle managers, or members of the board of directors of a corporation.

We now identify ten specific suggestions for building a project team, as listed in Exhibit 6.2 and discussed in what follows. Following these suggestions, together with those provided in the previous chapter, will likely lead to a strong and effective team operation.

Exhibit 6.2: Suggestions for Building a Project Team

1. Develop and maintain a personal plan for team building and operation.
2. Hold both periodic and special team meetings.
3. Clarify missions, goals, and roles.
4. Run the team in a participative, possibly consensual, manner.
5. Involve the team in situation analysis and problem solving.
6. Give credit to active, positive team members and contributions.
7. Assure team efficiency and productivity.
8. Obtain feedback from team members.
9. Integrate, coordinate, facilitate, and assure information flow.
10. Maintain effective communication.

6.3.1 Personal Plan

The team leader (PM, CSE, PC, and others) must always maintain an explicit level of consciousness about the team and how it is performing. This can take the form of an informal plan that reflects a continuous assessment of team behavior and operation as well as what might be done to bring about improvements. After each meeting of the team, the leader should evaluate what happened and determine whether some type of fine-tuning is necessary. Running a team should not be taken for granted. It requires constant adjustment to assure that the team, and the leader, have not fallen into bad habits and ineffective operation. The team has to be continuously stimulated to do its best.

6.3.2 Periodic and Special Team Meetings

The usual forum for team operation is a meeting of one type or another. Ground rules for running a meeting are considered in more detail later in this chapter. Here we simply note the requirement for both periodic as well as special meetings. Periodic meetings are necessary to maintain continuity and, depending on the project, can be held weekly, biweekly or monthly. They should be on the calendars of all participants so that other work pressures do not interfere. They should start on time and respect the fact that team members cannot spend all their time at meetings. Special team meetings are called in response to unique situations that may arise. Usually, they are necessary when an unusual and time-critical problem has unexpectedly surfaced, calling for a problem-solving response by the team.

6.3.3 Missions, Goals, and Roles

The team leader has the responsibility to clarify the overall mission of the team, the specific goals that are to be achieved, and the roles of the various team members. In a project context, these are often well known because, as a minimum, they are articulated implicitly in the project plan. The team may also be a "task force" within the project team, whose job it is to solve a particular problem. For example, such a team may be given the charter to maximize the effectiveness of interactions between the project team and the external interfaces with other entities or departments in the corporate enterprise (accounting, finance, contracts, human resources, graphics, etc.). Failure to be clear about missions, goals, and roles usually leads to confusion, thrashing, and less than a positive attitude from team members.

6.3.4 Participation and Consensual Operation

As a minimum, the team leader should adopt a style of participative management of a team. This implies full involvement of all team members in the process as well as in the products of the team. Those members who are laid-back should be encouraged to participate and bring their thoughts and ideas to the forefront. Lack of participation can be a signal of some type of dysfunction in the team; the leader should be particularly sensitive to the non-participating team member, following up with one-on-one conversations to see if there is some type of problem lurking in the background. The full-participation team is very likely to be a high achiever with solid relationships between team members and the leader.

In just about all participative team operations, it should be recognized that although there is strong participation in all processes, the team leader usually takes responsibility for all key decisions. This means that participation does not imply majority rule. In a project situation, the PM does not normally take a literal "note" and then automatically go with the majority. Indeed, the PM may disagree with the majority and therefore may make a decision that is a minority decision. Such are the vagaries of managing a project. In addition, all team members must be aware of and accept the prerogative of management in terms of the final decision, and once it is made, must use their best efforts to implement that decision. There is no room for the team member who undermines the team leader's decision, whether or not he or she agrees with it. If the team member cannot ultimately support a PM's decision, the next step is to leave the project team.

Participative operation, however, is not the same as consensual operation. More precisely, the latter can be interpreted as team agreement on a course of action, but without an explicit vote of the team members [6.1]. Consensual operation of a team is very desirable but not always possible. Most decisions, for a good team, turn out to be consensual. However, the leader should reserve the right, for a particularly contentious situation, to both take a vote (show of hands) and to make a decision that is contrary to the majority viewpoint.

6.3.5 Situation Analysis and Problem Solving

In Chapter 4, we introduced the notion and specific steps of situation analysis. Its purpose is to examine difficult situations that inevitably arise during the course of a project and, through a team dynamic, develop a solution. The team, in many ways, is the centerpiece of both situation analysis and problem solving. The PM or the CSE may have well-developed ideas as to the nature of a problem as well as its solution, but it is critical to obtain the ideas and inputs from other key team members who may be able to suggest answers that may not have been considered. In short, many people working constructively on a problem usually leads to better problem solving both immediately and over the long run.

6.3.6 Give Credit

Give credit to team members that make a contribution; this is a very important part of team building. This can be achieved in many ways—from a pat on shoulder to a formal acknowledgment at a team meeting. This works wonders because everyone has a basic need to feel appreciated. If this can be done in public, so much the better.

6.3.7 Assure Efficiency and Productivity

A PM with excellent team-building skills will nonetheless fail unless the team is being productive. This means making progress in accordance with the project plan as well as being efficient in the use of everyone's time at all team meetings. It also means that the PM must know when it is time to stop beating a problem to death and move on to a solution or to the next issue or, indeed, to call a meeting to a close. Experience suggests that many team members become frustrated with long meetings that drone on and on and would prefer crisp, effective interactions that support a sense of accomplishment for all that are present.

6.3.8 Obtain Feedback

Feedback about the team and its operation should be sought from all team members, preferably through one-on-one sessions. Team members with problems with the team may be wary about expressing their views to all. By means of a supportive private discussion, team members can provide feedback and will understand that their inputs are being valued and serve as a contribution to the team. Some may have an issue with another team member that may be addressed through such private feedback sessions. It is yet another way for the PM to convey the message that this person is valued as an individual and as part of the team.

6.3.9 Integrate, Coordinate, and Facilitate

The PM and CSE do not have all the answers, nor do they possess all of the wisdom on a project. Adopting a position as integrator, coordinator, and facilitator normally pays great dividends in terms of overall team effectiveness. This may appear to be a passive role, but actually it can be exercised in a rather active manner. It also goes along with the notion of empowerment. The job of the team leader is not to do all the work, but rather to assure that the best efforts of all team members are brought out.

6.3.10 Communicate

The previous section and Exhibit 6.1 summarize a dozen key points with respect to communication. Remember, even if the PM is well-skilled at all the nuts and bolts of project management, the project is likely to fail, and a team not likely to be built, without effective communications. This is perhaps the most important single message in all of the fine art of management.

6.3.11 Additional Points on Team Building

Teams are formed in contexts other than that of a formal project as described here. There are other types of teams, such as quality circle teams, integrated product teams, concurrent engineering teams, task force teams, and others. The key point is that the earlier discussion applies to all these various types of teams, and that the required skills are more or less the same. Learning the basics of team building, therefore, is worthwhile whether or not you now serve as a PM or CSE. Sooner or later you will be part of a team and perhaps sooner than you think you will be asked to build and run a team effort.

We also note that in the presentation of the subject of leadership in the previous chapter, the third most important attribute of a leader was cooperating, sharing, team building, and team playing. To be a leader, then, is to know how to build a team. Conversely, and categorically, to not know how to build a team is to not be a leader.

There is an endless supply of literature on the critical subject of team building, including entire books, courses, videotapes, and magazine articles. For the reader interested in pursuing this matter beyond the suggested essential steps cited in Exhibit 6.2, reference is made here to the *Harvard Business Review* and several additional sources [6.2, 6.3, 6.4, 6.5]. Effort put into building a strong, effective, and productive team will pay large dividends for both the Project Manager and the Chief Systems Engineer.

6.4 TEAM BUSTERS

It is important to pause for a moment on the subject of team building and acknowledge the existence of a potential counterforce to team building, namely, the "team buster." This is a person, nominally a member of the project team, who works hard, either consciously or otherwise, in destroying the team that the PM and CSE are trying to build. We address here some ways to deal with such a person.

One can spot a team buster by the following types of behavior patterns, which may be manifest singly or in various combinations:

- Questions the authority of the PM and CSE at every turn.
- Challenges the management and technical approach of the PM and CSE.
- Does not follow the agreed-on decisions.
- Consistently "goes over the head" of his or her boss.
- Tries to monopolize meeting agendas.
- Attempts to embarrass or challenge the boss in front of others.
- If a manager, and you are not part of his or her team, by definition you are doing a poor and misguided job.

- Forces his or her people to clear all actions, and even conversations, strictly through the "chain of command."
- Tries to create a "we" and "they" mentality, whereby everyone who reports to the team buster is a "we" and everyone else is in the "they" group.

In point of fact, when severe enough, the team buster can only really play on one team, his or her own. The team buster undermines all attempts at team building by his or her supervisor, and is a detriment to the team. Often, this type of person can have significant talents. Otherwise, he or she would not have survived to the current level of responsibility. The team buster can also be viewed as a "bully," a person who is predisposed to bullying other people to get what he or she wants. In effect, aggressive nonteam playing is a manifestation of bullying behavior.

There is only one real way to deal with a team buster. Loosely speaking, it is the "three strikes and you're out" solution. The first time the team buster exhibits the type of behavior cited before, the PM or CSE must have a private conversation with the team buster that makes it clear that the behavior has to change. This point must be made forcefully and emphatically. All attempts at cajoling and persuading are likely to fail, as they are ultimately viewed as signs of weakness and indecision. At the second infraction, another private discussion is called for, now making the point that if the behavior continues, the person will be removed from the job. This, of course, has to be supported in advance by the next in the chain of command, and might imply an exit from the company. When the next infraction occurs, and it usually does, all the pieces have to be in place in order to take the action of removal of the team buster from the project, and possibly from the organization at large. Again, it should be done in private, but with the appropriate member of the human resources department present. This is also an option after the second infraction.

In some cases, assistance in dealing with a team buster may be provided by human resources. This can take the form of counseling and other methods for creating awareness of the severity of the situation at hand. Again, if one is truly dealing with a team buster, the chances are that this attempt at changing behavior will not work.

The final point, with respect to a true team buster, is that the PM or CSE must act decisively and clearly, and show no signs of wavering. The behavior of the team buster poisons the team and creates all manner of havoc. There are times when reconciliation is impossible, as it usually is with the true team buster. The best solution is separation, as difficult as that may appear to be. The PM and CSE should resist blaming themselves for being unable to reform the team buster and move on to the more productive activities of team building.

6.5   CONFLICT MANAGEMENT

Conflict can be considered an inevitable part of running a project, especially a large one. It is not necessarily and always negative. It often can be turned into a positive, growing, and learning experience. In this section, we explore how conflict may be approached and managed, and also some of the styles that people adopt in attempting to deal with conflict. As with the other interpersonal relationship issues in this book, we acknowledge that there is great deal that cannot be covered here and attempt to focus on the key elements and essentials that must be known to both the Project Manager (PM) and the Chief Systems Engineer (CSE).

6.5.1   Areas of Conflict

Studies have shown the areas in which conflict tends to arise most frequently. Apparently, these areas have also changed with time. As an example, the following list shows conflict areas and their rank in studies in 1986 and in 1976 [6.5]:

Conflict Area	Rank in 1976	Rank in 1986
Schedules	1	1
Costs	6	2
Priorities	2	3
Staffing	3	4
Technical opinions	4	5
Personality	7	6
Procedures	5	7

It is interesting to note the persistence of schedules over the decade shown as the number 1 area in which there is conflict. Costs jumped into the number 2 position, with overall priorities staying within the top three. Thus, we see schedule and cost as critical items over which conflict occurs. Technical opinions is about midrange in both lists, and personality conflicts are present but toward the bottom of the list.

Conflicts regarding impersonal issues (schedules, costs, etc.) can be easier to deal with than personal issues. In principle, the former deal with different perceptions of objective facts. Personal issues are less than objective, and people will be at odds with one another simply because they do not like or approve of one another. From the point of view of the PM or the CSE, conflicts and conflict management should be considered part of the job. The question is how to deal with it when it does occur and what are one's individual propensities toward coping with conflict situations. These are referred to in the literature as conflict resolution styles, and are examined in what follows.

6.5.2 Styles

People approach conflict in different ways and these can be identified and measured. As an example, conflict styles may be articulated as [6.5]:

- Competing (forcing)
- Compromising (sharing)
- Avoiding (withdrawal)
- Accommodating (smoothing)
- Collaborating (problem solving)

In addition, there exists a measurement instrument known as the Thomas-Kilmann Conflict Mode Instrument [6.6] whereby one can measure an individual's tendency toward adopting one or another mode of conflict management. The reader with a interest in knowing more about his or her own tendency is urged to contact Thomas and Kilmann and take their conflict mode measurement "test."

Competing (forcing) is an approach whereby power is used to resolve a conflict. This may be done in a variety of ways. The most obvious is to utilize the dominant position as a supervisor in order to force resolution. In effect, "We will do it this way because I am the boss." This may temporarily resolve the conflict, but it may not persuade or convince anyone to change positions. The power may be applied directly or even subtly, but competing or forcing is not a long-term and reliable way to resolve conflicts. In certain situations, it may exacerbate the conflict and cause people to respond in kind when they have greater power leverage.

The *compromising* or sharing style involves trying to find a position that is acceptable to all parties. It is a classical "negotiation" stance and can often lead to an effective resolution. Unfortunately, the results may be acceptable in terms of human relations but may be wrong for the project. As an example, if a conflict occurs with respect to estimation of the time it might take to perform a given set of activities, a compromise solution might be to accept the mean value between the estimated values. This argues for "beauty" instead of "truth," and may hurt the project by failing to get to underlying facts that might be important. Some researchers in the area of conflict resolution have also called this approach the "lose-lose" solution because the combatants each lose a little in order to come to a resolution. This approach might work well in international negotiations, but has its shortcomings in a project context.

The *avoiding* or withdrawal approach simply refuses to come to terms with the conflict and face it squarely. Under these conditions, of course, the conflict remains and festers like a bad sore. No resolution occurs, and a poor model of behavior is established. The conflict may go underground for a while, but because its essence is not dealt with, it does not really go away.

Many novice managers adopt this mode of behavior because they are unsure as to their position, power, and skill in contentious situations. Some do not see alternative modes of behavior that lie between the extremes of "fight" or "fly" and therefore prefer to fly. It is not a recommended way of resolving conflicts because it really "pretends" that the conflict does not exist or, if it does, is not in need of action.

The *accommodating* or smoothing solution acknowledges the conflict but plays down its severity or possible impact. This approach is sometimes referred to as suppression because its ultimate purpose is to dampen the conflict and reduce its potential effects. It can be a good approach when the conflict cannot be dealt with at the moment it occurs. For example, if two members of the team flare up in conflict at a meeting, it may be entirely reasonable to suppress such a conflict, thus preventing progress on the meeting's agenda. In short, accommodating may be a good temporary solution but it does not really resolve the conflict. It is recommended only when the situation at hand does not provide sufficient time to tackle the conflict in a more fundamental manner.

A *collaborating* or problem-solving style recognizes that the combatants have a right to state their different views and that all views are accepted as valid. In this mode, there is encouragement to bring all views and perspectives to the forefront so that they can be explored in detail. Reasons "why" are elicited so that there is a clarification as to the issues and positions. If handled correctly, this will usually lead to a better understanding between combatants and a willingness to go beyond the surface conflict to its deeper roots and rationale. Listening is encouraged so that the participants can learn how to accept other positions with grace and equanimity. The objective of this approach is not only to collaborate, but also to truly solve the immediate problem. It may indeed have the ultimate effect of teaching people how to resolve conflicts in a productive manner. This, of course, is the recommended conflict resolution mode and, when skillfully applied, can support the long-term effectiveness of the project team.

We do not expect, in this short discussion, to delve deeply into a subject as complex as the human behavior aspects of conflict and its resolution. The basic point is that all of us have natural tendencies to handle conflict in different ways. If you can identify your own tendencies in relation to the preceding alternative modes, you may have a new way of looking at and approaching the difficult problem of handling conflict. Many people are good in conflict situations as long as they are not one of the combatants. In general, it is a good idea to try to see alternatives when you are a part of the conflict and can take a step back in an attempt to move into a less personal problem-solving mode. Backing down from a previously held position is not the end of the world. Indeed, it may actually represent the dawning of a new acceptance of the wisdom you have gained. Giving up old styles of combatting and competing may help you avoid ulcers and burnout.

6.6 MEETINGS

In Exhibit 6.2, meetings were suggested as one of the primary mechanisms for team building. Meetings, of course, create the opportunity for the team to "do its thing" in terms of real information exchange and problem solving. They are the operating crucibles in which the dynamics of team interaction are played out. For a healthy team, they are a thing of beauty. For an unhealthy team, they may bring out and encourage further dysfunctional behavior. The next time you go to a meeting, observe the team dynamics with a critical eye to see if your team is operating on all cylinders.

We cite here a number of ideas for establishing and carrying out meetings, as shown in Exhibit 6.3.

Exhibit 6.3: Ideas for Managing Meetings

1. Make clear the purpose of the meeting.
2. Decide if the meeting is periodic or special.
3. Establish an agenda.
4. Determine who should attend.
5. Fix the length of time for the meeting.
6. Make notes on expectations:
   - Problem discussion
   - Problem solving
   - Information exchange and sharing
   - Other
7. Elicit ideas and alternatives.
8. Define action items.
9. Have someone take minutes.
10. Determine, if necessary, when the next meeting is to occur.

The PM or CSE should have a clear idea as to the purpose of the meeting and be able to convey that purpose to the team, either in written form prior to the meeting, and certainly as the first item of the meeting. This provides focus for the meeting and avoids straying to a variety of possibly irrelevant subjects. Suggestions for extending the meeting's purpose are acceptable but are at the discretion of the leader.

All parties should know if the meeting is part of the stream of periodic meetings or is a special meeting to handle a more critical issue. In this context, when unforeseen problems arise, the leader should feel free to call a special meeting to use the team for situation analysis (see Chapter 4) and problem solving.

If possible, the team leader should establish and distribute a written agenda. This may not always be possible, as with emergencies, but it is helpful to know the scope of what the leader plans to deal with in advance. This allows team members to think about the meeting beforehand and also to bring appropriate materials to the meeting. It helps in the overall flow of the meeting and avoids a scene in which everyone is waiting for one person to retrieve important data for distribution to everyone.

The leader should give prior thought as to who should attend the meeting; they might include people who do not normally attend project meetings, such as folks from human resources, contracts, and so forth. The leader should be careful not to inadvertently exclude people who should be at the meeting or who are normally part of such deliberations. Excluding key players from meetings by not thinking may well damage the relationship with such people. No one likes to be excluded from important project considerations and problem-solving sessions.

When the meeting is announced, both a start and end time should be established. Team members are busy people who have other commitments and need to know when they can fit all these obligations into their time-pressured days. The team leader should be thoughtful about this item and not set up a pattern whereby all meetings tend to overrun by significant amounts. If you need four hours, take them, and let everyone know that the meeting will be a long one. But it should be over at the end of four hours, or earlier. Demonstrating time discipline and respect for the time of others is part of the job of the team leader.

Prior to and as preparation for the meeting, the leader should make notes on expectations for the meeting. This includes such items as discussion of the main points of a problem (schedule, cost, performance, etc.), approaches to solving the problem (e.g., alternatives), what information has to be brought to and out in the meeting (e.g., cost reports and master schedule), and anything else that appears to be relevant. These notes are not part of the agenda but are private scribblings that the leader can refer to during the meeting.

The team leader should make sure to elicit ideas from all participants. This helps not only to build the strength of the team, but also assures the broadest range of inputs from the participants. Even the quietest member of the team may have the right solution for the problem at hand. All inputs should be respected and listened to very carefully. Special attention should be paid to bringing new alternatives to the table in an attempt to define all the available options for team consideration and eventual decision and action.

The meeting should not be concluded without a clear recapitulation of all action items that flowed from the meeting. This includes actions decided on early in the meeting that may have been forgotten or overlooked. Everyone should take notes on the action items for which they are responsible. This can also be recorded on a blackboard or a whiteboard that provides immediate hard copy.

Minutes should be taken of the meeting, but these should be brief and also recap, as a minimum, all action items. Action items should carry information not only on what is to be done, but also by whom and when. This type of permanent but short record of meetings adds discipline to the process and also serves as a way of resolving potential conflicts about what was done and what conclusions were reached. This should be viewed as a way of facilitating information exchange rather than "papering the file." The leader should also consider sending selected minutes to his or her boss for particularly important situations and subjects.

Finally, and before the meeting adjourns, the time for follow-up meetings, if necessary, should be determined. This allows all members to check their calendars immediately, and the best times for the next meetings can be chosen in real time.

Running effective and efficient meetings is an integral and critical part of managing a project. They should not be approached without a consciousness of their importance and what one expects to accomplish. They serve many purposes, not the least of which is to establish a productive team dynamic. As one might expect, therefore, they have been the subject of considerable attention in the literature over the years. As an example, a book on making meetings work [6.7] suggests an approach called the New Interaction Method. This method, simply put, focuses on roles and responsibilities of four key players at any meeting, namely:

- The manager/chairperson
- The facilitator
- The recorder
- The group member

Basically, the book supports the notion that the preceding roles will create a dynamic that keeps the meeting on course and that they are crucial functions in any meeting. In addition, a variety of helpful hints are provided, including such subjects as finding win/win solutions, establishing a good agenda, working the issue of room size, and several others. The reader with a special interest in meetings is urged to consider the referenced book as well as others on this important topic.

6.7 PRESENTATIONS

The matter of preparing and giving presentations arises in at least two team contexts:

- Presentations made by team members within the team
- Presentations made by team members to persons (e.g., customers) outside the team

In the latter case, the team often meets to develop the presentation and also to "dry run" its presentation to others.

A set of eight essentials in terms of preparing and giving presentations is provided in Exhibit 6.4. The assumption here is that most presentations are made by utilizing slides or viewgraphs of some type.

Exhibit 6.4: Eight Ground Rules for Presentations

1. Know your audience.
2. Tell them:
   - What they're going to see
   - What they're seeing
   - What they saw (summary)
3. Maintain eye contact with key people in the audience.
4. Do not read each viewgraph/slide; paraphrase the main ideas.
5. Avoid slides that:
   - Are too cluttered
   - Talk down to the audience
   - Are too ostentatious
   - Cannot be read by everyone in the room
6. Leave enough time for each slide's message to sink in.
7. Be careful about interruptions.
8. Generally, hand out hard copy at the end, not at the beginning.

A critical part of constructing any presentation is to know your audience in advance. If you are able to do this, you will also be on track in terms of targeting key areas of interest. Try not to make assumptions about the audience when a phone call will give you some data to work with in this regard. Many presentations go astray from the beginning when it is realized that the primary focus is off center and the audience forces a change up front or is clearly impatient with material they already know.

A well-known and accepted ground rule is that the audience needs constant reminding of where you are, where you're going, and where you've been. This may sound like overkill, but keep in mind that the audience is often completely cold on the material being presented. These reminders give a sense of unity to the presentation and help the listener to integrate what is being said. The summary is especially important in pulling together the main thoughts, themes, and points that have been made.

Eye contact with particular people in the audience is extremely important. As a minimum, one should target the key players in the audience and make sure that eye contact is established. In general, the presenter should scan the audience and talk to *everyone* in it. One should try to avoid talking in the direction of the slides, looking at the ceiling, presenting to members of your own team if they are intermixed with a customer group and talking to only one person in the audience. The basic idea is to get your message across to every single person that has not previously heard the presentation.

A definite "no-no" is to read every word on every slide. Leave time for the audience to read the slides for themselves. Point to the key phrases (with a pointer, if possible) to direct the eye of the observer. Then paraphrase an important point or focus on only a few key words on which you can elaborate. Do not go through the slides at lightning speed because you will frustrate and lose your audience. Each slide has several messages and these need to be conveyed in an easy and flowing manner.

In terms of the slides themselves, there are lots of options. You can have a lot of information on each, but then need to go through the slides rather slowly. Slides should be readable, in general, unless you are trying to create a general impression without all the details in the slides. Avoid too much clutter that cannot be understood. Each slide should be "designed" so that the messages jump out rather than having to be dug out. Slides should not be too showy because that is likely to turn off at least some members of the audience. They also should be of a size that they can be read by everyone in the room. This means that you need to anticipate the size of the room and the number of people in the audience. Under no circumstances should you talk down to the audience as if they are dummies if they do not instantly understand everything you are conveying.

The presenter has to leave enough time for the audience to read the slide as it is directed to do so. More time is better than not enough. Even short periods of silence are acceptable because people often cannot process what they are reading and what they are hearing at the same time. Depending on the slide design and the method of presentation, a target might be two to three minutes per slide, assuming no interruptions. At three minutes per slide, a ten-slide presentation takes about half an hour. The idea is to have the messages sink in, not to meet a deadline. Dry runs with members of your team will help you to fine-tune a presentation. The PM should not have a novice presenter give an important briefing to a customer, for example, without a serious dry run.

Allowing interruptions is a matter that is somewhat controversial. In general, it is best to establish better audience contact by allowing interruptions in the form of questions from the audience. However, these should not be allowed to turn the presentation into a free-for-all. Losing control over the briefing is extremely undesirable and will likely damage the credibility of the presenter. By extension, this will damage the project team. A good presenter normally allows a modest number of questions but is able to draw people back to the presentation without losing control. This comes with practice and assistance from those who have this type of know-how. The PM and CSE must have mastered these types of skills. It is definitely all right to terminate further questioning and bring the audience back to the main thrust of the presentation.

The matter of handing out hard copy also has advocates on both sides. This author favors delivery of hard copy of your slides at the end of the presentation. People might be given three-by-five index cards to make notes on questions they might have during the presentation. The problem with the audience having hard copy in advance is that some people will leave you and go on the slide journey themselves. They can be ahead of you or backtrack to earlier slides. In either case, they're not likely to be listening to you. Under group pressure from the audience, however, it is difficult not to hand out the slide package when requested to do so.

Giving presentations represents a special skill that has to be mastered by the PM, CSE, and other key members of the project team. It is worth the time to take this activity very seriously, especially if the presentation is made to a customer or a large audience. The PM and CSE have the responsibility to maximize the positive impact of all presentations. Usually, this involves dry runs and supportive coaching. Paying a lot of attention to these matters will pay worthwhile dividends.

6.8 PROPOSALS

Proposals to prospective customers can take many forms, depending on the type of enterprise making the proposal. At one extreme is the oral presentation during which various points are made regarding the services and products to be provided. For this mode of proposal delivery, essentially all the suggestions made previously apply directly. At the other extreme is the full-blown written proposal that documents all the features of the products and services as well as the organization that is making the proposal. Such a proposal is often in response to a formal request for proposal (REP) provided by the customer. RFPs usually specify what the customer needs and also the evaluation criteria to be used in order to make judgments about the proposal. Proposals are considered by many to be a vital part of the lifeblood of an organization. They are also viewed at times as a fine art that can be mastered by some but not by others.

Many people believe that new business is won or lost by the amount of effort that goes into preproposal activities. Put another way, the formal written proposal is viewed as a necessary but insufficient condition for success with respect to winning a new contract. These views are widely supported. For this reason, we examine the matter of proposals in two parts. The first deals with suggestions as to what to do during the preproposal stage, and the second addresses the formal written proposal itself.

6.8.1 The Preproposal Phase

As indicated before, many proposals are won or lost as a function of the preproposal activity or lack thereof. For large contracts, this activity is literally a well-developed campaign with its own rather sophisticated plan. Because there is so much at stake, the preproposal work is dealt with as if it were a separate project that culminated in the customer's formal release of an RFP.

A summary of key points relative to the preproposal phase is provided in Exhibit 6.5. This is a minimal set of activities that should be undertaken in order to be successful with a proposal.

Exhibit 6.5: Recommended Preproposal Activities

1. Visit with the prospective customer to understand his or her needs and requirements.
2. Write think pieces and "white papers."
3. Present your company's capabilities.
4. Investigate alliances and teaming possibilities.
5. Identify and activate the proposal writing team.
6. Obtain internal corporate support.

There is no substitute for truly understanding what it is that the customer really wants. The best way to do this is to visit with the customer and talk through his or her various needs and desires. Find out what the customer is looking for—what the key issues are from the perspective of the customer. This notion applies as well when there are several customers, which is usually the case. In other words, the customer often consists of several people, all of whom are looking for one or more pieces of the puzzle known as customer requirements. If you have not spent the time necessary to visit across the table with your potential customer, you may have a "no-bid" staring you in the face.

In response to what you find out in your face-to-face meetings with your customer, seriously consider writing one or more think pieces or white papers. These are short (five- to ten-page) descriptions of your approach to the problem, as expressed by the customer. They are very specific as to what the key issues are, how you would develop a solution, and what the final product or service might look like. You are also requesting feedback from the customer as to whether or not you are on the right track. Therefore, after the customer has read these inputs, follow-up meetings are suggested in order to fine-tune your approach. You are also demonstrating your interest and responsiveness and, to the extent possible, preselling your approach to the customer. This spade work will not be lost on the customer when it comes time to evaluate your formal proposal. You will also be in a better position than your competitors because you will have had the benefit of this interaction with the customer.

As part of your meetings with the customer, you should also present the broad as well as specific capabilities of your company. This can take the form of a stand-up slide presentation in front of several people in your customer's organization. Such presentations involve corporate strengths such as personnel, facilities (e.g., special laboratories or test chambers), history of work with other clients on similar requirements, software already written, and so forth. In addition to "telling" your customer what you have to offer, try to "demonstrate" special capabilities. These can be expressed by software packages that you have developed that apply directly to the problem at hand. Instead of talking about the software, bring in a computer and illustrate the use of the software, in real time, on the screen. "Showing" is always better than simply "telling."

As you gain an understanding of the customer needs and requirements, you may find that your company does not possess all the required skills, capabilities, and experience with the knowledge domain of the customer. Both strategic and tactical alliances and teaming should be considered to make your team a winner. Many large-scale contracts cannot be won without teaming, thus requiring filling in all the gaps in capability and understanding by members of the team. You may be the prime contractor in such an arrangement or you may ultimately decide that a subcontractor position is the better part of valor. In the latter case, you must then try to determine who has the best chance of winning, and then do all of the preceding in attempting to sell your company to the prime.

Early identification and activation of at least part of the proposal team is mandatory. A proposal manager should be designated, along with at least a few key players of the proposal team. They need to be part of the preproposal campaign and play a role in preparing all written materials. In this manner, the learning curve is minimized when it comes time to actually respond to an RFP. All players are up-to-speed and understand what has transpired prior to the formal proposal preparation. Failure to make this investment will often lead to a disconnect between the preproposal and proposal phases, resulting in a poor written proposal.

Last, but by no means least, support has to be obtained from various people and departments in your organization. This includes marketing, contracts, human resources, matrixed managers, and your own line management. Although you may not have made a firm "bid decision," you have to give an early alert to all these other players so that they understand the program you are seeking and its unique requirements. For example, you may have to find certain key personnel to handle special technical areas. The human resources department and various matrixed managers may be able to help you in this regard. The contracts department needs an early alert so that it can review the terms and conditions of work done previously with this customer as well as get ready for the preparation of extensive cost estimates. Finally, you need support from your boss and other levels of line management in terms of concurrence with the bid as well as putting forth the resources necessary to prepare a winning proposal. This includes funding, but it also means making the right people available at the right time to contribute to a winning proposal.

6.8.2 The Proposal Phase

The proposal phase is entered when the RFP is formally issued and your company has made a decision to bid the contract in response to that RFP. How to write a winning proposal is a subject that itself represents a niche market for companies, so that what is recommended here is but an overview of key points in this regard. However, ten such points that are distilled from many years of experience are listed in Exhibit 6.6.

Exhibit 6.6: Ten Key Points in Writing a Winning Proposal

1. Develop a winning strategy.
2. Document winning themes.
3. Outline the proposal in detail, to the one-half page level.
4. Develop a detailed proposal schedule.
5. Discuss and confirm the preceding with your proposal team as well as management.
6. Interact efficiently with all support departments.
7. Use the best writers.
8. Confirm that all risk areas have been minimized.
9. Review and make changes in the "Red Team" context. ("Red Teams" are explained later in this section.)
10. Assure proposal readability and appearance.

Developing a winning strategy is more easily said than done. The issue here is complex, dealing with each and every key area in the RFP and how your approach can be distinguished from the competition. This is known as finding "discriminators" that place your proposal ahead of those of your competitors. Discriminators have to be found with respect to your technical proposal, management proposal, and cost proposal, with emphasis on the first and last. If you are operating in the domain of "best-value" contracting, you do not necessarily need to be the lowest bidder, but the overall value of your proposal must be as high as possible. Discriminators should be determined through a team approach, eliciting ideas from all members of the proposal team. Clear discriminators often can be found in cases for which your company has actually performed work sufficiently similar, so that the effort required is substantially less than that of your competitors. This can be embodied in software, for example, that is reusable and transferrable from a previous project to the one at hand. The bottom line is that if you cannot be positively distinguished from your competitors, you are not likely to win the contract in question.

Given that the team is able to find the necessary discriminators, they then should be converted into written themes that basically and convincingly answer the question: Why we should win this contract? These need to be mapped against the evaluation criteria of the RFP so that scoring is maximized. In other words, for each and every element of the evaluation criteria, it is desirable to document a theme that will score high. Scoring high involves bringing forth your discriminators in these areas and convincing yourself (and ultimately, your customer) that you have the best approach possible.

Next, the proposal has to be outlined, in detail, following precisely the ground rules provided in the RFP. Often, the RFP defines the proposal pieces (i.e., technical, management, and cost proposals) and their sections. These must be meticulously followed. At times, a dilemma is presented whereby the evaluation criteria are not easily mapped against the prescribed proposal outlines. In general, both should be accommodated. It must be evident to the evaluator as to where to find the material that is responsive to the evaluation criteria. If the evaluator has to struggle to find such material, a low score is inevitable. The mapping of the mandatory outline and the evaluation factors takes imagination and can be the most important determinant of a winning or losing proposal. After this is done, the outline should be divided into sections and subsections, with a page count for the smallest subsection. Page counts should be designated to the one-half page. Then the entire proposal should be reviewed for balance of coverage in relation to the evaluation criteria.

The proposal schedule should define all milestones and should be viewed as inviolate. Times have to be set aside for detailed reviews and rewrites. Many firms have three designated periods for rewrites, operating under the proposition that it is basically impossible to write a complex proposal without extensive rewriting. Times, of course, have to be allocated to the physical production of the proposal and delivery to the customer. In most government contracting, being even one minute late in terms of customer delivery will mean that your proposal will not be accepted. (This author has had this type of unpleasant experience twice in a period of thirty years.)

Given all of this, the next step is to brief the proposal team as well as management on the basic strategy for winning. All key players have to understand that strategy and concur with it. Alternative suggestions are acceptable and indeed accepted if they represent improvements. This is a basic sign-off on the approach and a commitment to implementing it in the best possible manner. All questions should be answered so that there is no misunderstanding of the approach.

Effective and efficient interactions with all supporting departments and people are mandatory to have a winning proposal. Contracts must handle the representations and certifications (reps and certs) smoothly and in consonance with the technical and management proposals. They must also provide cost spreadsheets that can be reviewed and concurred with by the proposal manager. Costing is often iterated numerous times, so that automating this process is essential. Matrixed managers must confirm the availability of key personnel. Human resource people must stoke up the machinery for hiring or establishing a hiring plan. The production department (including text and graphics) should be aware of and be able to handle the peak load represented by a major proposal. The proposal manager must assure that all these pieces fall into place smoothly and with a lot of lead time.

The best writers should be assigned to the proposal. Many technically talented people are not able to write coherent proposals. Many imaginative people decide to apply their imagination by not following the rules established for the proposal. At the same time, there are usually some people who are very talented from a technical point of view and who are also wonderful writers. These folks should be assigned to writing the proposal. The basic rule is to use your best technical *and* writing talents to write all proposals, wherever possible. In addition, use many charts, figures, exhibits, and diagrams to explain your approach to your customer.

Most projects involve some type of risk (schedule, cost, performance) and all potential risks have to be examined in detail as part of the proposal process. Some companies, of course, decide to "no-bid" when unacceptable risk levels are present. However, risk still needs to be double-checked and mitigated in the proposal itself. How you propose to minimize risk should be addressed in the proposal, so that you and the customer both know what the basic plan is. Two examples are worth noting. One involves the case in which the customer can review and reject documentation and reports indefinitely. Acceptance criteria have to be as explicit as possible so that this risk is mitigated. Another risk involves the warranty of software from "latent and patent" defects. Boundaries and criteria need to be established so that software improvements do not go on forever. In general, risk mitigation involves checking each and every documented requirement and setting forth a "buy-off" plan that is acceptable to your company as well as the customer.

Most large proposal writing efforts employ "Red Teams," which are groups of senior people who review and evaluate proposals. These teams engage in a formal review, simulating the process of review that is expected by the customer. Such reviews not only score the proposal, section by section, they suggest specific ways and means of making necessary improvements. These reviews are often brutal in their candor, and the proposal manager must know how to assure that they are constructively expressed. Invariably, Red Team inputs lead to extensive rewrites that must be built into the proposal schedule. Dispensing with Red Team reviews should be done with caution and is likely to be a serious mistake for large proposals.

Finally, the schedule should allow for a more than cursory review of the final product. This review usually involves overall appearance and readability. The objective is zero defects, meaning that there should be no typos or pieces in the wrong place. The proposal should not only *be* good, it should *look* good. However, it should not be ostentatious so that boundaries of good taste are crossed. Some customers react very negatively to showy and expensive proposals. Here again, know your customer.

With respect to the last point made and how the proposal looks, this author experienced what might be called a four-sigma case in proposal preparation. A proposal was sent out to a customer years ago but the xeroxing of the proposal was defective and resulted in the lack of fixing of the toner on various pages. A sweep of the hand on a page resulted in the immediate disappearance of all text—a type of invisible ink if you will. Fortunately, this was discovered very early, and the customer allowed us to substitute good copies in place of the defective ones. This is a once-in-a-lifetime experience, but it underscores the possible intrusion of Murphy's Law. Take the time necessary for a final review that assures your proposal looks the way you want it to look in terms of overall presentation to your customer. A poor-looking proposal with lots of typos, in spite of the excellence of its technical content, is not likely to be a winner.

6.9 A NOTE ON MOTIVATION AND INCENTIVES

Many Project Managers are eventually faced with the problem of keeping the members of their team motivated. They struggle with finding ways to throw money at the team as an incentive to sustain motivation. Basically, money is important but it is not the essence of motivation, at least from the experience of this author. Equitable treatment in terms of monetary rewards is necessary, but it is far from a sufficient condition for motivating members of your team.

There is a large amount of information in the literature regarding motivation and how it may be achieved. As an example, one text on the subject of management [6.8] examines, in a rather well-designed manner, such research results and theories as:

- Theory X and Theory Y
- Theory Z
- Argyris' maturity theory
- Self-fulfilling prophecy
- Various needs theories (including Maslow's)
- Equity theory
- Reinforcement theory
- Expectancy theory
- Various reward systems

The reader with a special interest in these matters can review the referenced text [6.8] and others on the overall subjects of motivation and incentives.

Experience in the worlds of project management and systems engineering has led to the following key elements in motivating members of a team:

1. Having interesting and challenging work
2. Treating people fairly and equitably
3. Establishing team affiliation with a highly functional team

4. Showing appreciation and recognition
5. Celebrating team successes

Experience shows that interesting and challenging work is the most critical factor in keeping team members excited and enthusiastic about what they are doing, day after day. The PM and CSE, therefore, must understand all members of the team and pay attention to how to maintain a flow of good work to them. This requires a deep sense of what is challenging to each person, which in turn means sufficient contact to obtain feedback about work assignments. Indeed, if challenging work is not a strong motivator, you may have to think about replacing such a person or moving him or her to an environment that is not as vital as a systems engineering project.

Beyond interesting and challenging work assignments, people on the team have to feel that they are being treated fairly. This means that the PM and CSE must be sensitive to these needs and be careful not to show favoritism that is unrelated to performance. All team members have equal standing in terms of being treated with respect and listening to what they may have to contribute, even though some will naturally contribute more than others. Monetary rewards must be perceived to be equitable, even though they may not be extremely high. Most people will respond well to a situation that is fair and equitable, and will be upset when this is not the case.

Given that your company provides adequate incentives in the form of salary, benefits, and bonuses, the addition of the previously cited actions normally keeps members of the team highly motivated. If your company policy is to keep the monetary incentives below average, you will very likely lose your people, especially the good ones. The point is that monetary rewards are not as important as some would think, assuming however that they are not substandard. Throwing money at people as a substitute for the preceding actions is basically the wrong approach. The right approach is to assure adequate but not outrageous levels of salary, benefits, and bonuses, with special attention to the five items listed before. Productive and exciting connections between people are ultimately more satisfying than a few more dollars in the monthly check.

A third suggestion is to assure that everyone is an accepted member of the team and that the team is highly functional. This type of team affiliation can be like having a second family, but in a work situation. The PM and CSE must keep a watchful eye to assure that some team members do not gang up on or ostracize other team members, and intercede where necessary. Having "we's and they's" within the team itself should be avoided whenever it damages the team's performance. A healthy team with full participation by all will go a long way toward creating an exciting and productive work environment.

Team leaders must always show appreciation and recognition for a job well done. As suggested earlier, this can take many forms, from a pat on the back to a cash bonus. People who are doing some special work on behalf of the team need to have positive feedback and the knowledge that the boss appreciates what they are doing. The PM and CSE must respond to this need on a continual basis, not just at raise time. Conversely, withholding positive feedback is likely to be a demotivator for most people.

Finally, and somewhat related to the preceding, the team should always celebrate its successes. When something good happens, it's time to have a celebration, which can take the modest form of a small party or after-hours pizza and beer. When something extraordinary happens, like a technical breakthrough or winning an important new contract, these celebrations should be augmented by a bonus in addition to the monthly paycheck. The bonus should be awarded as close in time to the exciting event as possible. Public recognition is almost always a preferred way to celebrate success and show appreciation for achievement beyond the call of duty. Very modest but public rewards, such as a certificate for a dinner for two at a good restaurant, go a long way to demonstrating that the boss recognizes both effort and achievement.

The bottom line is that a relatively small number of nonmonetary incentives and actions can be extremely effective in supporting team motivation. On the other hand, failure to pay attention to these nonmonetary actions will be a demotivator that cannot be made up by cash awards and raises. People who are not appreciated ultimately will seek an environment where they are able to get the psychic rewards they feel they deserve. Finally, and inevitably, the best performers will normally be the first ones to leave.

6.10 ANOTHER TEAM-RELATED PERSPECTIVE

Some further insights into teams and team building may be gained by examining related areas in the U. S. Department of Defense (DoD). When teams are discussed, it is virtually always in the context of *Integrated Product Teams* (IPTs), which are defined as [6.9]:

> A multidisciplinary group of people who are collectively responsible for delivering a defined product or process. The IPT is composed of people who plan, execute and implement life-cycle decisions for the system being acquired. It includes empowered representatives (stakeholders) from all of the functional areas involved with the product—all who have a stake in the success of the program, such as design, manufacturing, test and evaluation (T&E), and logistics personnel, and, especially, the customer.

Further, the Secretary of Defense "directed that the Department perform as many acquisition functions as possible, including oversight and review, using IPTs." Broad principles under which the IPTs are supposed to operate include:

1. Assurance of open discussions with no information withheld
2. Team members that are qualified and empowered

3. Participation that is proactive, consistent, and success-oriented
4. Continuous "up-the-line" communications
5. Reasoned disagreement when there is a difference of views
6. Important issues raised and resolved as early as possible

These are basically self-explanatory and strongly suggest a key item in the operation of these teams, namely, that much power is vested in the team members so that they can accomplish their goals. They also must be strong individuals who have learned very well how to be part of a team.

Aspects that are considered important are revealed by a brief glance at the key topics that make up the Teaming Guidebook [6.10]. These topics are:

- Team charters
- Team member selection
- Kick-off meeting
- Facilitator
- Communication
- Team decision making
- Conflict resolution
- Team evaluations and ratings
- Team awards
- Team training
- Team management
- Leadership
- Team problems

Important questions to consider when establishing a team include [6.10]:

1. Does the team purpose set the stage for subsequent actions by the team?
2. Is there a cogent and viable reason for the existence of the team?
3. How long is the team expected to be in existence?
4. Who are the customers of the team?
5. What are the team's products expected to be?
6. To whom does the team report?
7. Are all of the above items dealt with appropriately in the team's charter?

Within the DoD, IPTs and Integrated Process and Product Development (IPPD) are more-or-less inseparable. IPTs facilitate IPPD; IPPD cannot be completely executed without the contributions of IPTs. Both are considered to be long poles in the tent that houses the overall system acquisition process.

The relevant quotes are that IPPD "is a widely defined management technique normally implemented by Integrated Product Teams (IPTs)" and also that the "concepts of IPPD and IPTs shall be applied throughout the acquisition process to the maximum extent practicable" [6.9].

Integrated Process and Product Development (IPPD) is defined as [6.9]:

A management technique that simultaneously integrates all essential acquisition activities through the use of multidisciplinary teams to optimize the design, manufacturing and supportability processes. The IPPD facilitates meeting cost and performance objectives from product concept through production, including field support.

In explaining the functions of IPPD, business processes are also included along with design, manufacturing, and support. Further, modeling and simulation, as well as best commercial practices, are employed in order to arrive at the best processes and products. Additional principles of IPPD may be gleaned from the following list:

- Customer focus
- Concurrent development of products and processes
- Early and continuous life-cycle planning
- Proactive identification and management of risk
- Maximum flexibility for optimization and use of contractor approaches.

Part of the way that customer focus is maintained is by including the customer in the IPT, or a subset thereof, and listening carefully to the continuous feedback that the customer is providing. The focus on products and processes transcends the process reengineering concept by also assuring that the product is acceptable. Careful attention to planning for the entire life cycle means that fewer critical downstream milestones and activities will be missed. Risk management is critical to any and all serious engineering projects, and will be one of the important elements of systems engineering. Finally, flexibility is maintained by allowing for alternative contractor approaches, ones that fit the problem at hand and give the contractor some leeway to employ its experience and creativity to arrive at a cost-effective solution for the customer.

So we see in the DoD considerable attention to teams, especially IPTs, and we also note the use of these IPTs to accomplish Integrated Process and Product Development (IPPD). The highest levels within the DoD have committed themselves to this approach, and they believe that in doing so they will achieve a fundamental positive change in the way they are acquiring goods and services.

6.11 GROUP PROCESSES

Systems engineering is often embedded in a project, and both are discussed in considerable detail in this book. Indeed, project management and systems engineering management are the two top-level subjects in this text. Within the former, there has been, and continues to be, enormous emphasis on the building of teams. Phrases such as *integrated product teams* (*IPTs*), *high-performance teams* (*HPTs*), *project engineering teams* (*PETs*), and others are used extensively to examine one or the other aspect of a team. The literature has been almost overwhelming on the matter of teams, including those that do not work very well. The label is ultimately not what counts; what counts is the behavior in terms of facilitating highly functional analysis and decision making.

Another relevant topic related to teams is that of *groups* and the behaviors of groups for a variety of purposes that may go beyond the classical project and systems engineering environments and situations. In this section we explore some of the work done in regard to how groups can operate and how these operations can be improved. Very brief overviews of group situations are cited, drawn from an earlier book by this author [6.11].

6.11.1 Delphi Method

The Delphi method is often employed when one is attempting to reach a consensus from several experts, with controlled interactions between the participants. This method basically has survived over more than fifty years of adaptive practices.

6.11.2 GroupThink

GroupThink is a group decision-making process that often can lead to poor decisions. Reason? People do not speak up when they have a dissenting view. Specific steps need to be taken to encourage, and ultimately assure, full participation from all group members.

6.11.3 Thinking Hats

The Thinking Hats method, devised by Edward de Bono [6.12], has each group member put on a hat with a color that represents a different direction of thinking. This author has tried this process and found that it was well worth the attempt.

6.11.4 Advocacy versus Inquiry

In the advocacy versus inquiry group process, a specific inquiry approach is used rather than the more common advocacy situation, in the belief that advocacy has not been as productive as many would like. In advocacy, strong and highly verbal participants sometimes try to dominate the process as well as the ultimate decision.

6.11.5 SAST

The Strategic Assumption Surfacing and Testing (SAST) process forces assumptions to be made explicit so that they can be debated and analyzed. A final critical "synthesis" step sums up and articulates agreed-upon courses of action.

6.11.6 Team Syntegrity

The Team Syntegrity process was formulated by Stafford Beer, a well-known and influential cybernetician. A number of participants (e.g., fifty) are divided into subgroups that address a set of topics (e.g., 12), all of which are discussed and analyzed in detail. Teams present their results at a closing session.

6.11.7 Facilitation

The facilitation process, as implied, focuses on all aspects of facilitation, including a "zen" philosophy, toolkits, and training. Part of the central theme and goal is achieving true synergy.

6.11.8 Self-Directed Work Teams

Under the self-directed work team approach, teams are given the charter to devise a new product or service with very close to no supervision. Various successes with implementing this notion, along with the benefits, have been documented [6.11].

6.11.9 Synectics

Synectics embodies well-defined processes whereby groups are able to achieve high levels of creativity. Methods include the advantageous use of metaphors, analogies, expert leadership, and freewheeling imagination.

6.11.10 Collaboration

Collaboration is a group process in which two or more people establish a highly productive and interactive arrangement with respect to problem solving. Members of such groups support each other by sharing and examining ideas in a back-and-forth manner. Most of the time, they are able to achieve together what neither could do on his or her own.

* * *

In this chapter we have explored some ways in which teams and groups have behaved and have suggested possibilities for improvements. In the end, project management and systems engineering management are mostly about people, sometimes doing it right or failing to do so. Management is an art *and* a science, and is more easily talked about than actually achieved in an often imperfect and changing environment. Highly functional teams and groups are literally a joy to behold as well as learn from.

QUESTIONS/EXERCISES

6.1 Evaluate your boss against the list of essentials in Exhibit 6.1. Use a college scoring system (i.e., A = 4, B = 3, C = 2, D = 1, F = 0). Calculate a composite numeric score. Repeat this same exercise with respect to yourself.

6.2 Identify and discuss three additional suggestions for building a project team, beyond those listed in Exhibit 6.2.

6.3 Recall a "team buster" you have encountered. How did this person behave? What was done, or might have been done, to deal with such a person?

6.4 What conflict resolution style might be attributed to your boss? What do you think your style is? What does this suggest with respect to your relationship with your boss?

6.5 Cite and discuss three ways in that you have been frustrated by the manner in which your boss runs meetings.

6.6 Identify three motivation/incentive approaches used in your organization. Have they been effective? Why?

6.7 Find a definition of an Integrated Product Team (IPT) in the literature and compare it with the definition provided in the text, as set forth by the U.S. Department of Defense.

6.8 Find a definition of Integrated Process and Product Development (IPPD) in the literature and compare it with the definition provided in the text, as set forth by the U.S. Department of Defense.

6.9 You have probably attended many meetings over the years. Identify and briefly discuss five areas of focus for the meetings that you have led or might lead in the future.

6.10 Which three of the group processes cited in this chapter are easiest for you to relate to? Why? If you were designing a group problem-solving process, what three areas would you emphasize? Why?

REFERENCES

6.1 Peck, M. Scott (1993). *A World Waiting to Be Born*. New York: Bantam Books.

6.2 Meredith, T. R., and S. J. Mantel (1985). *Project Management—A Managerial Approach*. New York: John Wiley.

6.3 Archibald, R. D. (1976). *Managing High Technology Programs and Projects*. New York: John Wiley.

6.4 Kerzner, H. (2000). *Project Management: A Systems Approach to Planning, Scheduling and Controlling*, 7th edition. New York: John Wiley.

6.5 Kezsbom, D. S., D. L. Schilling, and K. A. Edward (1989). *Dynamic Project Management*. New York: John Wiley.

6.6 Thomas, K. W., and R. H. Kilmann (1974). *Thomas-Kilmann Conflict Mode Instrument*. Tuxedo, NY: XICOM, Inc.

6.7 Doyle, M., and D. Straus (1982). *How To Make Meetings Work*. New York: Jove Books.

6.8 Mondy, R. W., and S. R. Premeaux (1980). *Management—Concepts, Practices and Skills*, 6th edition. Needham Heights, MA: Allyn and Bacon.

6.9 *DoD Integrated Product and Process Development Handbook* (1998). Washington, DC: Office of the Under Secretary of Defense (Acquisition and Technology), August. See also http://web2.deskbook.osd.mil

6.10 *The Art of Teaming Guidebook* (1999). Published by the U.S. Army Materiel Command, Integrated Product and Process Management Working Group, June. See also http://web2.deskbook.osd.mil

6.11 Eisner, H. (2005). *Managing Complex Systems—Thinking Outside the Box*. Hoboken, NJ: John Wiley & Sons.

6.12 de Bono, E. (1985). *Six Thinking Hats*. Boston, MA: Little, Brown.

PART III
SYSTEMS ENGINEERING AND MANAGEMENT

7

THE THIRTY ELEMENTS OF SYSTEMS ENGINEERING

7.1 OVERVIEW OF THE SYSTEMS APPROACH AND ENGINEERING PROCESS

The first two chapters of this book provided an overview of some of the systems engineering notions that are dealt with in more depth in this chapter as well as the following four chapters. In particular, Figure 1.1 and the supporting text provided an overview of the *systems approach*. To reiterate the key features of this important perspective, the systems approach:

1. Establishes a systematic and repeatable process
2. Emphasizes interoperability and harmonious system operation
3. Provides a cost-effective solution to the customer's problem
4. Assures the full consideration of alternatives
5. Utilizes iterations as a means of refinement and convergence
6. Leads to satisfaction of all user and customer requirements
7. Creates a robust system

The three main aspects of this approach, in the context of developing a real-world system, involve:

1. Architecture design
2. Subsystem design
3. System construction

Architecture design deals with the top-level considerations and the fundamental design choices that have to be made in constructing a system. After this is completed, one then addresses the more detailed design of each subsystem. After these two steps are complete, it is then possible to actually construct the subsystems, composed of hardware, software, and human components.

In Figure 2.2, an overview of the systems engineering process and its management was illustrated. This expanded the preceding three aspects of the systems approach by adding two additional elements:

1. Other system and subsystem considerations
2. Project control

This gave more substance to the systems approach, and also introduced many more detailed elements of systems engineering and its management. In this chapter, we describe the fundamental elements of systems engineering and its management in the form of thirty specific tasks to be accomplished by the systems engineering team. Before addressing the thirty key elements, we briefly examine two other approaches in order to place these elements in other contexts.

7.2 TWO SYSTEMS ENGINEERING PERSPECTIVES

Over the years, there have been many descriptions of systems engineering, its internal processes, and its management [7.1]. These have ranged from the first book on systems engineering [7.2], to the first systems engineering standard [7.3], to systems engineering management guides [7.4], to a variety of more current texts on systems engineering [7.5, 7.6, 7.7, 7.8, 7.9, 7.10, 7.11]. In most cases, the systems engineering process has been described in some type of flow or activity chart that has embodied the elements of systems engineering, similar to the chart shown in Figure 2.2. Here we provide a brief overview of two such representations, one drawn from Military Standard 499B and the other from a series of processes defined by a government agency, the National Aeronautics and Space Administration (NASA).

7.2.1 Mil-Std-499B Revisited

This draft standard, discussed in Chapter 2, is rather specific about what it is that constitutes the systems engineering process. Referring to Figure 7.1, we see the four main features of this process, as represented in the standard:

1. Requirements analysis
2. Functional analysis/allocation

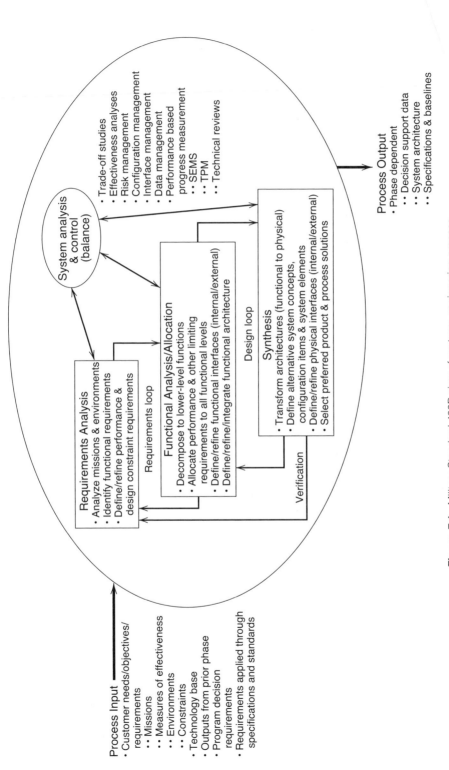

Figure 7.1. Military Standard 499B view of systems engineering process.

3. Synthesis
4. System analysis and control

This process requires a variety of inputs, delineated in the figure as "process input." The outputs are also shown under the heading of "process output." The view of systems engineering and its management in this text, however, is considerably broader. As indicated before, it goes beyond architectural design of the system and into detailed subsystem design as well as the physical construction, operation, and maintenance of the system. This broader view can be expressed as a set of some thirty interrelated elements and is the central focus of this chapter. In addition, this view also includes the management of the systems engineering process that is normally the purview of the Chief Systems Engineer (CSE).

7.2.2   The NASA Mission Design Process

NASA's Engineering Management Council has produced an engineering guide to conceptual design, mission analysis, and definition phases of a system [7.12]. NASA tends to build systems by going through a series of formal phases, from Prephase A to Phases C and D. The previously cited engineering guide defines the first three phases as:

1. The conceptual design process: Prephase A
2. The mission analysis process: Phase A
3. The definition process: Phase B

These are well-known by the industry that supports these phases that define the "front-end" portion of the system's life cycle. Table 7.1 provides a summary of the mission design activities, reviews, and products for each phase; Table 7.2 shows examples of the required study resources for the phases [7.12].

Two points are especially significant with respect to NASA's approach. The first is that NASA has defined the flow of activities associated with each phase. As examples, Figures 7.2 and 7.3 show these processes for Phase A and Phase B, respectively. These are the systems engineering processes for these early life-cycle phases. The second point is that NASA has implicitly given significant attention to the notion of *mission analysis*, which has often been overlooked as an integral part of, and indeed key element of, systems engineering. The latter is viewed as an important contribution to a more effective way of ultimately dealing with the significant subject of missions as well as requirements analysis and engineering.

TABLE 7.1 Summary of Mission Design Activities, Reviews, and Products

	Prephase A	Phase A	Phase B
Input to Each Phase	Mission Needs and Objectives	Preliminary Mission Requirements Document Prephase A Study Report Evaluation Criteria	Phase A Study Report Mission Requirements Document Operations Concept Document Project Initiation Agreement Evaluation Criteria Mission Requirements Request (Preliminary)
Activities	Prephase A Study Preliminary Mission Requirements Definition Top-Level Trade Studies ROM Costs and Schedule Feasibility Assessment	System and Subsystem Trade Studies Analysis of Performance Requirements Identification of Advanced Technology/Long Lead Items Risk Analysis End-to-End System Life-Cycle Cost as a Trade Parameter Cost and Schedule Development Top-Down Selection of Study Concepts Operational Concepts	Revalidation of Mission Requirements and System Operations Concept System Decomposition, Requirements Flow-Down, and Verification Risk Analysis System and Subsystem Studies and Trades Development of the Work Breakdown Structure and Dictionary Updating Cost, Schedule, and Life-Cycle Cost System and Subsystem Concept Selection and Validation

(Con

TABLE 7.1 (*Continued*)

	Prephase A	Phase A	Phase B
Reviews	Peer Reviews Mission Concept Review	Peer Reviews Mission Design Review	Peer Reviews System Requirements Review System Design Review Nonadvocate Review Nonadvocate Review Package Phase C/D Request for Proposal Package(s)
Products	Prephase A Study Report Preliminary Mission Requirements Document Evaluation Criteria Science Definition Team Report ROM Cost Estimate	Phase A Study Report Mission Requirements Document (preliminary) Operations Concept Document Project Initiation Agreement Phase B Study Plan Technology Development Plan Top-Level System and Mission Architecture Preliminary Cost and Manpower Estimates	Baseline Systems Description Technology Development/Risk Mitigation Design and Technical Documents Management and Control Plans Externally Required Documents and Agreements Mission Cost and Manpower Estimates Operations Plan

TABLE 7.2 Examples of Required Study Resources

	Duration	Funding	Staffing
Prephase A	2 to 4 months	$0 to < $100,000	Study Manager/Systems Engineer Project Scientist Part-time help from others
Phase A	6 months to 1 year	1 to 2% of total system cost	Study Manager Systems Engineer Project Scientist Resource Specialists Discipline Engineers and Specialists
Phase B	1 to 2 years	4 to 8% of total system cost	Project Manager Systems Engineer Project Scientist Resource Specialists Discipline Engineers and Specialists

7.3 THE THIRTY ELEMENTS OF SYSTEMS ENGINEERING

We come now to a variety of tasks that form the central core of systems engineering and its management. In this book, these tasks are called the thirty elements of systems engineering, and are listed in Exhibit 7.1.

Exhibit 7.1: The Thirty Elements of Systems Engineering

1. Needs/goals/objectives
2. Mission engineering
3. Requirements analysis/allocation
4. Functional analysis/decomposition
5. Architecture design/synthesis
6. Alternatives analysis/evaluation
7. Technical performance measurement (TPM)
8. Life-cycle costing (LCC)
9. Risk analysis
10. Concurrent engineering
11. Specification development
12. Hardware/software/human engineering
13. Interface control
14. Computer tool evaluation and utilization

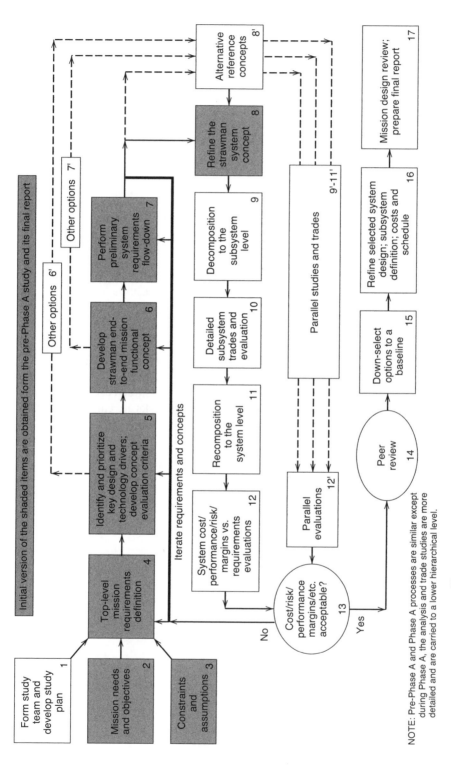

Initial version of the shaded items are obtained form the pre-Phase A study and its final report

Figure 7.2. Representative top-level flow for a mission analysis study—Phase A [7.12].

NOTE: Pre-Phase A and Phase A processes are similar except during Phase A, the analysis and trade studies are more detailed and are carried to a lower hierarchical level.

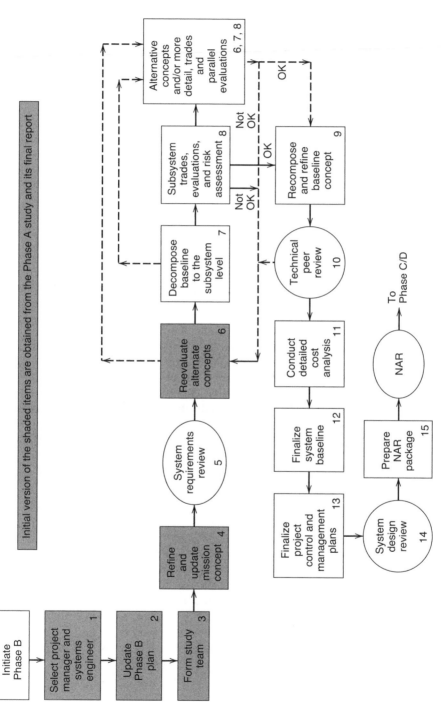

Initial version of the shaded items are obtained from the Phase A study and its final report

Figure 7.3. Representative top-level flow for a definition study—Phase B [7.12].

15. Technical data management and documentation
16. Integrated logistics support (ILS)
17. Reliability, maintainability, availability (RMA)
18. Integration
19. Verification and validation
20. Test and evaluation
21. Quality assurance and management
22. Configuration management
23. Specialty engineering
24. Preplanned product improvement (P3I)
25. Training
26. Production and deployment
27. Operations and maintenance (O&M)
28. Operations evaluation/reengineering
29. System disposal
30. Systems engineering management (planning, organizing, directing, and monitoring)

These thirty elements span the overall systems engineering process over the full life cycle of a system. It is an inclusive rather than exclusive interpretation of systems engineering and its management. It is the "union" of the set of tasks and activities that have previously been defined as necessary over a system's life cycle. It is not confined only to the systems architecting or synthesis phase and extends to the reengineering of a system, or portions thereof, after it has been in its operations and maintenance phase, possibly for many years. Reengineering has emerged as a critical activity because we cannot, as a general philosophy, afford to scrap many old systems and engage in completely new starts. As we have "legacy" software, we also have legacy systems. As we add functionality to older systems, we are not able to completely throw away these older core systems.

With respect to these thirty elements, not only must the CSE master each and every one of them, he or she must also understand the interrelationship between these elements. This is an enormously challenging assignment that requires both a broad and deep commitment to this discipline as well as its supporting knowledge base. Although there are numerous persons involved in some element of the process of systems engineering, a considerably fewer number achieve the level of Chief Systems Engineer (CSE). Such a position is often, but by no means always, a prerequisite to the position of Project Manager (PM), especially for a large program or project.

We now briefly examine each of the thirty elements of systems engineering and its management. The following four chapters provide additional depth for the more important elements. In effect, the following discussion provides a broad overview of all core elements of systems engineering and its management.

7.3.1 Needs, Goals, and Objectives

This element is the same as that cited as the first element of a project plan, as described in Chapter 3. Therefore, all that should be done under this element is to confirm that the statements of needs, goals, and objectives are correct. Because these are provided by the system user or acquisition agent, the system developer should go back to these sources for the purpose of assuring that the needs, goals, and objectives are still current and appropriately stated. Examples of such statements are given in Chapter 3.

7.3.2 Mission Engineering

Mission engineering has often been overlooked as a central element of systems engineering. In general terms, it involves the detailed articulation and analysis of the intended mission of the system that is being engineered. The main purposes of mission engineering are

- To verify that the system has legitimate missions to be executed that are not being carried out by other systems
- To provide a technical basis for the full definition of requirements for the system

With respect to the first point, mission analysis confirms that there is a real need for the system. It forces the very early consideration of the possibility that, for example, minor upgrades of existing systems will not suffice. Many system developments have ultimately failed because this important step was not fully explored. With the complexity and expense associated with our large-scale systems, we must be sure that a real need exists for the new system.

However, mission engineering also serves another critical function by providing a technical framework for defining the requirements for the system. Requirements, in general, flow from the need for the system and the mission that it is designed to accomplish. Thus, the definition of the system requirements is closely related to the mission engineering element. In the systems engineering world, poor requirements almost always lead to major schedule, cost, and performance problems downstream. Flexibility is needed in the handling of and possible updating of requirements as we go through other elements of the systems engineering process. The reason is simple—the systems engineering process is itself a learning process and we find that certain requirements have to be changed as we achieve a more complete understanding of the system.

7.3.3 Requirements Analysis and Allocation

Requirements analysis and allocation is a set of activities that review an existing set of requirements, derive new requirements, and then allocate these requirements to the functional elements of the system. In the latter sense, the allocation process cannot occur without the next element in the systems engineering process, namely, the full definition of functions and subfunctions.

Although this element of systems engineering is covered in detail in the next chapter, we cite here some of the important aspects of requirements analysis and allocation, which are:

- Checking requirements for
  - Completeness
  - Accuracy
  - Compatibility
  - Consistency
  - Traceability
  - Appropriateness
  - Level (e.g., mandatory vs. optional)
- Developing a set of derived requirements
- Placing the requirements in an automated context
- Allocating requirements to the functionally decomposed representation of the system
- Recommending changes in requirements where and when the preceding aspects suggest that such changes are desirable or essential

Because requirements analysis and allocation is such an important aspect of systems engineering and the ultimate success or failure of a system development, we deal with it in considerably more detail in the next chapter.

7.3.4 Functional Analysis and Decomposition

At the top level, systems are normally described by the functions they are to perform in distinction to specific subsystems and components. For example, the functions of a command, control, and communications system may be described by the following selected functions:

- Command
- Communications
- Computation
- Control
- Detection
- Guidance

- Identification
- Surveillance
- Tracking

By maintaining a focus on function, rather than the manner in which the function is to be executed in hardware, software, and human components, we allow the systems engineer to consider a host of alternative ways of implementing a given function. We explicitly separate the "what" is to be done from the "how" it should be done. We consciously want to avoid leaping to a premature conclusion regarding a specific way to implement a given function. Selecting such specific ways is represented by the tasks of detailed design.

The basic steps under this element of systems engineering involve the following:

- Definition of the top-level functions of the system
- Decomposition of the preceding into lower-level functions and subfunctions
- Allocation of generic information and data flows among and between functions and subfunctions
- From the requirements analysis and allocation element, assuring that all requirements are allocated to the functions and subfunctions

The preceding steps are specifically related to the system architecting process and also set the stage for subsystem design once the architecture is developed. As such, they are considered further in Chapter 9.

7.3.5 Architecture Design and Synthesis

Developing a basic architecture is the centerpiece of the total systems engineering process. It is fundamentally a synthesis procedure that normally requires

- The formulation of alternative system architectures
- The analysis of the postulated architectures to verify that they satisfy system requirements

Architecting is performed at the top level, dealing with functional descriptions rather than detailed subsystem design features. Various methods are available to facilitate architecting, and these are discussed later in Chapter 9, with case examples in the Appendix.

A good example related to the issue of architecting has emerged rather clearly in the computer information system world. Years ago, computer systems largely involved a dominant mainframe computer and thus the information system had a highly centralized architecture. As minicomputers, workstations, and microcomputers came into use, architectures evolved into more decentralized configurations. In today's world, we see networks of client-server configurations as a preferred architecture for many types of computer information systems. In broad terms, matters of the degree of centralization or decentralization represent architectural alternatives.

Architectures can also involve fundamental technology considerations and choices. An example is the basic selection of a time-division multiplexed system versus a frequency-division multiplexed system. These are very basic choices that "drive" the remainder of the system design. Once a selection is made, further architecting must be compatible with this basic approach.

The world of architecture and engineering (A&E) firms demonstrates, through analogy, another way of understanding notions of architecting. In designing an airport, for example, the architect portion of the firm does the basic top-level design and architecture for the airport. Once this is complete, the engineers take over (civil, structural, mechanical, electrical, etc.) to convert the basic architecture into a physical system. However, modern concepts of concurrent engineering would suggest that the engineering team should be represented in the "front-end" architecting processes. Concurrent engineering is viewed as one of the key thirty elements of systems engineering.

Finally, we note that this element of systems engineering, as herein defined, is devoted to the formulation of alternative architectures, all of which are deemed to satisfy, in differing degrees, the system requirements. This element *does not* include the selection of a preferred architecture. Such a selection is subsumed in the following element, alternatives analysis and evaluation.

7.3.6 Alternatives Analysis and Evaluation

This element accepts the results of the previously cited element as an input, namely, the definition of alternative architectures for the overall system. The essence of this element is to carry out a comprehensive analysis and evaluation of the alternatives, resulting in a preferred system architecture. However, this cannot be performed without several other elements, as shown in Figure 7.4 Key inputs are required from the following other elements:

- Technical performance measurement (TPM)
- Life-cycle costing (LCC)
- Risk analysis
- Concurrent engineering
- Systems engineering management

Secondary inputs are necessary as well from

- Integrated logistics support (ILS)
- Reliability, maintainability, availability (RMA)

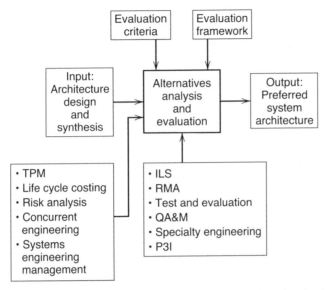

Figure 7.4. Inputs to and output from alternatives analysis and evaluation.

- Test and evaluation (T&E)
- Quality assurance and management (QA&M)
- Specialty engineering
- Preplanned product improvement (P3I)

Also shown in Figure 7.4 are two additional aspects of this element—the definition of evaluation criteria and an evaluation framework within which to carry out the assessment of alternatives.

At the completion of this element, the systems engineering team has produced a preferred system architecture together with the analysis that supports the selection. In short, it is not enough to simply select a preferred architecture; it is also necessary to explain why and how that architecture was selected, both to oneself as well as to the customer. Many system development efforts have gone astray as the preferred architecture was arrived at through some type of leap of faith without a clear demonstration of its superiority in relation to other alternatives. This has been particularly true when the contractor has chosen a proprietary or "closed" system architecture.

7.3.7 Technical Performance Measurement

Technical performance measurement (TPM) is a crucial element in the overall process of systems engineering. It is the underlying basis for evaluating the performance of the architecture alternatives. It also serves as a key ingredient in selecting more detailed design parameters in the form of hardware, software and human components (element 12 in Exhibit 7.1).

Examples of technical performance measures (TPMs) that might be used for different types of systems include the following:

- For a radar system:
  - Range to target
  - Probability of detection
  - False alarm probability
- For a transportation system:
  - Trip time
  - Capacity (for passenger and cargo transfer)
  - Quality of service
- For a communications system:
  - Bit error rate (BER)
  - Message error rate
  - Quality of service
  - Speed of service
  - Capacity (number of channels for various classes of communication such as data, voice, video, etc.)

These TPMs represent only a small number of possible measures that must be considered. Some additional measures are represented by the parameters upon which these TPMs are dependent. Such parameters, called technical performance parameters (TPPs), can be illustrated by the following:

- Power required
- Size of target
- Frequency of operation
- Size of antenna
- Bandwidth
- Signal-to-noise ratio
- Size of vehicle
- Load (external demand) on the system

The last is especially important because system performance generally degrades as the load on a system increases. Understanding this relationship, in quantitative terms, is a critical aspect of technical performance measurement.

In today's world, technical performance measurement is usually supported by one or more computer-based models and simulations. Thus, this capability is directly related to element 14 of Exhibit 7.1, namely, computer tool evaluation and utilization. The successful systems engineering team has a variety of such tools in place so that system performance can be evaluated as quickly and completely as possible.

7.3.8 Life-Cycle Costing

Clearly, a critical part of selecting among alternative architectures is the cost of each architecture and the systems that they imply. We adopt the perspective that costs must be considered on a life-cycle basis. Thus, the three main categories of cost are:

1. Research, development, test, and evaluation (RDT&E)
2. Acquisition or procurement
3. Operations and maintenance (O&M)

The above three cost categories may be further broken down into subordinate cost elements, as shown in Exhibit 7.2.

Exhibit 7.2: Breakdown of Major Cost Categories into Cost Elements

1. Research, Development, Test, and Evaluation
   1.1 Research and development
        Preliminary studies
        Design engineering
        Hardware
        Software
        Other personnel costs
   1.2 Test and evaluation
        Test planning
        Test hardware
        Test software
        Test operations
        Test evaluation
        Other personnel costs
2. Procurement
   2.1 Installations
        New construction
        Modification and renovation
   2.2 Equipment (hardware and software)
        Primary mission
        Mission support
        Other specialized

   2.3 Stocks
          Initial stock—primary mission
          Initial stock—support mission
          Spares—primary and support
   2.4 Initial training
          Training and support personnel
          Training materials and equipment
          Training facilities
   2.5 Other procurement (e.g., transportation) costs
3. Operations and Maintenance (O&M)
   3.1 Equipment replacement (hardware and software)
          Primary mission
          Mission support
          Other specialized
   3.2 Maintenance
          Primary mission
          Mission support
          Other specialized
   3.3 Training
          Training and support personnel
          Training materials and equipment
          Training facilities
   3.4 Salaries (operators)
          System operators
          Other operational support
   3.5 Material
          Expendables
          Other support material
   3.6 Other operations and maintenance (e.g., transportation)
   3.7 Disposal of system
          Dismantling of system
          Disposal of parts

   Using the above information, we may then construct the three *dimensions* of a Life-Cycle Cost Model (LCCM). The first dimension is the list of all the cost elements, as delineated in Exhibit 7.2. These may be visualized as the rows of a spreadsheet. Since the model applies to a system's life cycle, the next dimension is time, measured in years from the beginning of work on the system. This dimension is constructed as the columns of the spreadsheet. With these two dimensions, we can readily appreciate how data intensive the LCCM will be. For example, with about three dozen cost elements and perhaps a twenty-year life for the system, we have a total of $(36)(20) = 720$ cells that make up the spreadsheet. Each one needs to be estimated with as much fidelity as possible. If we then sum across any row, we are able to see the overall life-cycle cost of one of the elements, for example, the total RDT&E costs. If we sum down one of the columns, we can easily calculate the cost in any particular year. This is a necessary step in most budgeting processes. Finally, we recognize that systems are ultimately broken down into major subsystems. Thus, we may expand the two dimensions of the LCCM into a third dimension, namely, the subsystems of the overall system. Various "sheets" of the LCCM spreadsheet represent the subsystem costs, and then these are added into a summary "sheet" that represents the final LCCM for the system. To summarize, the three *dimensions* of an LCCM, constructed in the manner herein described, become:

1. The cost elements of the system
2. The years of useful life for the system
3. The subsystems of the overall system

During the early days of design, the subsystems may be thought of as the functions that are to be implemented in the system.

During architecture assessment, we usually only have a first-order estimate of these costs. However, they generally suffice to discriminate between the defined alternatives. As we move on to subsystem design, we develop costs more precisely, leading to more complete and, we hope, accurate cost estimates. Often, we also utilize various cost-estimating relationships (CERs) to develop the necessary cost estimates. An example is the COnstructive COst MOdel (COCOMO) [7.13], which is used in its various forms as a way of estimating costs and time to completion for software projects.

Cost estimating relationships (CERs) provide for the systems engineering team a way of estimating and calculating system costs in a generally top-down fashion. Instead of a process by which all the subordinate costs are estimated for, as an example, a radar system (bottom-up approach), we are able to estimate the costs as a function of only a few key variables for the system in question. This concept is further illustrated in Exhibit 7.3.

Exhibit 7.3: Illustrative Cost Estimating Relationships (CERs) [7.1]

The Cost of a. …	Can Be Considered a Function of. …
Radar system	Output power
	Frequency of operation
	Bandwidth or pulse width
	Weight
Missile booster	Weight
	T of propellant
Satellite terminal	Output power
	Antenna size
	Frequency of operation
	Receiver sensitivity

Aircraft engine	Thrust
	Bypass ratio
Software effort	Delivered source instructions
	Type of development effort
	Experience of team
Radio equipment	Number of channels
	Frequency
	Power
Dish antenna	Size of antenna
	Frequency of operation

The essential reason for using the above types of CERs, therefore, is that it is the best we are able to do at an early stage of design (e.g., the conceptual or architecting stages). If the estimating is being done for a new, clean-sheet-of-paper system, and by the users, it is especially important to have CER data so that budgets can be prepared for the system. Later on, it is useful to have these estimates to compare what the user believes the system should cost to what the system developer might believe it will cost. These concepts are traceable to the Department of Defense in the McNamara days, when he set up a group of "Whiz Kids" whose job, in part, was to be able to independently (from the contractor community) estimate the costs of new systems.

In general, we seek a cost-effective architecture and system, as illustrated in Figure 7.5. This figure shows three alternatives: a low-cost, low-effectiveness system; a high-cost, high-effectiveness system, and a system in between in

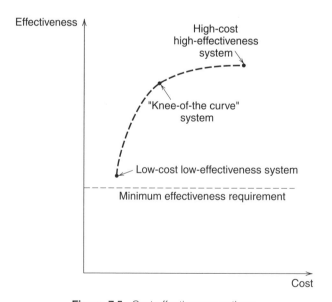

Figure 7.5. Cost-effectiveness notions.

terms of cost and effectiveness. The latter, hopefully defined at the "knee-of-the-curve," may turn out to be our preferred system. However, the low-cost, low-effectiveness system might be our preferred choice as long as it satisfies all the system requirements. This first-order representation, however, must be examined in greater detail in order to select a preferred system. This important topic is covered in Chapter 9 and again in the Appendix.

7.3.9 Risk Analysis

With the many problems that we have experienced in developing large-scale systems, we have included the subject of risk analysis as a formal element of systems engineering. In broad terms, risk may be explored in four key areas from the perspective of the CSE: (1) schedule risk, (2) cost risk, (3) performance risk, and (4) societal risk.

Schedule risk is prevalent, of course, under tight deadlines that must be met and that may have penalties associated with not meeting critical milestones. In such a situation, the critical path and all near-critical paths must be examined in detail to assure that schedule risk is minimized. Cost risk may be dominant when new development efforts are required along with the introduction of new technologies. Pushing the state of the art often results in considerable cost risk and the CSE and his or her team must examine all cost uncertainty areas extremely closely. Another factor is the type of contract under which the system is being procured. The discussion in Chapter 4 provides some further insight into this latter issue.

The most difficult area is performance risk, and it generally leads to both schedule and cost problems. Performance risk can be minimized by assigning your best systems engineering team and incorporating as many commercial-off-the-shelf (COTS) and nondevelopmental item (NDI) components as possible.

Finally, there is the matter of societal risk. This is the risk to the public that could result from the deployment of the system in question. Examples might include nuclear power plants, chemical processing plants, systems that might release toxic materials into our air and water, and the like. It is a clear responsibility of the systems engineering team to acknowledge such risks and build a system that minimizes potential hazards to the public. Issues of public safety are now squarely a feature of building large-scale systems and therefore fall upon the systems engineering team to consider in detail. Other administrative risks, as cited in Chapter 3, remain generally within the purview of the Project Manager.

In essentially all aspects of risk assessment, there are two key factors that play a central role. These are:

1. The likelihood that a high-risk event or set of events will occur
2. The consequences, given the occurrence of the foregoing event(s)

TABLE 7.3   Risk Table or Matrix

	Low Probability	Medium Probability	High Probability
High Consequence	High–Low	High–Medium	High–High
Medium Consequence	Medium–Low	Medium–Medium	Medium–High
Low Consequence	Low–Low	Low–Medium	Low–High

The most serious risks, of course, are represented when both factors are high. Therefore, a careful analysis has to ferret out such possible situations and then try to reduce both factors. If the consequence factor cannot be significantly changed (e.g., the catastrophic failure of a manned space mission), then system modifications should be made that reduce the probability factor.

The above notions of high-risk likelihood and consequence can be incorporated into a two-way table or matrix that has been the basis for a variety of risk analyses. These analyses partition the likelihood into high, medium, and low, and do the same for the consequences. These are then explored by reference to Table 7.3.

If we construct such a risk table, we can readily conclude that the situations represented by the High-High cell are the ones to which we should pay most attention in terms of what can be done to mitigate the risk. The next two troublesome cells are, of course, shown as the High–Medium and the Medium-High cases. Finally, if the previous cases have been satisfactorily dealt with, it may be possible to move to the Medium–Medium case. There is normally considerable discretion with respect to the assignment of numerical values to probabilities as well as the consequences. As an example, for the former, one might consider the ranges listed below:

- *High Probability*: From 0.7 to 1.0 (as close to unity as measurable)
- *Medium Probability*: From 0.4 to 0.7
- *Low Probability*: From 0.01 to 0.4

Consequences are often converted into dollar figures, unless other loss measures are more appropriate. This, however, is an area that can be quite controversial, especially when human lives are at stake with respect to negative consequences.

The manner in which the overall subject of risk analysis is approached can be quite varied. This may be illustrated by briefly looking at two agencies, the Department of Defense (DoD) and the National Aeronautics and Space Administration (NASA). The former, for example, defines *risk management* as [7.14]:

Risk management is a program management responsibility and is the act or practice of controlling the risk drivers that adversely affect the program. It includes the process of identifying, analyzing, and tracking risk drivers, as-sessing the likelihood of their occurrence and their consequences, defining risk-handling plans, and performing continuous assessments to determine how risks change during the life of the program.

A sample format for a Risk Management Plan includes the following sections:

1. Introduction
2. Program summary
3. Definitions
4. Risk management strategy and approach
5. Organization
6. Risk management process and procedures

Special attention is paid to *technical* risk management, since it is often the case that technical risk "is a significant driver of all other program risks" [7.14]. The three primary approaches to technical risk management involve a detailed analysis of:

1. Critical processes
2. Deliverable products (usually identifiable from the work breakdown structure), and
3. Integrated processes and products

A brief look at some of NASA's documentation with respect to risk man-agement reveals a serious attempt to define the life-cycle risk management elements [7.15], namely:

1. ISO 9001 compliance
2. Industry best practices: design and engineering
3. Key characteristic and critical process management
4. Industry best practices: manufacturing
5. Test/verification/validation
6. Supply chain integration and management
7. Implementation assurance
8. Independent assessment

We note the elements of best practices in industry and a topic known as verification and validation. This latter area was listed as one of the thirty elements of systems engineering and is discussed in a later section of this chapter.

7.3.10 Concurrent Engineering

Concurrent engineering is defined as "a systematic approach to the integrated, concurrent design of products and their related processes, including manufacturing and support. This approach is intended to cause the developers, from the outset, to consider all elements of the product life cycle from concept through disposal, including quality, cost, schedule, and user requirements" [7.8, 7.16]. From this definition, concurrent engineering seems to be similar to systems engineering. However, we view concurrent engineering as a subset (element of) systems engineering for two basic reasons:

1. It lacks the formal definition and historical discipline of systems engineering.
2. Its main thrust and significance flow from the notion of engineering concurrency, and therefore it is a contribution to, rather than a replacement for, systems engineering.

The essential perspective of concurrent engineering is that all parties that have a role to play in a given system design should be involved with that system throughout its life cycle. The sequential or serial process of moving from development to test to production to operations, and so forth, is augmented by adding all significant life-cycle disciplines and personnel to all life-cycle phases. Therefore, for example, even though a system may be in its early conceptual, mission analysis, or definition phases (see Section 7.2.2 for NASA phase descriptions), personnel from manufacturing should be involved in the process to assure that the system is designed for optimum manufacturability. This basic idea holds for all personnel who have a clear stake in the product or system. It also allows for parallel developments whenever possible that have the general positive effect of compressing schedules.

The implications of these notions of concurrent engineering have been that systems engineering teams must also contain integrated product teams (IPTs) that focus on integrating all life-cycle considerations. The systems engineering process, with its element of concurrent engineering, must explicitly take account of downstream integration, test, production, manufacturing, quality, operations, maintenance, logistics, and all other relevant aspects of a system's life cycle. Concurrent engineering forces this discipline as a vital part of systems engineering. Thus, it is herein defined as an essential element of systems engineering.

7.3.11 Specification Development

In principle, after the preferred system architecture has been selected, we require a detailed specification to continue on with the process of subsystem design. This is done by developing a set of detailed specifications for these subsystems, usually considered in three categories: hardware, software, and

human. In many system developments, there are actually levels of specifications so that a top-level spec has to be developed to define system architectures.
This element of systems engineering covers the rather important task of writing a series of specifications at whatever levels are needed. Indeed, a military
standard exists to help define the numerous specification requirements for
military systems [7.17].

7.3.12 Hardware, Software, and Human Engineering

As suggested earlier, this element of systems engineering deals with detailed
subsystem design in the three dimensions of hardware, software, and human.
In general, deciding what should be implemented in hardware versus software
is not an easy choice. Software often provides additional flexibility, but can
be more difficult to manage and implement. Human engineering functions
are normally relegated to those of operator and maintainer of the system, and
the system requirements often specify these functions.

In general, this element refers to the design and construction of subsystems
as well as the components that make up these subsystems. As with the architecting process, design choices involve the postulation of alternatives and the
selection of the best alternative to satisfy the stated or derived requirements.
As we move more toward the use of COTS and NDI for hardware, and reuse
for software, we are able to define and choose subsystems and components
more easily. This, however, is a nontrivial task and depends on the experience and design expertise of hardware, software, and human engineering
personnel. A short list of some typical subsystem design areas includes:

- Power supplies
- Multiplexers
- Modulators–demodulators
- Transmitters
- Receivers
- Servomechanisms
- Graphical user interfaces (GUIs)
- Large-scale displays

7.3.13 Interface Control

Experience indicates that many of the difficulties in building large systems
show themselves in the interfaces between the subsystems and between the
system and external systems with which it must interoperate. For this essential
reason, we include interface control as a key element of systems engineering.
In the hardware arena, we have come a long way in specifying such interfaces
and assuring interface compatibility. In the case of software, we are further

behind and have to develop new methods of specification for software modules that are candidates for reuse. The basic tasks that have to be accomplished under this element include:

- Definition of all interfaces
- Identification of the nature of all interfaces (physical, data, electrical, mechanical, etc.)
- Assurance of interface compatibility at all defined interfaces
- Strict control of the interface processes (design, construction, etc.)

This is a difficult bookkeeping and control problem that may involve tracking thousands of critical interfaces with the possibility that many interfaces may creep in that are not found until testing (such as cross-coupling of mechanical or electrical forces and fields). Also, many systems engineers simply do not give enough attention to this issue, and find that they pay an unexpected price downstream.

7.3.14   Computer Tool Evaluation and Utilization

Commercially available computer tools that support systems engineering have evolved so quickly, and can be obtained at such low prices, that it has become essential that all systems engineering teams have a set of such tools immediately at hand to carry out the thirty elements of systems engineering. This includes the set of tools known as CASE (computer-aided software engineering) tools that are used for software engineering, a subset of systems engineering. Under this element of systems engineering, the team evaluates and then uses the set of computer aids appropriate to the system being developed.

A listing of some thirty categories of systems engineering support tools is provided in an early systems engineering text [7.1]. As a practical matter, because there are so many tools to choose from, a formal evaluation is necessary in relation to the system requirements and the domain knowledge attendant to that system.

The maturity of computer tools and their widespread use have also spawned the notion of system and software engineering environments (SEEs). Such environments incorporate integrated sets of tools that support both systems and software engineering [7.18, 7.19]. The systems engineering team that has itself reached a significant level of maturity will establish a systems engineering environment that will allow it to execute the thirty elements of systems engineering quickly [7.20] and efficiently. It will also put such a team in an excellent position relative to its competitors.

7.3.15   Technical Data Management and Documentation

A complex systems engineering effort generates a large amount of technical information. For large systems, these data can be overwhelming, and come in many forms such as text, graphics, large-scale drawings, correspondence, requirements and specifications, and others. If these data are not organized and managed, the systems engineering team will soon be lost in mounds of paper that are not readily retrievable. Here again, with the power afforded by microcomputers and workstations, it is easier to manage large amounts of technical data on behalf of the entire project and the systems engineering team. A commercially available database management system (DBMS) becomes the most likely candidate as the centerpiece of technical data management, with other specialized software playing a supporting role. The reader with a special interest in this area may examine in detail the wide variety of other tools that might be used to assist in the task of technical data management.

The second part of this element is effective and efficient production of all required system documentation. Of course, this must be well-organized so that all files are retrievable by the project team. Word processors as well as other software (e.g., graphics) support this subelement, as do report generators of other types of software. Of special significance are ways and means of handling large-size drawings (e.g., blueprints) so that they can be stored and retrieved in an automated environment. All documentation should be available on magnetic disks to facilitate reproduction and usage.

7.3.16 Integrated Logistics Support (ILS)

Integrated logistics support (ILS) is defined as "a disciplined, unified and iterative approach to the management and technical activities necessary to (a) integrate support considerations into system and equipment design; (b) develop support requirements that are related consistently to readiness objectives, to design, and to each other; (c) acquire the required support; and (d) provide the required support during the operational phase at minimum cost" [7.21]. This definition was derived from a Department of Defense (DoD) Directive for ILS that was published in 1983 and is relevant and appropriate today. A more recent DoD Instruction [7.22], also cited in Chapter 2, establishes ILS policies and procedures for ensuring that (1) support considerations are effectively integrated into the system design, and (2) required support structure elements are acquired concurrently with the system so that the system will be both supportable and supported when fielded. This latter instruction emphasizes ILS procedures in the following areas:

a. Readiness objectives

b. Integrated logistics support plan (ILSP)

c. Computer resources support

d. Planning factors

e. Logistics support analysis

f. Manpower, personnel, training, and safety

g. Accelerated acquisition strategies

   h. Interim contractor support
   i. Depot maintenance support
   j. Spares acquisition integrated with production
   k. Postproduction support
   l. Logistics resources
  m. Milestone decision reviews

The ten specific elements of ILS are further defined as:

1. Maintenance planning
2. Manpower and personnel
3. Supply support
4. Support equipment
5. Technical data
6. Training and training support
7. Computer resources support
8. Facilities
9. Packaging, handling, storage, and transportation
10. Design interface

Clearly, given its defined scope, integrated logistics support is a critical element of systems engineering.

7.3.17   Reliability, Maintainability, Availability (RMA)

The RMA element focuses specifically on matters of the operating life and ease of maintenance of a system. In broad terms, reliability and availability are defined as follows:

Reliability. The probability that a system successfully operates to time "t."
Availability. The probability that a system is available to operate when called on to do so for a particular mission or application.

Maintainability does not have a generally accepted quantitative definition, but it is interpreted as the degree to which a system has been constructed so as to facilitate maintenance at an affordable cost.

There is clearly an important relationship between RMA and ILS. A system must be designed to meet specific reliability, availability, and maintainability requirements, as they are stated in the requirements and specifications documents. However, all three are supported by ILS considerations. For example, when a system or subsystem or component fails, a spare is usually required to put the system back on line. It is the ILS engineer who determines the level of sparing required and assures that there is a system in place to provide these spares in minimum time.

7.3.18 Integration

Integration refers to the set of activities that bring together smaller units of a system into larger units. Such units, often called configuration items (CIs) and components, are generally manifest in hardware and software. As these units are completed, they are tested to assure that they are operating correctly. If so, they are then interconnected (integrated) with other tested units to build larger and larger subsystems. As problems of incompatibility show themselves, they are solved, one at a time. Often, this physical integration is the first time that there is a real opportunity to see if units will "play together." If the interface control element (element number 13 in Exhibit 7.1) is performed with diligence, many such potential problems can be anticipated and averted.

This process of unit integration is a natural part of building any system. It can go astray when not enough time is allocated in the master schedule to account for the situations in which unexpected interoperability problems occur, despite all efforts to avoid them. The process of rapid prototyping has been introduced into some system development efforts, in part, to deal with such issues as early as possible.

7.3.19 Verification and Validation

These two parts of systems engineering may be defined as follows [7.1].

Verification. The confirmation that products and processes of each development phase fulfill the requirements for that phase, and interoperate with the results of an earlier phase.

Validation. The confirmation that requirements are correct and that all products and processes, when taken in combination, satisfy all system-level mission needs.

It will be recalled, from the beginning of this chapter (i.e., Figure 7.1), that the four key elements that make up the systems engineering process, according to the Military Standard 499B as well as other approaches, are:

1. Requirements analysis
2. Functional analysis/allocation
3. Synthesis
4. System analysis and control

From the perspective of the Institute of Electrical and Electronics Engineers' (IEEE) systems engineering standard [7.23], a fifth element needs to be added to the above four elements, namely, verification and validation (V&V). This is an important step, coming as it does from this well-respected and influential source. The IEEE definitions of verification and validation, which may be compared with the above definitions, are these:

Verification. The "process of determining whether or not the products of a given phase of development fulfill the requirements established during the previous phase."
Validation. The "process of evaluating a configuration item, subsystem, or system to ensure compliance with system requirements."

Although the concepts of verification and validation (V&V) may well have been established during the early days of missiles and space technology, they have become part and parcel of today's world of systems engineering. They can be carried out as an integral part of a given contractor's systems engineering activities, or they may be executed by a third party, in which case they would likely be called independent verification and validation (IV&V).

V&V are to be a part of each and every phase of an overall development effort, focusing on the special needs and requirements of each phase. For example, within an agency such as NASA, V&V would be scheduled to deal with each of the key phases [7.24], as below:

- *Phases A/B*. Preliminary analysis and definition
- *Phase C*. Design
- *Phase D*. Development

Further, some of the important methods that could be used to perform V&V include [7.24]:

1. Testing
2. Analysis
3. Demonstration
4. Similarity
5. Inspection
6. Simulation and modeling
7. Validation of records

V&V has also been specialized to the development of software whenever this aspect of a program has been considered to be of especially high risk, since the main purpose of V&V is to try to reduce the risk in a program. In such a case, the life cycle of interest may be narrowed, typically dealing with the software phases cited below:

- Requirements
- Design
- Development and coding
- Integration and test

A more complete set of software phases, including variations that depend upon the software development model that is selected, is cited in Chapter 10, which deals with various aspects of software engineering. Further, an excellent discussion of V&V in the context of software development is provided in John Wiley's *Encyclopedia of Software Engineering* [7.25].

7.3.20 Test and Evaluation

Test and evaluation (T&E) normally refers to physical confirmation of the performance of an overall system. It is carried out in at least two key contexts: (1) completion of the full-scale development of a system, and (2) placing the system in an operational environment. The former is called development test and evaluation (DT&E), and the latter is operational test and evaluation (OT&E). For many systems, OT&E can be exceedingly difficult because the operational environment for the system cannot be adequately represented or simulated. Examples include many military weapon systems for which it would be prohibitively expensive to truly test the system in terms of stated operational requirements. In such cases, attempts are made to combine real-world test data with models in order to verify performance. An example is the rather large-scale National Test Bed (NTB) established for the strategic defense initiative (SDI).

The PM and the CSE must pay a great deal of attention to the systems engineering element of T&E because these are the arenas in which there is a "sell-off" of the system to the customer. It is the stage at which there is overall confirmation that the system meets the given user requirements. If failures occur during T&E, it may be extremely expensive to go back and make the changes necessary to improve performance. It is also usually very time-consuming. As described in the quality assurance element of systems engineering, trying to test quality into a system is basically the wrong approach. In that sense, a system that moves into DT&E and OT&E with a large number of defects still remaining in the system is very likely to become a major disaster.

As the SEMP (systems engineering management plan) is the key document that describes what is to be done to execute and manage the systems engineering elements, the TEMP (test and evaluation master plan) documents the T&E activities. For the reasons cited earlier, the best systems engineers should be involved in structuring all T&E concepts and plans as well as the interpretation of results.

As with many of the other elements of systems engineering, test and evaluation is undergoing continuous review, evaluation, and change. This is to be expected since test and evaluation (T&E) is truly one of the most important aspects of bringing a system from development into operation. An example is a report by a Defense Science Board Task Force [7.26] that presented these six findings:

1. T&E should focus on how to support the acquisition process.
2. T&E planning needs to start early in the acquisition cycle.
3. There is distrust between the development and test communities.
4. Contractor, development, and operational testing have overlapping functions.
5. It is essential that we have independence of test data evaluation.
6. The response to perceived test "failures" is often counterproductive.

Test and evaluation is also seen now as a key element of what is being called the DoD transformation [7.27]. This transformation depends greatly on being able to bring appropriate levels and types of information together at the right time and place. This also involves getting T&E activities into the planning and development processes much earlier in the system life cycle.

7.3.21 Quality Assurance and Management

Quality assurance and management (QA&M) is an element of systems engineering that addresses all matters related to product and service quality. As a discipline, quality assurance and management has undergone extensive changes in the last 15 years or so, mainly through the introduction of Total Quality Management (TQM). Some of the principles of TQM can be articulated as:

- Strict conformance with specifications
- Continuous improvement
- A focus on process
- Attention to customer needs
- Empowerment of people to implement
- Management support of its introduction and use

As TQM has developed, various tools or methods have evolved that support it, including statistical process control (SPC) and quality function deployment (QFD). These and other procedures have served as practical ways to enhance quality.

Military Standard 499B does not address matters of QA&M in detail. Specifically, the comment made in that standard is as follows:

The government and contractor shall insure the comprehensive application of the systems engineering process to integrate quality factors throughout all technical elements and activities of the program and to:
a. capture customer requirements and translate them into detailed design requirements that can be implemented in a consistent manner
b. deliver products that meet operational requirements under specified operational conditions

If the system is being developed in a corporate environment that has embraced TQM or some variant thereof, the PM and CSE should bring selected procedures and practices into the systems engineering process as a part of the QA&M element. If there is no corporate quality initiative or structure, the PM and CSE should develop a QA&M program for the project. This is a workable approach, but is clearly inferior to having a corporatewide quality program into which a given project fits.

7.3.22 Configuration Management

Configuration management (CM) is largely a bookkeeping and control activity that involves the following subelements [7.1]:

- Identification
- Control
- Auditing
- Status accounting
- Traceability

Although most CM concepts are simple, they are often carried out poorly or not at all. This is a serious mistake, especially for a large system. Failure to keep track of and control the status of all system configurations leads to backtracking to try to understand just exactly what version of the system the team is dealing with. This problem shows up especially in software development in which there is the potential for making changes quickly and easily.

A key concept in system development that relates to the element of CM is that of system "baselining." When a satisfactory version of a part of the system has been constructed and tested, it may be given a baseline name and number. From a CM point of view, that version of the system cannot be changed without going through a formal configuration control board (CCB) review and action. The CSE is thereby able to maintain control over the system and assure that it is built logically, incrementally, and under positive change management.

7.3.23 Specialty Engineering

Specialty engineering refers to a set of engineering topics that have to be explored on some, but not all, systems engineering efforts. In other words, some systems involve these special disciplines and some do not. Examples of specialty engineering areas include:

- Electromagnetic compatibility and interference
- Safety
- Physical security
- Computer security
- Communications security
- Demand forecasting
- Object-oriented design
- Value engineering

A sampling of technical domains that might support engineering specialty areas on a particular project are

- Network engineering
- Aerodynamics
- Thermal analysis
- Structural analysis
- Mechanical design
- Hypersonics
- Nuclear engineering
- Artificial intelligence

As one can see, the list of potential specialty engineering areas can be rather long. Also, many of these areas are supported by specific computer tools that facilitate the analysis and design processes.

7.3.24 Preplanned Product Improvement (P3I)

Preplanned product improvement considers the ways and means to enhance the system beyond the scope of the current contractual arrangement. Such improvements generally enhance performance, and more than satisfy the requirements, as currently stated. Examples of such improvements include:

- Extending the range of a radar system
- Decreasing the response time of a transaction processing system
- Increasing the storage capability of a database management system
- Adding built-in test equipment (BITE)
- Increasing the speed of a network

Improvements under P3I consideration represent candidates for follow-on efforts that the customer might consider in future versions of the system. By dealing with them during the current effort, migration paths to improved systems are identified early. With the rapid speed of technology advances, it is necessary to provide a clear and low-cost path to downstream system improvements.

P3I is normally a part of all commercial products, although it may not be called by this name. Whether your firm is making razor blades or software, plans for product improvement must be on the drawing board at all times in order to be competitive in a changing world.

7.3.25 Training

Training in the context of the thirty systems engineering elements generally refers to the training of system operators and maintainers. Training for these personnel must be designed and delivered in formal well-developed programs. Such programs also have to be scheduled during the operations and maintenance (O&M) phase of the system, often extending fifteen to twenty years downstream. Training, from the DoD perspective, has to be considered as part of the ILS discipline (see Section 7.3.16). Training specialists should be part of the systems engineering team.

7.3.26 Production and Deployment

Production refers to the phase of a project during which one or more installable systems are being produced for the customer. In many procurements, the user or acquisition agent, as a matter of policy, recompetes the production phase for a given system. Many contractors specialize in this production phase, with developed expertise in the "build-to-print" arena. Organizing and implementing a production or manufacturing capability is itself an area in which systems engineering procedures and practices can be brought to bear. It also can be an exceedingly complex set of activities because it deals with the fundamental capability of an enterprise to replicate a system that has already been built and accepted by the customer. For additional information regarding this subject, the reader should refer to the large available literature on production and manufacturing. The final points to be made in this respect are that (a) producibility issues have to be dealt with as part of systems engineering (note that manufacturing analysis is a subject cited in Mil-Std-499B), and (b) systems engineering processes need to be applied to the actual production and manufacturing of the system.

As the system is being produced, some units of the system may be moving into the installation/deployment phase. This is a rather straightforward sub-element, requiring excellent documentation on how to install the system in its operational environment. Such environments range from rather benign situations, such as an office building, to more stressful ones such as aboard a ship or a submarine. In some cases, special analysis has to be carried out so as to assure field operability. This applies, for example, to siting a radar in a particular ground-based location. At times, unanticipated problems arise during installation for a system that was operating without difficulties in a more controlled environment. Most companies have specialized installation teams that are familiar with the types of problems that surface during system deployment.

7.3.27   Operations and Maintenance (O&M)

This part of the systems engineering process refers to the rather long period of time during which the system is operational in the field. From a systems engineering perspective, emphasis should be placed on the continuous measurement of the system's performance. Measurement procedures range from simple manual data sheets to automated sensors that record operational status more or less continuously. In the commercial world, companies try to maintain contact with consumers through hot lines and reports of satisfaction (or not) from customers. It is important for the systems engineering team to explicitly consider how to install and sustain a performance measurement program during the O&M phase.

7.3.28   Operations Evaluation and Reengineering

If the preceding element is carried out correctly, a database is built as to how well the system is performing over a long period of time. This places in evidence ways and means of evaluating system performance and points toward specific improvement areas for reengineering of the system. Because our installed base of systems is extremely large, operations evaluation and reengineering is a rather serious issue. The latter has also been receiving a great deal of attention under the topic of business process reengineering (BPR).

7.3.29   System Disposal

It is clear that once a system has come to the end of its usefulness, as agreed upon by the users and sponsors, it is necessary to dispose of it. For systems that are 100% software, this is a relatively simple matter. Of course, the software must reside upon host machines, so there is a question as to whether or not the hardware remains or is to be retired. Whenever there is hardware that is no longer useful as part of the system, specific plans and procedures must be developed for disposal, followed by careful implementation.

Beyond the more-or-less standard methods of reusing, reselling, salvaging, or junking various hardware that is no longer needed, is the fact that many systems that need to be disposed of fall in the broad category of hazardous waste materials. In such cases, they must be disposed of in accordance with applicable standards so as not to present a human or environmental risk, either current or future. In the same vein, disposal should be consistent with the principles and practices of *sustainable development*, a topic that is discussed in somewhat greater detail in Chapter 12.

Matters of system disposal can be quite varied as well as unique. For example, in March of 2001, the Russians had to decide what to do about disposing of the 143-ton *Mir*, their fifteen-year-old (launched in February 1986) spacecraft, which had apparently become too decrepit and expensive to operate. With the spacecraft traveling at about 132 miles above the earth, *Mir's* engines had to be fired by mission controllers on the ground in order to slow the vehicle and change its orbit from circular to elliptical. The plan called for most of *Mir* to burn up as it entered the earth's atmosphere. However, up to 27.5 tons of "debris" was expected to hit the earth south of Tahiti, east of New Zealand, and southeast of Chile's Easter Island. According to reports, the plunge through the atmosphere was fiery and spectacular, and also successful. Despite the great care taken by the Russians, the government of Chile was not happy about this space "dumping" so close to their country. Although it did not cause an international incident, one can see that disposing of 143 tons of what had become space junk turned out to be a nontrivial adventure.

7.3.30 Systems Engineering Management

This last element covers all the management activities that must be considered by the CSE and lead systems engineering managers. As such, it includes the four basic functions of the project manager, namely, planning, organizing, directing, and monitoring. These are applied to the previous twenty-nine elements of systems engineering. In this sense, it is explicitly recognized that the CSE has management responsibilities that are to be executed in the context of all the elements of systems engineering. Thus, in addition to being the chief technical person on a project, the CSE must have excellent skills in the fine art of management. Several specific areas of interest have been established and presented in some detail in Part II (Chapters 3 through 6). A further listing of problem areas with respect to systems engineering management is provided below in Exhibit 7.4.

Exhibit 7.4: Selected Problem Areas Within the Element of Systems Engineering Management

1. Requirements unclear, incorrect, and/or creeping
2. Unrealistic schedule/cost constraints
3. Risks and their mitigation not seriously considered
4. Newly formed design/development team
5. Little project status monitoring
6. No well-defined design/development process
7. Poor system decomposition and work assignments
8. Alternative architectures not considered

9. Ineffective performance measurement
10. Inadequate communications skills
11. Late staffing and insufficient skill levels
12. Inadequate reserves and incentives

The reader is encouraged to compare the above issues with experiences that he or she may have had in developing systems.

7.4 THE IMPORTANCE OF DOMAIN KNOWLEDGE IN SYSTEMS ENGINEERING

We complete this chapter with a brief note on the matter of domain knowledge and its relationship to systems engineering. Domain knowledge involves an understanding of the technical domain of the system that is being developed, examples of which include:

- An on-line transaction processing (OLTP) information system
- A rapid-transit system
- An air defense radar system
- A missile system
- An aircraft
- A supercomputer
- An automobile
- A chemical processing plant
- A nuclear power plant

Effective design and construction of any system involves *both* an effective systems engineering process and a deep understanding of the domain knowledge implicit in the system. Transit system engineers are not likely to have the domain knowledge associated with building a new aircraft, and vice versa. Nuclear engineers may not understand the intricacies of developing information systems and their software elements. The best systems engineers have both the systems engineering and the domain-knowledge expertise. It is not really possible to function with excellence as a systems engineer on a given program without having the appropriate domain-knowledge understanding. This does not mean that such a person cannot make an important contribution. However, he or she may lack a depth of understanding of the domain and its related science and technology. Accordingly, it is a good idea for a person with a basic understanding of systems engineering, who is to move on to the field of radar systems, for example, to be immersed in the rather complex world of radars and their underlying technologies. This notion applies as well to all technical disciplines and their related domain-knowledge bases.

QUESTIONS/EXERCISES

7.1 Identify and discuss three systems engineering activities that you believe should be added to the list of the thirty elements of systems engineering.

7.2 Contrast the NASA life-cycle phases in this chapter with the system acquisition phases in Exhibit 2.1.

7.3 Functionally decompose a personal computer system. Write two requirements statements for each of the decomposed elements.

7.4 Define three technical performance measures (TPMs) for
  a. a personal computer system
  b. an automobile
  c. an air defense system

7.5 Find and describe two examples of cost-estimating relationships (CERs) from the literature.

7.6 Identify and describe three specialty engineering areas other than those cited in this chapter.

7.7 Find a risk analysis approach in the literature and discuss its major features.

7.8 Define and discuss two systems engineering management issues, other than those cited in this chapter, that are likely to have a significant impact upon the success of a systems engineering program.

7.9 Are there any elements of systems engineering that you would add to the thirty described here? Why? Are there any elements you would delete? Why?

7.10 In describing the content of systems engineering, develop some type of hierarchical structure.

REFERENCES

7.1 Eisner, H. (1988). *Computer-Aided Systems Engineering*. Englewood Cliffs, NJ: Prentice Hall.
7.2 Goode, H. H., and R. E. Machol (1957). *System Engineering*. New York: McGraw-Hill.
7.3 *Engineering Management*, Military Standard 499A (1974). Washington, DC: U.S. Department of Defense.
7.4 *Systems Engineering Management Guide* (1983, 1990). Fort Belvior, VA: Defense Systems Management College.
7.5 Beam, W. R. (1990). *Systems Engineering—Architecture and Design*. New York: McGraw-Hill.
7.6 Sage, A. P. (1992). *Systems Engineering*. New York: John Wiley.

7.7 Blanchard, B. S., and W. J. Fabrycky (1990). *Systems Engineering and Analysis*. Englewood Cliffs, NJ: Prentice Hall.

7.8 Chapman, W. L., A. T. Bahill, and A. W. Wymore (1992). *Engineering Modeling and Design*. Boca Raton, FL: CRC Press.

7.9 Wymore, A. W. (1993). *Model-Based Systems Engineering*. Boca Raton, FL: CRC Press.

7.10 Sage, A. P., and W. B. Rouse, eds. (1999). *Handbook of Systems Engineering and Management*. New York: John Wiley.

7.11 Sage, A. P., and J. E. Armstrong, Jr. (2000). *Introduction to Systems Engineering*. New York: John Wiley.

7.12 *The NASA Mission Design Process* (1992). Washington, DC: National Aeronautics and Space Administration, Engineering Management Council.

7.13 Boehm, B. W. (1981). *Software Engineering Economics*. Englewood Cliffs, NJ: Prentice Hall; see also Boehm, B. W. (2000). *Software Cost Estimation with COCOMO II*. Upper Saddle River, NJ: Prentice Hall PTR.

7.14 Department of Defense (DoD) Website: web2.deskbook.osd.mil

7.15 *Life Cycle Risk Management Elements for NASA Programs* (1997). Washington DC: National Aeronautics and Space Administration Headquarters, June.

7.16 Carter, D. E., and B. S. Baker (1992). *Concurrent Engineering—The Product Development Environment for the 1990s*. Reading, MA: Addison-Wesley.

7.17 *Specification Practices*, Military Standard 490 (1972). Washington, DC: U.S. Department of Defense.

7.18 Comer, E. R. (1972). *Catalyst: Automating Systems Engineering in the 21st Century*. Paper from the Proceedings of the 2nd Annual International Symposium of the National Council on Systems Engineering (NCOSE), Seattle, July 20–22.

7.19 Charette, R. N. (1986). *Software Engineering Environments—Concepts and Technology*. New York: McGraw-Hill.

7.20 Eisner, H., J. Marciniak, R. McMillan, and W. Pragluski (1993). *RCASSE: Rapid Computer-Aided System of Systems (S2) Engineering*. Paper from the Proceedings of the 3rd Annual International Symposium of the National Council on Systems Engineering (NCOSE), Arlington, VA, July 26–28.

7.21 Blanchard, B. S. (1986). *Logistics Engineering and Management*. Englewood Cliffs, NJ: Prentice Hall.

7.22 *Defense Acquisition Management Policies and Procedures*, DoD Instruction 5000.2. Washington, DC: U.S. Department of Defense.

7.23 *Standard for Systems Engineering*, IEEE P1220 (1994). Piscataway, NJ: IEEE Standards Department.

7.24 Shishko, R. (1995). *NASA Systems Engineering Handbook*, SP-6105. Washington, DC: National Aeronautics and Space Administration (NASA) Headquarters, June.

7.25 Marciniak, J., editor-in-chief (1994). *Encyclopedia of Software Engineering*. New York: John Wiley.

7.26 *Report of the Defense Science Board Task Force on Test and Evaluation* (1999). Washington, DC: Office of the Under Secretary of Defense for Acquisition and Technology. September.

7.27 Stenbit, J. (2002). "Test and Evaluation: A Key Component of DoD Transformation," *ITEA Journal of Test and Evaluation* (June/July), 6–8.

8

REQUIREMENTS ANALYSIS AND ALLOCATION

8.1 INTRODUCTION

Requirements analysis and allocation (RAA) has become a critically important element of systems engineering. All of the later elements of systems engineering are carried out with one dominant question in mind: Are we in fact satisfying the system requirements in the best possible (most cost-effective) manner? Thus, as we architect, build, and test the system, we continuously return to the requirements to be sure that we are completely satisfying them. All systems engineering activities thus can be said to revolve around the stated requirements, as defined by the user or acquisition agent for the system.

Requirements are defined during the very early stages of the system life cycle. However, they are often incomplete and inappropriately stated. This leads to difficulties in architecting and designing a system as well as controversies between the system developer and the user. This can be exacerbated because requirements are not easily changed, even when found to be inadequate. This flows from the contract mechanisms and rules of competition that often surround the system acquisition process. Attempts to improve or reconcile requirements take time and cost money, thus putting additional pressure on the systems engineering team in terms of meeting stringent time and cost constraints. Poor requirements are perhaps the most mentioned issue when examining reasons why systems engineering and development efforts ultimately result in lack of performance as well as cost and schedule overruns.

Requirements may represent problem areas even when dealing with upgrades to an existing system. In such a case, the context is often that it has not been adequately proven that there is a set of validated operational requirements. As an example, a report from the General Accounting Office (GAO) to the Acting Secretary of the Air Force claimed that certain planned upgrades were not based on validated requirements [8.1], with the following statement:

> Upgrade of System 8 ground station software and hardware is premature because the Air Force has not yet validated operational requirements. Validated requirements are needed to verify the need for planned DSP ground station upgrades. DOD Instruction 5000.2 and Air Force Regulation 57-1 require that an operational requirements document, identifying minimum acceptable performance characteristics, be prepared for all major weapon systems such as DSP. An operational requirements document identifies the minimum acceptable performance required to satisfy mission needs and is used to establish test criteria for operational test and evaluation.

From this, it follows that the prudent user, acquisition agent, Project Manager (PM), and Chief Systems Engineer (CSE) will pay a great deal of attention to the matter of requirements. Therefore, RAA becomes a central focus for the systems engineering effort.

8.2 DEPARTMENT OF DEFENSE (DOD) PERSPECTIVES

Looking at Military Standard 499B [8.2], we find requirements analysis to be one of the four key elements of the systems engineering process (see Figure 7.1), represented in a short form by the key words:

- Define/derive/refine performance requirements (what item must do and how well)
- The (systems engineering) process shall define and analyze performance and functional requirements, define and design system elements to satisfy those requirements, and establish the final configuration.

The full description of the element of requirements analysis is reproduced in Exhibit 8.1.

Exhibit 8.1: Requirements Analysis as Stated in Mil-Std-499B

User requirements/objectives shall be defined/refined and integrated in terms of quantifiable characteristics and tasks that item solutions must satisfy. Technical requirements shall be developed concurrently for all functions and subfunctions based on the system life cycle, and iteratively to provide progressively more detailed performance requirements definition. For each requirement, the absoluteness, relative priority and relationship to other derived requirements shall be identified. Impacts of identified user requirements/objectives and derived requirements in terms of mission (tasks), environments, constraints and measures of effectiveness shall be analyzed as the basis for defining and deriving performance requirements. These impacts shall be continually examined for validity, consistency, desirability, and attainability with respect to technology availability, physical and human resources, human performance capabilities, life cycle costs, schedule and other identified constraints. The output of this analysis will define technical performance requirements and either verify existing requirements or develop new requirements that are more appropriate for each item.

We note especially that the systems engineering team is to "either verify existing requirements or develop new requirements that are more appropriate for each item." Although this clearly makes good sense, the insertion of new requirements, as alluded to before, can be an extremely controversial issue. Strictly new requirements developed by the systems engineering team cannot be accepted until they are confirmed as acceptable by the user and system acquisition agent. Again, this can be a difficult process that is impeded by the general reluctance to change requirements, often necessitating an approved change in a formal contractual document and relationship.

In our definition of RAA, we include the matter of requirements allocation. This inclusive definition means that requirements, in addition to being analyzed, are allocated to system functions and subfunctions. Thus, the definition in this book leads to the explicit interaction between RAA and functional analysis/allocation (see the list of thirty systems engineering elements in Exhibit 7.1).

Another DoD perspective related to RAA can be found in the so-called DoD 5000 series. In DoD Directive 5000.1 [8.3], reference is made to evolutionary requirements definition. A key statement in this regard is:

> Once approved as a new start acquisition program, operational performance requirements for the concept(s) selected shall be progressively evolved from broad operational capability needs to system-specific performance requirements (e.g., for range, speed, weight, payload, reliability, maintainability, availability, interoperability).

In addition, the same directive identifies the three major decision-making support systems that affect the acquisition processes, namely, the

- Requirements Generation System (RGS)
- Acquisition Management System (AMS)
- Planning, Programming, and Budgeting System (PPBS)

Thus, the DoD sees the matter of generating requirements as a major factor in its overall acquisition approach. The front end of the RGS is developing a mission needs statement that defines projected needs in broad operational terms. Examples cited are

- The need to impede the advance of large armored formations 200 kilo meters beyond the front lines
- The need to neutralize advances in submarine quieting made by potential adversaries

The Mission Need Statement flow is depicted in Figure 8.1. As shown, when the Mission Need Statement is approved, the system enters Milestone 0, allowing for the initiation of concept studies. Again, all of this is defined as part of the Requirements Generation System within the DoD.

In DoD Instruction 5000.2 [8.3], there is a section on Evolutionary Requirements Definition. The stated purpose of that section is to "establish the basis for the determination, evolution, documentation, and validation of mission needs and system performance requirements." The focus is on establishing system performance objectives and minimum acceptable requirements, and the need for documentation in an Operational Requirements Document (ORD). Another point of emphasis is the progressive refinement of the initial broad objectives and the minimum acceptable requirements. Thus, there is a carry-forward of the notion of defining and then evolving a set of minimum acceptable requirements for the system.

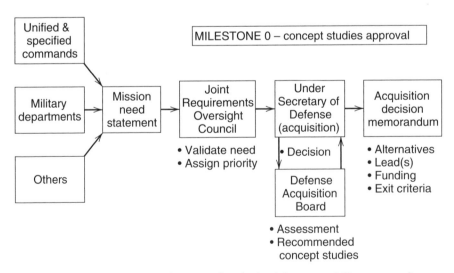

Figure 8.1. Mission Need Statement flow (major defense acquisition programs).

8.3   A NATIONAL AERONAUTICS AND SPACE ADMINISTRATION (NASA) PERSPECTIVE

A NASA perspective with respect to handling requirements is well represented in its engineering guide to the mission design process [8.4]. NASA's view is that "the definition and tracking of requirements are among the most important aspects of the mission definition and system design process," which itself is defined in terms of Prephase A, Phase A, and Phase B activities, as alluded to in the previous chapter. Requirements are defined and tracked to produce functional, performance, operational, and interface requirements that are to be implemented and verified during the next system phase, which NASA defines as Phase C/D. The flow-down of requirements is shown in Figure 8.2. NASA

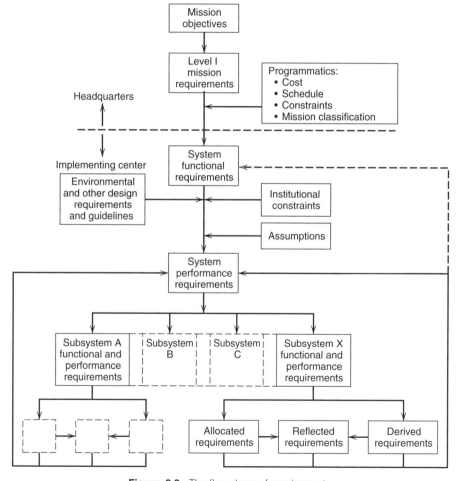

Figure 8.2. The flow-down of requirements.

encourages a "proactive iteration with the customer" as the *only* way that all parties can come to a true understanding of what should be done and what it takes to do the job. This is an enlightened view that is wholly supported by this author. Only by the developer and customer working together is it truly possible to successfully engineer a complex system. However, this flexible and joint relationship can be interfered with by an inflexible contracting officer and acquisition agent.

NASA also cites the need, with respect to requirements, for iteration, traceability, verification, and validation. This acknowledges that requirements must be analyzed and worked with, usually over long periods of time. They are not viewed as static, but rather as changeable in the best interest of all parties.

8.4   THE ORGANIZATION OF REQUIREMENTS STATEMENTS

We examine here some specific requirements documents and statements that have been used by selected government agencies in their procurement of systems.

8.4.1   Selected NASA Requirements Formats

A very large system that is being procured by the National Aeronautics and Space Administration (NASA) is called EOSDIS, the Earth Observing System Data and Information System. This system is likely, when considering all related costs, to be in the multibillion dollar range. The preliminary requirements for this system were documented in a rather voluminous NASA report [8.5]. EOSDIS was divided into three major segments, each of which was further delineated in terms of functions, as shown in Exhibit 8.2. We note the functional description as a key aspect of defining the system.

Exhibit 8.2: Functional Description of EOSDIS [8.6]

1. Flight Operations Segment
   Functions: 1.1 Mission Control
              1.2 Mission Planning and Scheduling
              1.3 Instrument Command Support
              1.4 Mission Operations
2. Science Data Processing Segment
   Functions: 2.1 Data Processing
              2.2 Data Archive
              2.3 Data Distribution
              2.4 Data Information Management
              2.5 User Support for Data Information
              2.6 User Support for Data Requests
              2.7 User Support for Data Acquisition and Processing Requests

3. Communications and System Management Segment

Functions: 3.1 Distribution of EOS Data and Information to EOSDIS Nodes

3.2 Distribution of Data Among Active Archives

3.3 Interface with External Networks

3.4 Network/Communications Management and Services

3.5 System Configuration Management

3.6 System/Site/Elements Processing Assignment and Scheduling

3.7 System Performance, Fault, and Security Management

3.8 Accounting and Billing

This functional description sets the stage for all further systems engineering activities and is central to the systems engineering process. NASA, in turn, establishes a set of requirements for the system for each and every function. At the top (system) level, however, NASA elected to organize the requirements as shown in Exhibit 8.3:

Exhibit 8.3: Organization of EOSDIS Systemwide Requirements [8.5]

5.0 EOSDIS SYSTEMWIDE REQUIREMENTS

   5.1 OPERATIONAL REQUIREMENTS

   5.2 FUNCTIONAL REQUIREMENTS

   5.3 PERFORMANCE REQUIREMENTS

   5.4 EXTERNAL INTERFACES

      5.4.1 EOS Project

      5.4.2 Space Station Information System (SSIS)

      5.4.3 Networks and NASA Institutional Elements

      5.4.4 Cooperating Institutions

      5.4.5 The EOSDIS User Community

      5.4.6 International Partners

   5.5 SECURITY

      5.5.1 Technical Security

         5.5.1.1 Computer Systems Selection Criteria

         5.5.1.2 EOSDIS Security Procedures

      5.5.2 Physical Security

      5.5.3 Contingency

         5.5.3.1 Restart/Recovery

         5.5.3.2 Software/Data

         5.5.3.3 People

         5.5.3.4 Equipment

   5.6 RELIABILITY, MAINTAINABILITY, AVAILABILITY (RMA)

      5.6.1 Reliability

      5.6.2 Maintainability

5.6.3 Operational Availability
5.6.4 EOSDIS System-Level RMA
5.6.5 Fault Detection Requirements

From Exhibit 8.3, two important points are

1. The system is described by *function*, and specific requirements are defined for each such function.
2. A set of top-level requirements is also defined, which applies at the system level.

An example of the system-level requirement for reliability is as follows:

The reliability for EOSDIS shall be measured by the Mean Time Between Failures (MTBF) for a system or component over the entire life cycle of the equipment. Failures are those equipment and software malfunctions which result in interruptions in service. Interruptions in service resulting from external factors beyond the control of EOSDIS . . . shall not be considered failures unless EOSDIS equipment or software contribute to the interruption.

Beyond this system wide reliability requirement, NASA invokes two military standards and one handbook:

- Mil-Std-785B for Reliability
- Mil-Hdbk-472 for Maintainability Prediction
- Mil-Std- 470A for Maintainability Status Reporting

The reader interested in NASA approaches to requirements definition may refer to the EOSDIS requirements specification [8.5] and similar requirements documents for other NASA programs [8.6, 8.7].

8.4.2   A DoD Requirements Format (I-CASE)

The DoD has long been interested in building an Integrated Computer-Aided Software Engineering (I-CASE) system. The statement of work for such a system, produced by the Air Force, has a specification that contains a statement of requirements for I-CASE [8.8]. The four-digit table of contents for these requirements is shown in Exhibit 8.4.

Exhibit 8.4: Requirements Outline for I-CASE

10.3 REQUIREMENTS
    10.3.1 Overview
    10.3.2 I-CASE Software Engineering Environment
        10.3.2.1 Standards

From Exhibit 8.4, we note two aspects of this statement:

- Section 10.3.2.3 identifies the requirements by functional area
- The emphasis on test and execution environments

With respect to the former, the requirements document deals extensively with the requirements for each functional area, confirming the general practice that requirements are keyed to the system as defined by its functions and subfunctions. The lead-in to the functional requirements section states as follows:

> The I-CASE SEE [software engineering environment] will consist of an information repository and COTS [commercial-off-the-shelf] software components that support the software development and maintenance process. The I-CASE SEE will provide an environment that supports software development and maintenance activities including business case analysis, software engineering, application migration, program/project management, configuration management, quality assurance, life cycle documentation, presentation production, requirements traceability, impact analysis, error reporting, security, and external system interface support.

8.4.3 A Coast Guard Example of Requirements

As another example of a statement of requirements, the U.S. Coast Guard (USCG) in the Department of Transportation produced a specification and statement of work (SOW) for mission-oriented information system engineering (MOISE) [8.9]. A three-digit definition of their requirements is shown in Exhibit 8.5. We note that this specification of requirements is largely related to the procurement of services and thus is different, in general, from the procurement of a system.

Exhibit 8.5: Example of Coast Guard Requirements Format

C.3 REQUIREMENTS
  C.3.1 MULTIPLE INFORMATION SYSTEM INTEGRATION
  C.3.2 INITIATION AND CONCEPT PHASES
  C.3.3 DEFINITION PHASE
  C.3.4 DESIGN PHASE
  C.3.5 DEVELOPMENT PHASE
  C.3.6 DEPLOYMENT PHASE
  C.3.7 OPERATIONS PHASE
  C.3.8 METRICS
  C.3.9 MULTIPHASE, CROSS-CUTTING SKILLS AND
      SERVICES
  C.3.10 CONTRACTOR ACQUIRED FEDERAL INFORMATION
      PROCESSING (FIP) RESOURCES
  C.3.11 SYSTEM DEVELOPMENT CENTER
  C.3.12 CONTRACTOR/USCG MANAGEMENT RELATIONSHIPS
  C.3.13 CONTRACTOR PERSONNEL REQUIREMENTS
  C.3.14 CONTRACTOR SECURITY REQUIREMENTS

8.5 SPECIFIC REQUIREMENTS STATEMENTS

This section identifies certain types of requirements statements and also presents a few examples of specific statements, drawn from government programs.

8.5.1 Types of Requirements

Requirements are sometimes stated in terms of their levels of applicability that, in effect, establish the importance of these requirements to the system user or acquisition agent. An example follows with respect to the weighting or significance of requirements statements [8.10]:

- The most important requirements are stated as "shall," to indicate mandatory requirements.
- The next most important requirements are stated as "shall, where practicable."

- The next most important requirements are stated as "preferred" or "should."
- The next most important requirements are stated as "may."

These may be simplified to three- [8.8] and two-level schemes, such as

- Minimum features
- Desirable features
- Highly desirable features
- Minimum or mandatory requirements
- Optional requirements

The preceding requirements represent various ways to designate the importance or weight of these requirements. This gives further guidance to the system developer as to the needs of the customer.

8.5.2 Specific Examples of Requirements Statements

A series of specific requirements statements drawn from various government programs follows:

- The contractor shall allocate the Earth Observing System Data and Operations System (EDOS) Requirements identified in the functional and performance specification and shall identify the methods and procedures necessary for verifying compliance with these requirements.
- EDOS shall provide the capability to support a processing rate of at least 50% above the average aggregate throughput rate for nonreal-time data.
- The Payload Data Services System (PDSS) shall:
  — Transmit up to eight real-time and eight nonreal-time data streams concurrently
  — Have an availability of greater than or equal to 0.95
  — Support a real-time S-band input rate of 192 kbps
- I-CASE shall have the capability to create or support
  — 100 databases
  — 2000 tables per database
  — Tables with 250 columns and one billion rows
  — Columns with 250 characters
  — At least 100 indexes per table
- Identify existing systems that have requirements similar to an Automated Highway System (AHS) in terms of public interaction, safety, reliability, and complexity; analyze these systems to derive "lessons learned" in their implementation that could be appropriate to AHS development

- The system performance boundaries are as follows:
  - Customer Data and Operations System (CDOS) shall protect all data to achieve a BER (bit error rate) of 10 E-12
  - The maximum interruption of data delivery services shall be ten seconds or less
  - Reinitialization of data delivery services shall require no more than two minutes
  - The processor utilization for CDOS shall not exceed 50%
  - The memory and storage utilization for CDOS shall not exceed 75%
- The EOSDIS ground system shall have an operational availability of 0.992 with an MTBF (mean time between failures) of 500 hours and an MDT (mean down time) of 4 hours
- Catalyst shall support the calculation of evaluation measures of merit for the alternatives through ranking, weighting, hierarchical weighting, and probabilistic weighting
- Catalyst shall automate generic systems engineering functions that are adaptable to the way an organization does business

8.6   ESSENTIAL STEPS OF REQUIREMENTS ANALYSIS

The following eight areas represent a minimum set of essential steps with respect to the analysis of requirements.

8.6.1   Automation of Requirements Analysis and Allocation (RAA)

There has been an increasing recognition of the need to automate the process of RAA. For a large and complex system, manual handling of RAA is basically obsolete. Early automated systems for RAA include such tools as:

- The Input-Output Requirements Language (IORL)
- The Software Requirements Engineering Methodology (SREM)
- The Problem Statement Language/Problem Statement Analyzer (PSL/PSA)

These tools are attributed to Teledyne Brown, TRW, and Meta Corporation, respectively.

More recent software packages have been reviewed and analyzed by the Air Force's Software Technology Support Center (STSC) [8.11], describing over 100 so-called Upper CASE tools for this purpose. These include tools for requirements specification and analysis, a dozen of which follow:

1. Teamwork/RT: Cadre Technologies
2. CARDTools: Cardtool Systems Corporation

3. Cohesion: DEC
4. Power Tools: Iconix Software Engineering
5. Software Through Pictures: IDE
6. Excelerator: Intersolv
7. IEW: Knowledgeware
8. Battlemap: McCabe & Associates
9. System Architect: Popkin Software & Systems
10. RTrace: Protocol
11. SES Workbench: Scientific and Engineering Software
12. IEF: Texas Instruments

Two very popular and capable packages not listed above are Telelogic's DOORS and Vitech's Core.

Many requirements documents are now calling for a requirements analysis tool in hand and in operation from the start of a new contract and systems engineering effort. The state of the art in systems engineering tools strongly supports the use of software aids to carry out requirements analysis and allocation.

8.6.2 Relationships and Traceability

A key step under the element of RAA is to explicitly identify relationships and traceability. Relationships are clearly important because the systems engineering team must always be aware of and track requirements interdependencies. For example, one section of a requirements document may refer to the pointing error of an instrument and another section may specify the allowable error for the stabilization and control system for a satellite carrying that instrument. The latter error is very likely to influence the former. The systems engineer must establish such relationships and, as a minimum, annotate the dependency.

Traceability has at least two meanings. The first is the explicit connection between all requirements, through all documentation in which such requirements are stated. The second is the traceability of a requirement through all the life-cycle elements of the systems engineering process. Using the earlier instrument pointing error as an example, the T&E (test and evaluation) element must include traceability back to this requirement. Similarly, the TPM (technical performance measurement) program should include this type of traceability. Thus, traceability involves, as a minimum:

- The interdependence between requirements
- The longitudinal tracking of requirements through the various elements of the systems engineering process as they are applied during the system's life cycle

8.6.3 Ambiguous Requirements

The systems engineering team also must look for requirements that are unclear or ambiguous. Such ambiguities sometimes are revealed when comparing different sections of a requirements document that may have been prepared by different people and the overall review of the integrity of the requirements has not brought forth these types of problems. As an example, in a requirements specification for a radar system, one section may refer to the target size as one square meter and another part of the spec may call out a target size of up to two square meters. Similarly, the range to target may be sixty miles, but if the level of detection at that range is not specified (as, for example, the probability of detection at that range), the requirement remains ambiguous. Essentially, any requirement that leaves doubt as to its precise meaning must be classified as ambiguous until the ambiguity or lack of clarity is resolved.

8.6.4 Incorrect Requirements

Certain requirements, when analyzed, can be shown to be clearly incorrect. As an example, suppose a requirement for the operational availability of a system is stated as follows:

- The system shall have an operational availability of 0.995 with an MTBF (mean time between failures) of 500 hours and an MDT (mean down time) of 6 hours.

If we use the standard relationship for availability as

$$\text{Availability} = \frac{\text{MTBF}}{\text{MTBF} + \text{MDT}} = \frac{500}{500 + 6} = 0.988$$

we obtain an availability calculated as 0.988, which is less than the specified value of 0.995. Thus, there is an error in the requirement as stated. This requires further discussion and clarification with the customer in order to resolve the problem. A failure to deal with what appear to be incorrect requirements places the systems engineering effort at risk. This may result in downstream schedule and cost overruns.

8.6.5 Incompatible Requirements

Incompatible requirements suggest that there is conflict or incompatibility between two or more stated requirements. This can happen, for example, when there is a designation of COTS software that may not run on a designated host computer. Software may be specified that runs on a workstation computer, but the basic architecture may call for a microcomputer solution. In such a case, the software and the computer requirements are essentially incompatible. This also may occur with external interface systems. The system in question may have to interoperate with an external system, but with the specified requirements, it may not do so. Many systems that are part of a larger system run into this type of potential incompatibility issue.

8.6.6 High-Risk Requirements

The systems engineer must go through the full list of requirements and identify, as early as possible, those requirements that represent high risk to the development effort. Typically, high-risk requirements push or extend the state of the art in one or more technology areas. Examples of such areas may include

- Integration of commercial-off-the-shelf (COTS) software with development software
- Advanced expert systems
- Voice-actuated computer systems
- Extremely high-power or high-frequency devices
- Multimedia systems
- Very high-capacity, dense storage systems
- Real-time command and control systems
- Forcing the team to adopt a development approach where a COTS approach would be sufficient

High-risk requirements also show up as risks in terms of meeting schedule and cost constraints. As the team experiences difficulties in solving the technical problems associated with high-risk developments, schedules tend to slip and costs begin to rise. This "domino effect" in trying to deal with high-risk requirements can easily result in failure of the project.

Where there is substantial risk associated with certain requirements, the systems engineering team has to focus attention on how to meet these requirements, or attempt to relax such requirements through negotiations with the customer, especially where performance is not significantly affected.

8.6.7 Low-Performance Requirements

Requirements that lead to low performance should also be identified by the systems engineering team. In today's world of rapidly changing technology and long development times, attention has to be paid to situations in which following the requirements leads to low performance. This may be observed in systems that contain large amounts of software and where the requirements emphasize the use of specific commercial software packages. In this situation, instead of specifying commercial software by name and version, the requirements should call for the best software available within the class of software desired. In this way, the developer can select the most recent version at the latest possible time, thereby obtaining the high-performance version instead of an obsolete version.

8.6.8 Derived Requirements

A critical part of requirements analysis is the construction of derived requirements. These are a set of requirements derived from the original requirements specification but are not actually a part of the specification itself. In other words, such requirements flow from the systems engineering analysis and must be formulated to proceed constructively with the design of the system.

An example can be seen by correcting the requirement for availability cited Section 8.5.2. This requirement is restated as:

- The overall system shall have an operational availability of 0.988 with an MTBF of 500 hours and an MDT of 6 hours.

The issue here is that requirements for the subsystems have to be derived from this overall system requirement in order to design the subsystems. Such derived requirements are not part of the requirements document; it is up to the systems engineering team to derive them. Thus, the team has to annotate all cases for which top-level system requirements must be analyzed in greater detail to develop a set of derived requirements. Such derived requirements are then "allocated" to the subsystems to give guidance to the subsystem design teams. More specific examples of how this might be done are explored in the next section.

8.7 DERIVED AND ALLOCATED REQUIREMENTS

Three situations that demonstrate the issue of deriving and allocating requirements are examined in this section.

8.7.1 The Weight of a Spacecraft and Its Subsystems

The weight of an overall spacecraft (communications, weather, surveillance satellites) is constrained by the launch vehicle that places such a satellite in orbit. The requirements specification therefore usually will identify the total maximum weight for the satellite. The systems engineering team must therefore derive from this top-level requirement a series of derived weight requirements for each of the subsystems making up the total satellite. Such subsystems, for example, might include:

- The satellite structure
- The stabilization and control subsystem
- The telemetry and data handling subsystem

- The thermal control subsystem
- The power supply subsystem
- The satellite payload subsystem

In turn, each of these subsystems might be examined to see if new derived weight requirements have to be determined. As an example, suppose the satellite payload subsystem consists of five remote measurement instruments, as in a weather satellite. It would be typical then to derive a weight requirement (maximum weight permissible) for each of the five instruments. In this general manner, requirements are derived from the original requirement and then allocated to the subsystems of a given system.

8.7.2 Deriving and Allocating Reliability Requirements

If we refer again to Section 8.6.8, we see a requirement for a mean time between failures (MTBF) of 500 hours for the overall system. Let us assume that the overall system has four major subsystems, as shown in Figure 8.3. The requirements document specifies only the overall system MTBF requirement of 500 hours and not the subordinate MTBF requirements for the subsystems. It therefore becomes a task of the systems engineer to derive and allocate MTBF requirements for the four subsystems. Based on analysis of the subsystem technical information and requirements, as well as data from previous programs and experience, the systems engineer might derive a set of MTBF requirements as illustrated in Figure 8.3. The four subsystem MTBF-derived requirements are

- *Subsystem 1:* MTBF = 2500 hours
- *Subsystem 2:* MTBF = 2000 hours
- *Subsystem 3:* MTBF = 1667 hours
- *Subsystem 4:* MTBF = 2000 hours

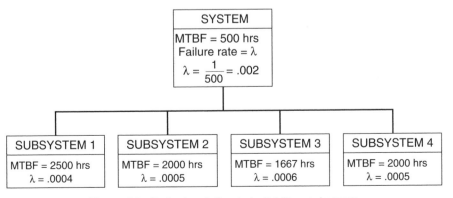

Figure 8.3. Derived and allocated reliability requirements.

Note that all of these MTBFs are larger than the system requirement of 500 hours. In order to develop this set of subsystem-derived requirements, however, the systems engineer must understand the basic relationship between MTBF and failure rate for a given system, given the assumption of an exponential failure law. The relationship is

$$\text{Mean time between failures (MTBF)} = \frac{1}{\text{failure rate}}$$

Also, the failure rates are additive under these circumstances, so that the overall system failure rate is the sum of the subsystem failure rates. The resultant MTBFs and failure rates for the system and its subsystems are shown in Figure 8.3. We leave it as an exercise at the end of the chapter to verify the compatibility between the stated MTBFs and failure rates.

Through this example, we see how requirements may be derived and also how they are allocated to lower levels of indenture for the system. We also note that this process cannot be accomplished without an understanding of the specific relationships between MTBFs and failure rates. This is true for all situations of derived requirements and allocation of these requirements, namely, that the algorithms that apply to the situation at hand must be known to the systems engineer.

8.7.3 Deriving and Allocating Errors in a System

We take now as a third example of deriving and allocating requirements that of sighting and pointing at a target in a shipboard environment. We assume that the stated requirement for the system is that the sighting to the target have a maximum permissible error of one-half of a degree, which is equivalent to 0.00873 radian. We further break down this error into three fundamental error sources:

1. Error in pointing by a human (operator error)
2. Error in the pointing instrument (instrument error)
3. Error in mounting the instrument to a platform (platform error)

The issue is then to derive pointing requirements (maximum errors) for these three error sources and the systems they represent.

This situation is depicted in Figure 8.4. Under a model that claims independent additive errors associated with the random variables that represent the error sources, the total root-mean-square (rms) error is related to the subordinate errors by the relationship:

$$\sigma_T^2 = \sigma_1^2 + \sigma_2^2 + \sigma_3^2$$

where σ is the rms error (standard deviation) and its square is the variance of any error distribution. The derived maximum errors, as determined by the

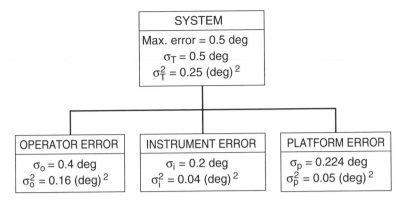

Figure 8.4. Derived and allocated pointing-error requirements.

systems engineer, are

Maximum human (operator) error $= \sigma_o = 0.4$ deg
Maximum instrument error $= \sigma_i = 0.2$ deg
Maximum platform error $= \sigma_p = 0.224$ deg

These then represent a consistent set of derived error requirements that may be utilized in the design of the various subsystems. Because one of these "subsystems" is the human operator, we have to verify that such a person is able to point the instrument within the designated error limits. If this is not possible, then there has to be a further rederivation of errors such that the overall pointing error requirement is satisfied. This may turn out to be a long and difficult process, but it is necessary to meet the overall system requirement.

We further note that the algorithm used in this error calculation is different from that used in both the weight example (Section 8.7.1) and the reliability example (Section 8.7.2). The error calculation called for a summation of the squares of the rms errors, equivalent to the sum of the variances of the independent error-source variables. The error, in this example, is also associated with a "one-sigma" value, another choice that might be made by the systems engineer. Here again, the nature of error algorithms must be part of the body of knowledge of the systems engineer in order to carry out this derivation and allocation in a correct manner.

8.8 OTHER REQUIREMENTS ISSUES

8.8.1 Six Additional Requirements Problem Areas

In addition to the eight essential steps of requirements analysis cited and discussed earlier in this chapter, we close this section with a brief examination of what might be called special requirement problems. Here are six such problem areas:

1. Requirements creep/volatility
2. Not verifiable/testable
3. Unable to prioritize
4. Pre-defined solution
5. Incomplete
6. Stakeholders not sufficiently involved

Creep and Volatility. Requirements creep and volatility cause numerous problems. However, if changing requirements actually point us in the direction of a better system, then we should welcome such changes. There are clearly ways to handle changing requirements, to the ultimate satisfaction of both developer and customer. All such changes should be made explicit, and if they are good for the customer, then the customer should be willing to pay for them. The developer needs to be steadfast in keeping track of the original requirements and agreed-upon requirements changes.

Cannot Verify or Test. In general, we have a problem area when it is not possible to test and verify when a requirement has been met or not met. Fuzzy and non-quantitative requirement statements lead us in that direction, and can become arenas for massive debate, with a lot at stake. To the extent possible, the developer should challenge these types of requirements as early as possible so that a satisfactory agreement can be made before the area blows up on all concerned parties.

Cannot Prioritize. We should recognize that not all requirements are equally important. Section 8.5.1 suggested different levels of requirements, and they may go a long way toward dealing with this issue. A more simple procedure is to have a two-level description of requirements: mandatory (M) and optional (O). This procedure also reinforces the architecting notions in this book whereby we can offer alternative architectures from the low end (satisfies only the mandatories), to the high-end (satisfies the mandatories and most of the optionals), and the knee-of-the-curve solution (satisfies all the mandatories and some cost-effective set of optionals). In addition, many requirement documents are written so that the developer can make a very good "guess" as to what the priorities tend to look like. For example, if there is a separate section on information security, a message is being conveyed simply by that structure.

Predefined Solution. Requirements can be written so that certain system architectures are a foregone conclusion. This may be all right, but it also may be counterproductive. For example, if the requirement for a communication system is stated as a frequency division multiplex approach, then other options that the developer might prefer are basically nonresponsive. It might be better if the customer asks for a choice between frequency division and time division multiplexing, supported by a complete trade-off analysis between the two. Once a customer provides a "predefined solution," it also has to accept responsibility if that solution turns out to be a bad one. This situation occurred several years ago, when the government mandated the use of "Ada" as the preferred programming language.

Incomplete Requirements. Incomplete requirements are a frequent problem area. The system developer, as the expert in any given field, should recognize when requirements as initially stated are incomplete. The well-known discipline of functional decomposition can aid the system developer here. Each and every function and subfunction should have at least one (and usually more than one) requirement. If no requirements are allocated to a subfunction, we have an incomplete requirement. The answer to this problem was embedded in the earlier discussion of essential steps. The developer, in each such case, should set forth a list of "derived" requirements and have them formally accepted by the customer. If it is difficult to obtain customer acceptance, this may be a good area for a mandated trade-off study.

Stakeholders Not Sufficiently Involved. The immediate customer may turn out not to be the final customer. All stakeholders (on the customer side) should have an opportunity to provide and critique the requirements defined for the system. These stakeholders should also be asked to formally "sign off" on the last version of the requirements documentation. If they are not able to, then perhaps there is a problem that needs to be debated and resolved. The goal is to have all stakeholders involved, at an appropriate level, and for a solid consensus to be reached.

8.8.2 An Illustrative Trade-off Study of Requirements

Barry Boehm, one of the leaders in both systems and software engineering, has been a strong advocate of closer interactions between systems and software engineering. In an important article [8.13], he conveys some observations about the unequal positions of systems and software engineers in the early stages of system design. Hardware, software, and systems engineers all need to have a position at the table when considering the design and architectural features of large-scale systems.

Boehm also discusses certain aspects of a contract between TRW and the government regarding a difficult information query system, dating back to the 1980s. The customer required a response time of less than one second. Driven primarily by that singular requirement, it was estimated that the system cost would be nearly $100 million. This was an alarming figure for everyone, so a detailed trade-off analysis was undertaken that revealed that if the response time requirement were changed to four seconds, there would be two important consequences: (1) this response time would satisfy the users about 90 percent of the time, and (2) system development costs would drop to about $30 million. So we have this astonishing result: a one-second response time leads to a $100 million system, and a four-second response time leads to a $30 million system, a "savings" of some $70 million, largely on the basis of one requirement.

Although this is only one "data point," it demonstrates the potential value of questioning one or more requirements on the basis of the impacts on the system cost and schedule. Treating the requirements as absolutely inviolate precludes these types of considerations, which can be a critical part of the process of real-world systems engineering.

8.8.3 Conclusion

The preceding discussion has emphasized the importance of requirements analysis and allocation and the problems associated with requirements. Although customer or user requirements are the touchstone for the systems engineering effort, it is also true that problems with requirements suggest that certain requirements be negotiated and changed. Living with requirements that are ambiguous or cannot be satisfied, in the long run, decreases the likelihood of success of the project and its systems engineering activities. For those requirements in that category, the PM and the CSE must enter a process of discussion and negotiation to resolve any and all problems. This means that despite the perspective that requirements are often taken as inviolate, problems with requirements must be solved.

To some extent, the matter of changing or updating requirements has been acknowledged. One such manifestation is in the so-called "spiral model" for software development. This model explicitly revisits the requirements several times to deal with problems in this domain. This is a step forward in the context of the overall subject of software development and engineering.

Another step that has been taken to deal with requirements issues is that of having a representative of the user or customer be part of the systems engineering group. In this way, a direct and immediate link is established with the customer, who is then able to respond to the issues that might be raised with respect to interpretations and changes of requirements. Closer linkages between the project and engineering activities and the customer is also a major step forward in trying to deal realistically with what has been a problem area for many years.

Finally, requirements relative to software have constituted large parts of our systems as they have leaned more heavily on software solutions. Therefore, software requirements specification and analysis have received a great deal of attention in recent years [8.12].

In summary, the systems engineering team must treat requirements analysis and allocation as an extremely important part of the systems engineering process. It must also be prepared to resolve difficulties with requirements and not necessarily accept poor requirements as fashioned in concrete. A full and open dialogue with the customer with respect to such issues helps to increase the chances of success of the project.

QUESTIONS/EXERCISES

8.1 Define two additional steps in the process of requirements analysis, in addition to the essential steps cited in this chapter.

8.2 Functionally decompose an automobile system. Define two requirements for each of the decomposed elements.

8.3 Identify and briefly describe two automated requirements analysis tools that are available from software vendors, other than those mentioned in this chapter.

8.4 Contrast the NASA perspective on handling requirements with the approach taken by the Department of Defense.

8.5 Verify the compatibility between the stated MTBFs and the failure rates in Figure 8.3. Given a system composed of four independent subsystems with MTBFs of 200, 250, 500, and 1000 hours, what is the MTBF of the overall system?

8.6 A system with three independent subsystems has a total failure rate of 0.0108. We also know that the MTBFs of the subsystems are in the ratio 2:3:4. What are the subsystem MTBFs and failure rates?

8.7 **a.** The total admissible error variance for a system is 70 and two of the three subsystem root-mean-square (rms) errors are 4 (value of x) and 6 (value of y). What is the largest acceptable integer value of the third error source (z) if all error random variables are additive and independent?

   b. Recalculate your answer if the error model is based on the relationship: total error $= 2x + y + z$

8.8 Systems can be designed such that errors correspond to multiples of standard deviation (sigma) values. If the resultant distribution were normal (Gaussian), interpret numerically the intended consequences of designing to the one-sigma, two-sigma, and three-sigma values.

8.9 Construct three additional examples of derived requirements.

8.10 Problems with requirements are almost always on our list of what went wrong in building a system. Select three problems, and discuss the reasons why we do not appear to be able to eliminate or mitigate them.

REFERENCES

8.1 *United States General Accounting Office (GAO) Report for the Acting Secretary of the Air Force*, Defense Support Program, GAO/NSIAD-93-148 (1993). Washington, DC: USGAO.

8.2 *Systems Engineering*, Military Standard 499B (Draft) (1991). Washington, DC: U.S. Department of Defense.

8.3 Department of Defense (DoD) Directive 5000.1 and Instruction 5000.2 (1991). Washington, DC, February 23.

8.4 *The NASA Mission Design Process* (1992). Washington, DC: National Aeronautics and Space Administration, NASA Engineering Management Council.

8.5 *Phase C/D Requirements Specification for the Earth Observing System Data and Information System (EOSDIS)* (1990). Greenbelt, MD: Goddard Space Flight Center.

8.6 *EDOS Contract Data Requirements List and Statement of Work* (1992). Greenbelt, MD: Goddard Space Flight Center.

8.7 *CDOS Requirements Specification, Level 3* (1990). Greenbelt, MD: Goddard Space Flight Center.

8.8 *Integrated Computer-Aided Software Engineering (I-CASE) Solicitation* (1992). Gunter AFB, AL: Department of the Air Force, Standard Systems Center.

8.9 *Mission Oriented Information System Engineering (MOISE) Solicitation*, Request for Comment (1992). Washington, DC: U.S. Coast Guard, Department of Transportation.

8.10 Eisner, H. (1988). *Computer-Aided Systems Engineering*. Englewood Cliffs, NJ: Prentice Hall.

8.11 *Requirements Analysis & Design Tool Report* (1992). Hill AFB, UT: U.S. Air Force, Software Technology Support Center.

8.12 Davis, A. (1990). *Software Requirements—Analysis and Specification*. Englewood Cliffs, NJ: Prentice Hall.

8.13 Boehm, B. (2000). "Unifying Software and Systems Engineering," *Computer Magazine* (March), 114–116.

9

SYSTEMS ARCHITECTING: PRINCIPLES

9.1 INTRODUCTION

Architecting a large-scale complex system is the centerpiece of systems engineering. Without the systems engineering process, such architecting is likely to be disorganized and unsatisfactory. With the systems engineering process and perspective, it is more likely that a sound system architecture will ultimately evolve.

In Chapter 1 (Figure 1.1), an overview of the systems approach was presented. This approach showed the following as key elements of architectural design:

- Requirements
- Functional design of alternatives
- Analysis of alternatives
- Evaluation criteria
- Formulation of a preferred system architecture

In Chapter 2 (Figure 2.2), the architectural design process was expanded, showing the following elements:

- Requirements
- Mission engineering
- Requirements analysis/allocation

- Functional analysis/allocation
- Architectural design/synthesis
- System analysis
- Life-cycle costing
- Risk analysis
- Other system/subsystem considerations
- Formulation of a preferred system architecture

In Chapter 7 (Figure 7.4), inputs to the important step of analyzing and evaluating alternatives were defined, with the output representing the preferred system architecture. In this chapter, we bring all these representations together into a coherent set of essential steps for developing a preferred system architecture. We also show some examples and approaches preferred by other investigators of this important process and issue.

9.2   A VIEW OF SYSTEMS ARCHITECTING

Any discussion of architecting large and complex systems would not be complete without reference to the extraordinary and landmark treatise by E. Rechtin on systems architecting [9.1]. Rechtin's view is that "the core of architecting is system conceptualization" and that there are four basic approaches to the process of architecting:

1. The normative (pronouncement) methodology
2. The rational (procedural) method
3. The argumentative approach
4. The heuristic approach

The approach delineated in this chapter is a combination of the rational and the heuristic approaches. The rational method is exemplified by the evaluation framework discussed later in this chapter; the heuristics lie mainly in how alternative systems are defined and also how a rating scheme is established in order to assess the strengths and weaknesses of these alternatives. In seeking the appropriate mix between these approaches, Rechtin suggests that "the scientist, engineer and architect follow the heuristic: *simplify, simplify, simplify.*"

Another key observation made by Rechtin is:

The essence of systems is relationships, interfaces, form, fit and function. The essence of architecting is structuring, simplification, compromise and balance.

The balance is achieved by the appropriate compromise between the following types of factors:

- System requirements
- Function
- Form
- Simplicity
- Affordability
- Complexity
- Environmental imperatives
- Human needs

These factors can be said to represent a high-level set of evaluation criteria for all types of systems.

In addition, Rechtin provides an interesting citation of notably successful systems as well as a rather long list of heuristics, chapter by chapter. The latter list is must reading for any and all Project Managers (PMs) and Chief Systems Engineers (CSEs).

9.3 A NATIONAL AERONAUTICS AND SPACE ADMINISTRATION (NASA) PERSPECTIVE

NASA defines an architecture as [9.2]:

> How functions are grouped together and interact with each other. Applies to the mission and to both inter- and intra-system, segment, element, and subsystem.

In their mission design activities, NASA emphasizes the following:

- Requirements
- Decision making and evaluation criteria
- Optimization and descope options
- Robustness and flexibility
- Cost and schedule as a trade parameter
- Developing the cost plan and schedule
- Risk assessment and mitigation
- Establishing design margins
- Analyses and trade studies
- Technical performance measurement

We will find some of these themes reiterated as we look at the formal process of system architecting.

As indicated in earlier chapters, NASA looks at mission design in terms of three early phases:

1. The conceptual design process: Prephase A
2. The mission analysis process: Phase A
3. The definition process: Phase B

If we examine closely the specifics for these phases, we see the following purpose for Phase A:

> The purpose of the Phase A study is to **refine** the mission and systems(s) requirements, determine a *baseline mission configuration and system architecture*, identify risks and risk mitigation strategies, identify the "best" candidates, and select one.

Thus, it is clear that systems architecting and selection of a preferred architecture are carried out during Phase A. With respect to the Phase A study final report, NASA requires that the topics listed in Exhibit 9.1 be covered.

Exhibit 9.1: NASA Phase A Study Document Topics

- Technology needs and development plan
- Refined and validated mission requirements
- Final feasibility assessment
- Disposal requirements
- Functional/operational description
- Hard ware/software distribution
- Design requirements
- Definition of top-level interfaces and responsibilities
- System/subsystem description
- Mission description
- Data handling requirements
- Launch vehicle requirements
- Mission operations
- Preliminary work breakdown structure
- Refined cost estimates and schedules
- Establishment of accountability for delivery of an end item and its performance
- Apportionment of technical resources, the distribution of margins, and allocation error budgets

- System-level block diagram, flight, and ground
- Maintenance and logistics requirements
- Top-level system architecture

This documentation must provide answers to the following types of questions:

1. Does the conceptual design meet the overall mission objectives?
2. Is the design technically feasible?
3. Is the level of risk acceptable?
4. Are the schedule and costs within the specified limits?
5. Do the study results show this option to be better than all others?

NASA is explicit regarding the use of "appropriate and predetermined weighted and prioritized evaluation criteria to select the best of the candidate designs that were evaluated." Weighted evaluation criteria for this purpose come up again later in this chapter as part of the recommended system architecting process.

9.4 ARCHITECTURE DESCRIPTIONS

9.4.1 NASA Descriptions

We indicated in Chapter 8, "Requirements Analysis and Allocation," that the requirements for a system are almost always delineated by system function. Therefore, more often than not, the system functions, at least at the top level, are an input to the systems engineering team.

In many cases, the architectural description of the system consists of the reiteration of the given system functions, along with the way in which those functions, and the subsystems representing those functions, are interconnected. As examples, Figures 9.1 and 9.2 show system architectures defined by NASA for the EOSDIS and the EDOS systems [9.3, 9.4]. Both figures are from the requirements specifications for the two systems, as defined by NASA. We note the implicit emphasis on interconnections between functions, extending as well to the physical locations of the functional capabilities. In some cases, the architecture description is supplemented by the identification of functional interfaces between a given system and other systems with which it must interoperate. This is illustrated by Figure 9.3 for a NASA system known as the PDSS (Payload Data Services System [9.5]). Types of data and information flow in terms of functional interfaces are emphasized in this description.

Although the previously cited functional architectures provide valuable information to the systems engineering team, the position taken here is that this information is only an input to the true architecting of a system. System

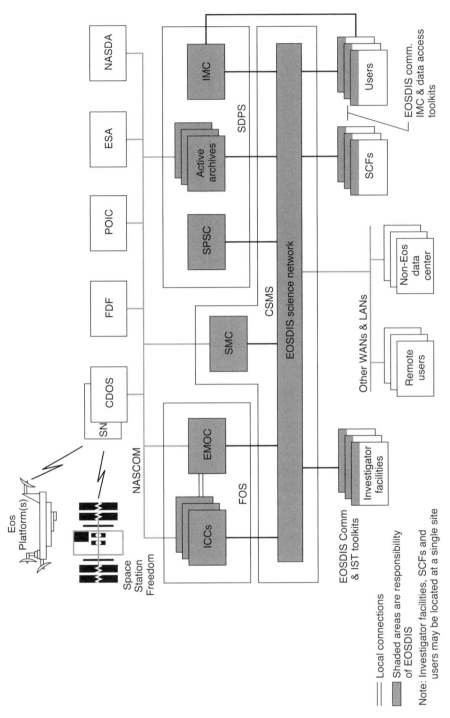

Figure 9.1. Illustrative architecture description—EOSDIS.

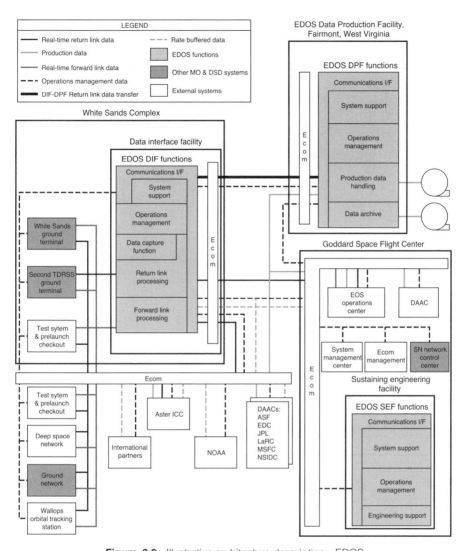

Figure 9.2. Illustrative architecture description—EDOS.

architecting takes several steps beyond this type of description, the key element of which is to define a technical approach for the implementation of each of the defined functions as well as an overall selection of the best approach (most cost-effective system) for the combination of functions. Such a selection defines the preferred system architecture. This position is made more concrete in the sections that follow in this chapter as well as in the Appendix.

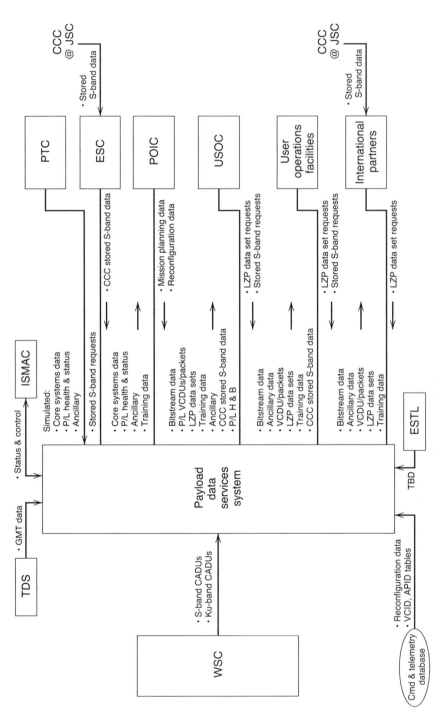

Figure 9.3. PDSS functional interfaces.

9.4.2 Department of Defense Descriptions

The Department of Defense (DoD) has formulated a framework for Command, Control, Communications, Computer, Intelligence, Surveillance, and Reconnaissance (C4ISR) architecture, development and integration [9.6]. This *C4ISR Architecture Framework* (now called DoDAF) is an attempt to:

- Ensure that architectures developed by different parts of the DoD are interrelated
- Set forth uniform methods for describing various types of information systems
- Provide guidance on describing architectures

The last point is of particular importance. In fact, the Framework is built largely upon the idea that there are three major perspectives on how to describe an architecture, namely:

- The operational view
- The systems view
- The technical view

These views can be thought of as giving us different types of information about the same architecture, rather than being different architectures themselves. The most useful architecture description is one that consists of multiple views, providing an "integrated" notion of the architecture. The more precise definitions of the above three views of an architecture are [9.6]:

The *operational architecture view* is a description of the tasks and activities, operational elements, and information flows required to accomplish or support a military operation

The *systems architecture view* is a description, including graphics, of systems and interconnections providing for, or supporting, warfighting functions

The *technical architecture view* is the minimal set of rules governing the arrangement, interaction, and interdependence of system parts or elements, whose purpose is to ensure that a conformant system satisfies a specified set of requirements

The Framework also provides a detailed explanation of the roles of each of these architectural views. Of special importance is the notion that an architecture description must provide explicit linkages between these various views. This leads to an integrated view as well as supporting the notion of interoperability, which has been a problem for DoD systems for many years.

The Framework also shows a six-step process for building an architecture as follows:

Step 1: Articulate the intended use of the architecture.

Step 2: Establish the scope, context, environment (and any other assumptions) of the architecture.

Step 3: Determine which characteristics the architecture needs to capture.

Step 4: Establish which architecture views and supporting products should be built.

Step 5: Build the needed products.

Step 6: Use the architecture for its intended purpose.

Steps 4 and 5 above suggest that each architectural view is supported by several products. That indeed is the case. In fact, there are a series of "Essential Framework Products" and also a set of "Supporting Framework Products." These products are intended to convey critical information about the architecture. The reader is referred to the reference for the *C4ISR Architecture Framework* [9.6] in order to gain a fuller understanding of the nature and structure of this work.

9.4.3   IEEE Descriptions

The discussion of the C4ISR Architecture Framework cites an architecture definition, attributed to the IEEE [9.6], as:

> A structure of components, their relationships, and the principles and guidelines governing their design and evolution over time

This definition is useful, but does not give the prospective architect a blueprint for how to architect a system. Once an architecture has been formulated, it is very likely to satisfy the broad definition cited above.

Yet another view of systems and their architectures is presented in the IEEE standard for systems engineering [9.7]. In Chapter 7, we noted that the IEEE expanded from four to five the key elements of the systems engineering process, yielding the following:

1. Requirements analysis
2. Functional analysis/allocation
3. Synthesis
4. System analysis and control
5. Verification and validation

The process ultimately produces a preferred system architecture, which the IEEE defines as:

> The composite of the functional, physical and foundation architectures, which form the basis for establishing a system design [9.7].

In addition, the definition points the reader to requirements traceability and allocation matrices that presumably connect the overall system design to various aspects of the above-cited three architectures. Although definitions of the above three subordinate architectures are provided in the standard, it is difficult to see how to construct these architectures without seeing an example or two. Nevertheless, one can accept the notion that "views" of the architecture can be associated with the ideas behind "functional, physical and foundation." These views, by immediate observation, are not the same as those developed by the DoD C3I (command, control, communications, and intelligence) world, as discussed above.

As we move with the IEEE into the software domain, we find a standard that focuses upon a recommended practice for architectural description [9.8]. This standard explores "the activities of the creation, analysis, and sustainment of architectures of software-intensive systems." The standard makes the very significant point that "concepts of architecture have not been consistently defined and applied within the life cycle of software-intensive systems." Instead, it argues that it is indeed possible to *describe* architectures. These descriptions can also be thought of as "views" of an architecture, even though we may not be completely clear as to how to develop an architecture. Selected views of an architecture may be illustrated, in the hardware case, by Figures 9.1 through 9.3 which have already been discussed. Thus, we are seeing a difference between having a "single accepted framework for codifying architectural thinking," and "views" or "descriptions" of an architecture. We attempt, in this chapter, to narrow this gap by setting forth a prescriptive method for developing an architecture, whether it be related to hardware, software, or a combination thereof.

9.4.4 Additional Selected Views

Here we provide more detailed information about *views* of architectures. We have introduced the original C4ISR (now consistently called DoDAF) approach to an architectural framework with the top-level orientation related to the (a) operational view, (b) systems view, and (c) technical view. These more detailed views have been called "products" and have been broken down into *essential* and *supporting* products. These products are shown in Table 9.1.

The annotated version of the supporting products can be found in Table 4.1 of the original referenced document.

A quite different set of views can be associated with the method of architecting suggested next in this chapter (section 9.5). Briefly, that method takes

TABLE 9.1   Essential and Supporting Products for Views [9.6]

a. Essential Views (with annotations)
   AV–1—Overview and Summary Information:
   Scope, purpose, intended users, environment depicted, analytical findings, if
      applicable
   AV–2—Integrated Dictionary:
   Definitions of all terms used in all products
   OV–1—High-Level Operational Graphic Concept
   High-level graphical description of operational concept (high-level
      organizations, missions, geographic configuration, connectivity, etc.)
   OV–2—Operational Node Connectivity Description:
   Operational nodes, activities performed at each node, connectivities and
      information flow between nodes
   OV–3—Operational Information Exchange Matrix:
   Information exchanged between nodes and the relevant attributes of that
      exchange, such as media, quality, quantity, and the level of interoperability
      required
   SV–1—System Interface Description:
   Identification of systems and system components and their interfaces, within
      and between nodes
   TV–1—Technical Architecture Profile:
   Extraction of standards that apply to the given architecture

b. Supporting Views (without annotation)
   OV-4—Command Relationships Chart
   OV-5—Activity Model
   OV6a—Operational Rules Model
   OV6b—Operational State Transition Description
   OV7—Logical Data Model
   SV2—Systems Communication Description
   SV3—Systems Matrix
   SV4—System Functionality Description
   SV5—Operational Activity to System Function Traceability Matrix
   SV6—System Information Exchange Matrix
   SV7—System Performance Parameters Matrix
   SV8—System Evolution Description
   SV9—System Technology Forecast
   SV10a—Systems Rules Model
   SV10b—Systems State Transition Description
   SV10c—System Event/Trace Description
   SV11—Physical Data Model
   TV2—Standards Technology Forecast

Note: These products are expected to be updated over time.
Key:
AV = All Views
OV = Operational View
SV = Systems View
TV = Technical View

TABLE 9.2   Additional Views Related to Cost-Effectiveness Architecting [9.9]

View 1. Requirements Satisfaction
View 2. Risk and Requirements
View 3. Interoperability
View 4. Cost by Function
View 5. Cost versus Requirements
View 6. Sensitivity to Changes in Criteria Weights
View 7. Effectiveness versus Risk
View 8. Effectiveness versus Human Factors
View 9. Effectiveness versus RMA
View 10. Effectiveness versus Residual Performance Factors

the approach that we are in search of a cost-effective architecture from among a set of alternatives. The top-level views in that regard are the:

- Synthesis view
- Analysis view
- Cost-effectiveness view

An additional set of ten views is shown in Table 9.2.

We see that each of these views is oriented toward a quantitative description of an architecture.

9.5   ESSENTIAL STEPS OF SYSTEM ARCHITECTING

Two earlier representations, Figures 1.1 and 2.2, have shown the development of an architecture as the first top-down design or synthesis of the system in question. It is an attempt to come to terms with the critical design choices for the system as a whole. Thus, for this author, the essence of architecting is defining these top-level design choices and placing them in a context that establishes a set of reasonable alternatives that can then be evaluated. Following this notion, consider the structure of Exhibit 9.2 [9.10].

Exhibit 9.2:   System Functions and Design Choices for a Communications System

Top-Level System Functions	Design Choices (Alternatives)
1. Multiplexing/demultiplexing	D11, D12
2. Modulation/demodulation	D21, D22
3. Switching and routing	D31, D32, D33
4. Encryption/decryption	D41, D42
5. Formatting/signal conversion	D51, D52, D53
6. Control and monitoring	D61, D62
7. Recording and playback	D71, D72
8. Satellite/terrestrial communications	D81, D82

This table shows a set of eight top-level functions of a communications system. For each of these functions, the system architect considers one or more design approaches that satisfy the requirements, as stated for the function. Thus, for the first function, the two design approaches are represented as D11 and D12, where D11 is the first design approach for function 1 and D12 is the second design approach for function 1. To generalize, DIJ is the Jth design approach for function I. Since we are operating at the top-level functional breakdown, these approaches are considered fundamental to the overall system design and represent true alternatives from which we will eventually construct a preferred architecture. As an example, for the mux/demux function number 1, D11 might be frequency division multiplexing whereas D12 could be time division multiplexing.

From the above discussion, we are able to immediately see the combinatorial nature of the architectural design problem. If all of the design combinations were admissible, then in principle there could be as many as $(2)(2)(3)(2)(3)(2)(2)(2) = 576$ combinations, each of which represents a single architectural choice for the system. When we introduce the fact that not all combinations are internally compatible or interoperable, the number of combinations (admissible alternatives) narrows, and usually in a rather dramatic way. On this basis, two critical tasks of the architect are to (a) set forth the various design approaches, and (b) look for ways to reduce the number of alternatives (combinations) to be considered. Given that the architect is able to do part (a) of the above, the approach delineated here reduces the number of alternatives by placing them in a cost-effectiveness context, specifically the one shown in Figure 7.6. From that perspective, a practical number of alternatives is constructed and then evaluated by the systems engineering team, led by the chief systems engineer (CSE).

If we return to the thirty elements of systems engineering, as defined in Chapter 7 (Exhibit 7.1), we can describe the essential steps of system architecting by means of elements 3 through 10, namely:

- Requirements analysis/allocation (element 3)
- Functional analysis/decomposition (element 4)
- Architecture design/synthesis (element 5)
- Alternatives analysis/evaluation (element 6)
- Technical performance measurement (element 7)
- Life-cycle costing (element 8)
- Risk analysis (element 9)
- Concurrent engineering (element 10)

This "short list" does not mean that some of the other thirty elements of systems engineering are completely neglected during the process of architecting; rather, it suggests that these above eight elements are critical and must be part of the process. Each of these essential steps is briefly described in the following sections.

9.5.1 Requirements Analysis/Allocation (RAA)

As discussed in some detail in the previous chapter, requirements analysis and allocation is an essential element in systems engineering. It is also critical in terms of architecting a system. The high-priority aspects of RAA have been delineated in Section 8.6. Here we emphasize, with respect to architecting, that requirements are allocated to the functionally decomposed blocks of the system. These allocated requirements come from two sources:

1. They are provided directly in the requirements document that comes from the system user or acquisition agent.
2. They are derived from the preceding by the system developer, thus "filling in the blanks" where necessary.

The result of RAA then establishes at least one functional requirement for each decomposed functional block.

9.5.2 Functional Analysis/Decomposition

The preceding step cannot be taken until the system has been decomposed into its functional blocks. This part of the process assures that functional decomposition has been carried out as a precursor to RAA. Other aspects of functional analysis and allocation have been described in Chapter 7, Section 7.3.4. We note also the primary position of functional analysis and allocation in Military Standard 499B (see Section 7.2.1).

9.5.3 Architecture Design/Synthesis

This step involves the synthesis of architectural alternatives. These alternatives flow from trying to define a set of choices that satisfy the requirements in each functional area. The conceptual problem that is attendant to this step can be described as follows. If there are a total of eight functional areas, and there are three design choices for each such functional area, then in principle there are a total of 3^8 or 6,561 (!) possible architectures. This, of course, makes no sense, so we try to narrow the field of possibilities through the addition of constraints.

One method for achieving this rather rapidly is depicted in Table 9.3. This table shows only three alternative architectures, mapped against the functional areas and the subordinate requirements in these areas. These three alternatives are:

1. A low-cost, minimum-effectiveness alternative
2. A "baseline" alternative
3. A high-performance, (high-cost) alternative

TABLE 9.3   Alternative Architectures versus Functions and Requirements

Functions and Requirements	Alternative Architectures		
	Low-Cost System	Baseline System	High-Performance System
Function 1			
Req. 1.1			
Req. 1.2			
Req. 1.3			
Function 2			
Req. 2.1			
Req. 2.2			
Function 3		$\begin{bmatrix} \text{How is each} \\ \text{system designed} \\ \text{to satisfy} \\ \text{the requirements?} \end{bmatrix}$	
Req. 3.1			
Req. 3.2			
Req. 3.3			
Req. 3.4			
Function N			
$\vdots$			
Req. N.1			
Req. N.2			
Req. N.3			
$\vdots$			
Req. N.M			

The entries in Table 9.3 describe how the alternative is designed to satisfy the key functional requirements. This notion was first presented in Figure 7.5. The idea is to bracket the reasonable alternatives by the low-cost and high-performance architectures to set the stage for evolving a cost-effective "baseline" alternative. All such alternatives satisfy the full set of requirements as defined by the customer. In this synthesis and top-level design process, there is a minimum of formal analysis and evaluation, which is the next step in the process. A greater number of alternatives may also be considered and evaluated by formal Taguchi and response surface methods that the reader may wish to explore but that are outside the scope of this presentation.

9.5.4   Alternatives Analysis/Evaluation

Given the alternative architectures as defined in the previous step, these alternatives are then analyzed and evaluated with the objective of deriving a preferred system architecture. We refer again for this process to Figure 7.4. This figure shows a series of inputs to the analysis and evaluation of alternatives, namely:

- Architecture design and synthesis
- The key elements of:
  - Technical performance measurement (TPM)
  - Life-cycle costing
  - Risk analysis
  - Concurrent engineering
  - Systems engineering management
- The ancillary elements of:
  - Integrated logistics support (ILS)
  - Reliability, Maintainability, Availability (RMA)
  - Test and evaluation (T&E)
  - Quality assurance and management (QA&M)
  - Specialty engineering
  - Preplanned product improvement (P3I)
- Evaluation criteria
- An evaluation framework

The figure also shows the output as the preferred system architecture.

The basic evaluation framework brings together the final form of the alternatives evaluation process, as shown in Table 9.4. This table shows three alternatives as columns and the evaluation criteria as rows. With such a matrix, we evaluate each alternative with respect to how well the evaluation criteria are satisfied. The evaluation criteria are weighted to capture the likely possibility that they are not all equally important. We chose to select these weighting factors so that they add to unity to normalize the numbers and simplify the evaluation. Weighting factors are developed by group decision-making processes [9.10] involving the systems engineering team. Customer inputs are highly desirable but usually difficult to obtain in quantitative terms.

The individual "cells" in the matrix contain two basic numbers. The first is the rating of the given alternative with respect to the stated evaluation criterion. The second is the product of the rating and the weight given to the evaluation criterion. This latter product is then summed down the column to place in evidence the final score for each alternative. This score is the basis for comparing the alternatives against each other.

There are many options available for the rating system that one might select. Three such options are:

1. A scale of 1 to 10
2. A Likert scale with the numeric values of 0, 2, 4, 6, 8, and 10
3. S college scoring system of A, B, C, D, and F, equivalent to numeric scores of 4, 3, 2, 1, and 0

TABLE 9.4 System Alternatives Evaluation Framework

| Evaluation Criteria | Criteria Weights (W) | Alternative Architectures | | | | | |
| | | System A | | System B | | System C | |
		Rating	$W \times R$	Rating	$W \times R$	Rating	$W \times R$
Criterion 1	0.08	7	0.56	8	0.64	9	0.72
Criterion 2	0.10	6	0.60	6	0.60	8	0.80
Criterion 3	0.13	6	0.78	7	0.91	7	0.91
Criterion 4	0.09	5	0.45	7	0.63	8	0.72
Criterion 5	0.12	8	0.96	9	1.08	10	1.20
Criterion 6	0.07	6	0.42	7	0.49	7	0.49
Criterion 7	0.11	7	0.77	6	0.66	5	0.55
Criterion 8	0.10	9	0.90	6	0.60	4	0.40
	1.00		5.44		5.61		5.79

Table 9.4 shows the first of these options and the resultant scores for the three alternatives. Each of the given scores can then be compared with the understanding that the maximum score for any given alternative is 10. The illustrative numbers in the table show preferences for Systems C, B, and A, in that order.

With this evaluation framework, one may then ask the question: Where do the evaluation criteria come from? These criteria can be derived from several sources simultaneously:

- The requirements document for the system
- Possible evaluation criteria defined by the customer
- The measures of merit and effectiveness (MOMs and MOEs) that may have been developed
- The technical performance measures (TPMs) for the system

In general, the TPMs (which are measurable) and other possibly subjective factors, as well as cost measures, may make up the full set of evaluation criteria. Examples of evaluation criteria for a communications system and a transportation system are listed in Exhibit 9.3. The broad factors cited by Rechtin (see Section 9.2) [9.1] may be said to apply to all systems, independent of the domain in which they are operative.

Exhibit 9.3: Illustrative Evaluation Criteria

A Communications System
- Availability
- Bandwidth
- Capacity

- Connectivity
- Expandability
- Grade of service
- Life-cycle costs
- Number of channels (by type of channel)
- Quality of service
- Reliability
- Response time
- Risk
- Security
- Speed of service
- Survivability
- Throughput

A Transportation System

- Capacity
- Capacity-to-demand ratio
- Comfort and convenience
- Economic impacts
- Environmental effects
- Frequency of service
- Growth capability
- Life-cycle costs
- Quality of service
- Reliability of service
- Risk
- Safety
- Security
- Speed
- Trip time

In a study of the future development of the U.S. airport network, the Transportation Research Board of the National Research Council addressed directly the matter of criteria for its evaluation and developed the following set of fourteen criteria:

1. Capacity
2. Safety
3. Cost

4. Competition
5. Flexibility
6. Time
7. Frequency
8. Reliability
9. Comfort and convenience
10. Congestion and pollution
11. Other environmental concerns
12. Compatibility
13. Funding
14. Management

So whether we are dealing with a very large system, such as the national airport network, or a smaller system, such as a radar, the issue of developing and using a coherent list of evaluation criteria remains approximately the same.

9.5.5 Technical Performance Measurement (TPM)

Technical performance measurement plays a prominent role in Military Standard 499B. Exhibit 9.4 lists some of the TPM references in that standard.

Exhibit 9.4: Selected TPM References in Mil-Std-499B

Section 4.6.8. Technical Performance Measurement (TPM). The contractor shall establish and implement TPM to evaluate the adequacy of architectures and designs as they evolve to satisfy the requirements and objectives selected for tracking. TPM shall be used to identify deficiencies that jeopardize the ability of the system to meet a performance requirement. Actions taken due to deficiencies are dependent on whether the technical parameter is a requirement or an objective. The level of detail and documentation shall be commensurate with the impact on cost, schedule, and performance of the technical program.

Section 4.6.8.1. Implementation of TPM. The contractor shall track the achievement to date for each technical parameter as the analyses, design and development activity progresses. In the event that the achievement to date value falls outside the tolerance band, the variation shall be determined by comparing the achievement to date against the corresponding value on the planned value profile. In the event progress in the technical effort supports identification of a different current estimate than previously predicted, a new profile and current estimate shall be developed. The current estimate shall be determined from the achievement to date and the remaining schedule and budget. Risk assessments and analysis shall be updated to reflect changes in the TPM profiles and current estimates, and impacts on related parameters.

Section 4.6.8.2. TPM on Requirements. For identified deficiencies, analyses shall be performed to determine the cause(s) and to assess the impact on higher level parameters, interface requirements, and system cost effectiveness. Alternative recovery plans shall be developed with cost, schedule, performance, and risk impacts fully explored. For performance in excess of requirements, the marginal cost benefit of the performance delta and opportunities for reallocation of requirements and resources shall be assessed and an appropriate course of action implemented.

Section 4.6.8.3. TPM on Objectives or Decision Criteria. The contractor shall perform TPM on objectives and decision criteria as contained in their SEMP (systems engineering management plan)

In general, TPMs are the key measures of technical performance of the system, as previously illustrated in Chapter 7 (Section 7.3.7). They allow the systems engineering team to carry out the evaluation ratings shown in Table 9.4. They deal with technical factors of the system and should be measurable in a technical sense. Their measurement may be exceedingly complex, requiring detailed modeling and simulation to compute the values of TPMs. TPMs, in general, depend on a series of other parameters known as technical performance parameters (TPPs). A method for relating TPPs to TPMs, developed by this author, is Parameter Dependency Diagramming (PDD) [9.10]. The PDD procedure is illustrated in Section 9.8.

9.5.6 Life-Cycle Costing

Life-cycle costs for the architectural alternatives may also be considered evaluation criteria. These costs may be broken down into subcategories, if considered desirable. Generally accepted subcategories include

- Research, development, test, and evaluation (RDT&E) costs
- Procurement or acquisition costs
- Operations and maintenance (O&M) costs

There are several different ways in which costs may be factored into the evaluation framework discussed before. Three such ways include

1. Costs as formal evaluation criteria
2. Costs in the context of effectiveness-to-cost ratios
3. Costs as a separate, but related, consideration

In the first approach, costs are listed among the various evaluation criteria and are weighted along with these other criteria. The weights reflect how important costs are in relation to the other criteria. In the second approach, formal effectiveness-to-cost ratios are computed in order to select one architecture over another. For the latter case, there is no integration of effectiveness with the cost metrics, at least in a formal sense.

There are times when cost considerations are elevated as well to a more global set of considerations, which themselves can be included as evaluation criteria. Such cost-related considerations include the extent to which the alternative:

- Supports overall economic growth
- Is amenable to financing
- Minimizes environmental costs
- Minimizes the dislocation of people, and so forth

As with the technical performance measures and their related evaluation criteria, costs must be estimated for the various alternatives under evaluation. This might imply the development of a first-order life-cycle cost model that can be refined during later stages of system development. For purposes of architecture comparison, a full life-cycle model is generally not necessary, or even feasible.

9.5.7 Risk Analysis

In Chapter 7, we noted the four key areas of risk attendant to engineering large-scale systems, namely:

1. Schedule risk
2. Cost risk
3. Performance risk
4. Societal risk

Risk must be considered one of the evaluation criteria for most complex systems. Measurement of risk is a nontrivial matter, but it has been considered in great detail in a variety of books and articles [9.10, 9.12, 9.13], ranging in scope from broad systems considerations to software development.

9.5.8 Concurrent Engineering

Concurrent engineering is viewed here as a crucial part of architecting a system because it brings together all the necessary technical and management skills needed to design, test, produce, and operate the system. This broad range of system considerations must be present to assure that some important features are not neglected. As an example, many design engineers assume that matters of production and producibility are easily resolvable. By having a production engineer as part of the concurrent engineering team and effort, we attempt to assure that such an assumption is valid. If it is not, then perhaps other alternatives will become more attractive. The basic idea is that the CSE must have the broadest range of technical and management inputs as part of the architecting process. Additional information regarding concurrent engineering can be found in Chapters 7 and 14.

9.5.9 Illustrative Example of Architecting Process

This section illustrates the core architecting process by using a concrete but simplified example to demonstrate the essential steps. A more detailed version of this example is also provided in the Appendix, Section A5.

The illustrative system is called a Severe Climates Anemometry System (SCAS) [9.11], and it contains the six major functions, and a set of subfunctions, as listed below:

1. Atmospheric Sensing
   1.1 Wind speed
   1.2 Wind direction
   1.3 Barometric pressure
2. Mechanical Service
   2.1 Instrument housing
   2.2 Instrument orientation
3. Environmental Service
   3.1 Ice control
4. Power Service
   4.1 Main supply
   4.2 Regulation/conditioning
   4.3 Backup power
5. Indoor/Outdoor Transmission
   5.1 Power
   5.2 Signal
   5.3 Physical linkages
6. Data Handling
   6.1 Collection
   6.2 Processing/storage
   6.3 Reporting, distribution, and display

As a next step, we are specifically interested in the fact that there are alternative design choices that we can make in order to implement each of the functions and subfunctions. We array our design choices in a tabulation that follows from Table 9.1, and in this case, becomes Table 9.5.

We note that we are building three architectures, one each for a low-cost system, a high-effectiveness system, and a baseline system. The latter is our attempt at trying to find the "knee-of-the-curve" system, as illustrated in Figure 7.5. The three architectures are constructed, in this case, at the sub-functional level. For the first subfunction of wind speed sensing (subfunction 1.1), the architect selected a commercial-off-the-shelf (COTS) pitot tube. In moving to the other two architectures, the architect added a hard-wired transducer, and then a radio transducer. This was the choice of the architect (or the architecture team) in order to add capability (effectiveness) in moving from left to right on the table.

For the case of wind speed direction (subfunction 1.2), the architect decided that all three alternatives should be based upon a simple shaft drive design choice. Further, the same three design choices for subfunction 1.1 were used in subfunction 1.3. Moving to subfunction 2.1, the architect went from an aluminum solution to a molded composite solution, to composites that are more compact and weigh less. Three different choices were made as well for the orientation/position subfunction 2.2. The process of selecting design choices for each subfunction continued until all functions and subfunctions were considered.

The important point is that this core architecting process *goes beyond the mere articulation of system functions and subfunctions.* It focuses upon the design (synthesis) process whereby alternative design choices are made at the subfunction level. Further, this is structured so as to produce only three alternatives, but ones that have a cost-effectiveness basis for ultimately choosing which of the three will turn out to be the preferred architecture.

In most cases, the architect thought that each architecture should have a different set of design choices. In two, the same design choice was selected across the board (i.e., for subfunctions 1.2 and 4.1) The selected choices were based upon the expertise of the architect and his or her subject knowledge in this particular domain. If a team is utilized to carry out this process, it is likely that better solutions will be forthcoming, since a team is able to draw upon the expertise of many instead of just one or two. In its very compact form, Table 9.5 can be thought of as the "synthesis" step in the top-level architecting of a system. Each subfunction is addressed, moving from left to right across the three alternative systems. Once the entire chart has been filled in with alternative design choices, the architecting team must scan each alternative from top to bottom in order to confirm that the selections, within a system, are compatible and interoperable. If they are not, then changes have to be made to assure harmonious intrasystem operation, that is, that the selected design choices within a system will "play together." If the team has formulated more than three design choices for one or more subfunctions, it

TABLE 9.5 Alternative System Architectures for Anemometry System

Functions	Subfunctions	Low Cost	Baseline	High Effectiveness
1. Atmospheric Sensing	1.1 Wind speed sensing	COTS pitot tube	COTS pitot with transducer	Add radio transducer
	1.2 Wind direction sensing	Simple shaft drive	Simple shaft drive	Simple shaft drive
	1.3 Pressure sensing	COTS pitot tube	COTS pitot with transducer	Add radio transducer
2. Mechanical Service	2.1 Instrument housing	Machined aluminum	Add molded composites	Less weight/compact
	2.2 Orientation/position	Wind-vaned COTS bear	Less tail boom length	High-precision bear/balancing
3. Environmental Service	3.1 Ice control	Analog feedback temperature control	Add digitized control	Add process & heat pipes
4. Power Service	4.1 Main power supply	Commercial 220/110 V COTS	Commercial 220/110 V	Commercial 220/110 V
	4.2 Power regulation/conditioning	Conditioners/rods	Add ground fault interrupter	Add lightning arrester
	4.3 Backup power	Battery-instruments	Gas generator with sensor	High-reliability diesel with switch
5. Indoor/Outdoor Transmission	5.1 Power transmission	Stranded wire harness	Stranded wire harness	Custom slip rings
	5.2 Signal transmission	Foil-shielded wire harness	Coaxial with slip rings	2-way radio, no wiring
	5.3 Physical linkages	Shaft/conduit, pressure tube	Add shielded transducer	Minimum shaft-physical support
6. Data Handling	6.1 Data collection	Potential and indoor Pneumatic cell	Magnetic position sensor	Optical position sensor
	6.2 Data processing/storage	Manual database entry	Automatic computer control	Automatic computer control
	6.3 Reporting, distribution, and display	Physical meters manual	GUI + modem access	DBMS + packet network

may then be appropriate to expand the number of alternative architectures to be considered.

The next step in the architecting process is to formally evaluate the three alternatives in terms of their costs and effectiveness. This step is the "analysis" part of the process, and it is clearly a critical one. Although the architect has attempted to construct low-cost, high-effectiveness, and "knee-of-the-curve" architectures, it is only by analyzing the three alternatives that we are able to see what the cost-effectiveness profiles turn out to be.

A simplified analysis procedure is illustrated in Table 9.6, following the construction suggested previously in Table 9.4. Five specific criteria are developed, namely:

- *Criterion 1*   Performance
- *Criterion 2*   Human factors
- *Criterion 3*   Maintainability
- *Criterion 4*   Risk
- *Criterion 5*   Other

Weights are developed for these criteria, and the three alternative architectures are evaluated on a scoring system of from 1 to 10. Each score is multiplied by the corresponding weight, and the resulting products are summed to obtain a total score for each of the alternatives. At the same time, the costs for these alternatives are determined. In this example, the results are:

Alternative	Score	Costs
Low-cost system	6.8	200K
Baseline system	8.1	250K
High-effectiveness system	8.4	400K

The question at this point becomes: which of the above three alternative architectures does the architect wish to put forth as the preferred architecture? If we treat cost as an independent variable (CAIV), we can plot the above scores (as a measure of effectiveness) against the costs to obtain the points shown in Figure 9.4. This graph illustrates the pattern shown in Figure 7.5. The low-cost system is verified to have both the lowest cost ($200K) and the lowest effectiveness. The high-effectiveness system has the highest cost ($400K), and also the highest effectiveness. The baseline system is in the middle, showing a significant effectiveness improvement over the low-cost system, but without a major increase in cost. Indeed, for this example, the baseline system "looks like" the knee-of-the-curve solution. The graph of Figure 9.4 is the third element in the architecting process, since it provides an important "view" of the cost-effectiveness results of the first two steps (i.e., synthesis and analysis). The alternative that the architect puts forth as

TABLE 9.6 Evaluation Framework for Architecting Illustrative System

Evaluation Criteria	Weights	Low Cost		Baseline		High Effectiveness	
		Score	Weight × Score	Score	Weight × Score	Score	Weight × Score
Performance	0.3	6	1.8	8	2.4	9	2.7
Human factors	0.2	7	1.4	8	1.6	9	1.8
Maintenance	0.2	7	1.4	9	1.8	9	1.8
Risk	0.2	8	1.6	8	1.6	6	1.2
Other	0.1	6	0.6	7	0.7	9	0.9
Sums	1.0		6.8		8.1		8.4
Costs of alternatives			200K		250K		400K

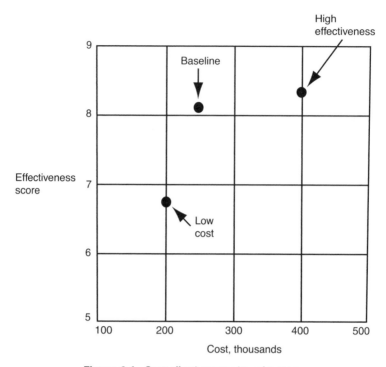

Figure 9.4. Cost-effectiveness view of system.

the preferred architecture now depends upon what the customer is ultimately looking for as well as the constraints under which he may be operating. Indeed, if you, the reader, imagine that you are the customer, you will be able to construct a case for any of the three as preferred, based upon the scenario that you envision (see Question/Exercise number 9.8).'

The example described here is a "boiled down" version of the one in the Appendix, with some modifications in the interest of simplifying the presentation. By referring to the Appendix, one will see that there are actually a set of subcriteria for each of the five major criteria, as listed below in Table 9.7. Other variations on these themes, as for example in the scoring system, may be employed by the architecting team to explore sensitivities, as suggested earlier. Also, as the team is able to spend the time required for a true and objective set of measurements of effectiveness, confidence in the architectural solution will grow, in most cases. If not, a change in the alternatives may be called for.

9.5.10 Interoperability

The method of architecting developed and described here provides some insights into how to deal *explicitly* with the matter of interoperability within a system. Looking back at the "synthesis" chart of Table 9.5, we see alternative

TABLE 9.7   Top-Level and Subordinate Evaluation Criteria for Anemometry System [9.11]

- Evaluation Criterion 1—Performance
  - 1.1 Vaning function/stability
  - 1.2 Average power consumption
  - 1.3 Impact resistance/robustness
  - 1.4 Speed of data processing
  - 1.5 Data availability
  - 1.6 System availability
  - 1.7 System reliability
  - 1.8 Useful life
- Evaluation Criterion 2—Human Factors
  - 2.1 Ease of use
  - 2.2 Operator safety
  - 2.3 Bystander safety
- Evaluation Criterion 3—Maintainability
  - 3.1 Frequency of scheduled maintenance
  - 3.2 Ease of maintenance
  - 3.3 Complexity of assembly
- Evaluation Criterion 4—Risk
  - 4.1 Cost risk
  - 4.2 Schedule risk
  - 4.3 Performance risk
  - 4.4 Technological risk
- Evaluation Criterion 5—Other
  - 5.1 Manufacturability
  - 5.2 Market potential/demand
  - 5.3 Appearance/aesthetic quality
  - 5.4 Expandability/upgradability

design choices being made for each function and subfunction. Generally, these selections are made by looking at a subfunction and moving left to right across the page to see how that subfunction might be instantiated for all three architectures (i.e., low cost, baseline, and high effectiveness). After this has been completed, we move down to the next subfunction and use the same process. By this procedure, eventually the entire chart is constructed, as shown in Table 9.5. Then we are in a position to check for interoperability between the design choices made for each of the suggested architectures. For example, we start with the low-cost architecture and compare the selected design choices with each other, top to bottom, and two at a time. For each pair, we ask: Are these two design choices interoperable with each other? We continue to do this for all pairs, which means that moving down one row, we have $n(n-1)/2$ such choices. For the chart in Table 9.5, $n = 15$, so that there are 105 questions of this type for each column. If all the answers show that the design choices are interoperable, then we move on to the next column. If some of the design choices are not interoperable, then other choices need to be made that are, in fact, interoperable. This is a systematic way of assuring interoperability throughout and explicitly as part of the synthesis process.

Having explored in this chapter a variety of ways to look at the matter of architecting a system, we are now in a position to formulate the following recommended short-form definitions of an architecture, a preferred architecture, as well as architecting:

> **Architecture.** An organized top-down selection and description of design choices for all the important system functions and subfunctions, placed in a context to assure interoperability and the satisfaction of system requirements
>
> **Preferred Architecture.** A choice among several architectures that is balanced, cost-effective, and most congruent with the stated requirements and what the customer is seeking, as tempered by program and/or system constraints
>
> **Architecting.** A process with the following simplified steps: (1) functional decomposition of the system, (2) construction of design choices for all important functions and subfunctions (synthesis), (3) evaluation of the resultant interoperable system alternatives (analysis), and (4) display of the results so as to facilitate the selection of a preferred, cost-effective architecture from among the constructed alternatives.

9.6 THE 95% SOLUTION

All alternative architectures referred to before are designed to satisfy all stated system requirements. That is, the low-cost, high-performance, and baseline architectures are developed such that all requirements are met. A "95% solution" is a term, coined by this author, to refer to a system architecture that, instead of satisfying *all* of the system requirements, is designed to meet only 95% of the requirements. The number 95, in this context, is more heuristic than it is precise. The notion flows from Rechtin's observation [9.1] that extreme requirements work against the balance necessary in architecting systems, "creating unexpected misfits and deficient performance." Indeed, Rechtin makes his point very strongly: "[E]xtreme requirements should remain under challenge throughout system design, implementation and operation."

On this basis, as well as trade-off statements made in various standards, another conceptual alternative architecture to the low-cost, high-performance, and baseline alternatives has been added, namely, the 95% solution, which satisfies 95% of the stated requirements. To achieve satisfaction of the additional 5%, extreme prices may have to be paid in terms of schedule and cost. Put another way, if the system developer can build a system that satisfies

95% of the requirements, and at the same time reduce both schedules and costs by significant amounts, then it is an obligation of the developer to make that alternative known to the customer. In such a situation, the 95% alternative would be added to the evaluation framework shown in Table 9.4 and that alternative would be compared to the other three alternatives in terms of the set of evaluation criteria. Special notations would have to be added to make it clear as to the precise requirements that have not been met as well as the level of performance that is achieved by the 95% solution in relation to these requirements. With the 95% solution, the developer is basically saying to the customer: If I had the freedom to back off from the stated requirements, here is the system architecture that I would recommend that would represent the best balance among requirements, performance, cost, and schedule. Although this approach flies in the face of many current system procurement practices, it also implicitly suggests that perhaps a reformation of some of these practices is in order.

9.7 TRADE-OFFS AND SENSITIVITY ANALYSES

Trade-off and sensitivity analyses are carried out to:

- Select approaches at the functional level in the architecting process
- Select specific design choices at the subsystem level
- Determine how sensitive the overall system selection is to changes in the weights and ratings given to the various architectural alternatives

9.7.1 Military Standard 499B Perspective

Military Standard 499B [9.4] cites the following rationale for what are called trade studies:

> *Trade Studies (Para. 4.6.1).* Desirable and practical trade-offs among stated user requirements, design, program schedule, functional and performance requirements, and life cycle costs shall be identified and executed. Trade-off studies shall be defined, conducted and documented at the various levels of functional or physical detail to support requirements, functional decomposition/ allocation, and design alternative decisions or, as specifically designated, to support the decision needs of the systems engineering process. The level of detail of a study shall be commensurate with cost, schedule, performance, and risk impacts.

We note that trade-offs involving user requirements and other system characteristics are definitively called for. From this perspective, requirements are not necessarily assumed to be fixed and inviolate.

In the important area of synthesis, the standard identifies the following areas as important in terms of trade studies:

- Establish system/configuration item (CI) configuration(s).
- Assist in selecting system concepts and designs.
- Support make or buy, process, rate, and location decisions.
- Examine proposed changes.
- Examine alternative technologies to satisfy functional/design require ments, including alternatives for moderate- to high-risk technologies.
- Evaluate environmental and cost impacts of materials and processes.
- Support decisions for new products and process developments versus nondevelopmental items (NDI) or commercial-off-the-shelf (COTS) products and processes.
- Evaluate alternative physical architectures to select preferred products and processes.
- Support materials selection.
- Select standard components, techniques, services, and facilities that reduce system life-cycle cost and meet system effectiveness requirements (force structure and infrastructure impacts that emphasize supportability, producibility, training, deployment, and interoperability must be considered).

9.7.2 A Radar Detection Trade-Off Example

To illustrate the matter of trade-off analysis from a technical point of view, we briefly discuss here an example of detection of a radar pulse signal. We assume a simple threshold detection scheme whereby a target is declared to be present when, at the sampling time, the threshold (T) is exceeded. If the threshold is not exceeded, the decision is that no target was present.

Under the general assumption of a pulsed system with additive independent Gaussian noise:

a. When a target is present, the threshold detector "sees" a signal-plus-noise Gaussian distribution with a mean value of voltage equal to V and a noise (power) variance equal to N.

b. When a target is not present, the threshold detector "sees" a noise Gaussian distribution with a mean value of zero and a noise (power) variance equal to N.

In situation (a), we are interested in the probability of detection, $P(d)$, that is, the probability that we will correctly detect a target when it is present. In (b), we wish to compute the false alarm probability, $P(fa)$, that is, the probability that when no target is present, we may incorrectly conclude that

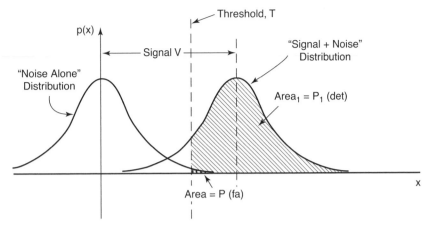

Figure 9.5. Signal Plus Noise and Noise Alone Density Functions.

there is a target. This can occur when, at the time of sampling, the noise alone is sufficiently large so as to exceed the set threshold.

9.7.2A   Density Functions

The overall situation may be depicted by the diagram of Figure 9.5. Here we see two Gaussian probability density functions. One is the "noise-alone" distribution, which has a mean value of zero since there is no signal present. It also has a standard deviation, sigma, whose square represents the noise power, N. The other distribution relates to the "signal plus noise" situation in which the noise adds to the signal pulse of magnitude V. Thus this latter distribution has a mean value of V and the same shape as the "noise-alone" distribution. We also see the threshold set at some value, T. We can visually see the detection probability $P(d)$ as the area under the "signal plus noise" density function to the right of the threshold. Also, the false alarm probability $P(fa)$ is the area under the "noise-alone" density function to the right of the threshold. In this case we erroneously decide that a signal is present when it is not. That is the basic definition of a false alarm.

Trade-offs between detection and false-alarm probabilities can occur when the threshold value (T), the pulse amplitude (V), and the noise power (N) are varied. The following three cases further explain this idea.

Case One: Increase Threshold; Pulse Amplitude and Noise Power Remain the Same

As we increase the detection threshold, less of the signal-plus-noise distribution remains to the right of the threshold value. Therefore, the detection probability decreases. This is an undesirable effect. However, less of the noise-alone distribution is to the right of the threshold, so the false-alarm probability also diminishes. This is a desirable consequence. Therefore, by the increase in threshold, we are "trading" to obtain better false-alarm performance [a lower $P(fd)$], but at the expense of detection performance [a lower $P(d)$]. A natural question is: Is there a threshold selection that allows us to meet *both* the detection and false-alarm probability requirements? By performing this trade-off analysis, that is, stepping the threshold through increasing and decreasing values, we determine the answer to this question.

Case Two: Increase Pulse Amplitude; Threshold and Noise Power Remain the Same

In general, the pulse amplitude is increased by increasing the power transmitted by the radar. This normally increases the cost of the radar system. The signal-to-noise ratio increases and the signal-plus-noise distribution has a larger mean value, but the same noise variance N. In this case, the detection probability increases for a target at the same range. Another way to look at this case is to say that for the same $P(d)$ as in Case One, we can see a target at a longer range. Pumping more power out of the transmitter results in an improved detection capability. But if the noise power remains the same, so will the false-alarm probability.

Case Three: Decrease Noise; Pulse Amplitude and Detector Threshold Remain the Same

Decreasing the noise may be achieved by designing a lower-noise front-end receiver. This increases cost and may also increase development time if we are pushing the state of the art. Less noise shows up as a decrease in the variance (N) of *both* the signal-plus-noise and the noise-alone distributions. This means that the detection probability increases *and* the false-alarm probability decreases! These are both desirable consequences. However, we must pay the price of the low-noise receiver and there are some natural limits as to how far the noise can be reduced.

We may also explore trade-offs that involve changing two of the preceding three key parameters at the same time. Such an exploration will reveal additional variations in the detection and false-alarm probabilities. In addition, other implementations and models may be considered (such as the integration of pulses), but the same basic notions of trade-offs remain. One is trying to find a *balanced* solution that satisfies user requirements. A more quantitative treatment of this particular example can be found in a variety of texts [9.9], as well as the discussion that follows.

9.7.2B   Quantitative Trade-Offs

Various qualitative relationships between the key variables, $P(d)$, $P(fa)$, N, T, and V were explored in the three cases cited above. We can expand these notions into quantitative terms by referring to a table of the normal (Gaussian)

TABLE 9.8 Changes in Detection and False Alarm Probabilities as the Threshold Changes

Threshold (volts)	Detection Probability	False Alarm Probability
3	.9938	.0668
4	.9772	.0228
5	.9332	.0062
6	.8413	.00135
7	.6915	.00025

distribution. The reader is asked to retrieve such a table from an appropriate source.

Specifically, and by way of illustration, we can see how the detection and false alarm probabilities change as the threshold is changed. Changing values in these probabilities represent a trade-off as a function of changes in the selected threshold. We will assume a noise power of 4 watts and a pulse voltage of 8 volts, and we will change the value of threshold from 3 to 7 volts, in steps of one volt. We will then see how the two probabilities change as this threshold is modified. Table 9.8 shows the results of these changes.

We thus can see the detection probability decreasing as the threshold increases, in quantitative terms. A smaller probability of detection is an undesirable feature. The false alarm probability decreases, however, as the threshold increases. This is a desirable consequence. Therefore, we are able to develop a trade-off relationship that shows how these two key probabilities are traded off against one another as a function of changes in the value of the threshold. An additional trade-off relationship can be seen by varying the noise and seeing what happens to the detection and false alarm probabilities. We will assume the threshold (T) to be 5 volts and the pulse amplitude (V) to be 8 volts, and we will start with a sigma equal to 3.5 rms (root-mean-square) units, and run it to sigma equal to 1, in increments of 0.5. The results are shown in Table 9.9.

We observe from these numbers that as the noise decreases, there is an increase in the detection probability and *also* a decrease in the false alarm rate. Both represent improvements in system performance. So we are able to

TABLE 9.9 Changes in Detection and False Alarm Probabilities as Noise Changes

RMS Noise (sigma)	Detection Probability	False Alarm Probability
3.5	.803	.0765
3.0	.8413	.0475
2.5	.8849	.0228
2.0	.9332	.0062
1.5	.9772	.0005
1.0	.99865	.000003

see, in numerical terms, how important decreases in noise can be. We get a double positive effect by decreasing noise in this scenario and model.

We can continue to explore such trade-offs, in numerical terms, by changing the pulse voltage, V, and calculating the effects on detection and false alarm probabilities. It is left as an exercise to confirm the numbers shown in Table 9.8 and Table 9.9. The reader is also urged to examine other trade-offs between the five key variables, N, V, T, $P(d)$, and $P(fa)$. Thus, this compact model can serve as a powerful tool for generating and displaying trade-off relationships.

9.7.3   Sensitivity to Criteria Weights

The evaluation framework shown in Table 9.4 reveals that the results are obviously sensitive to:

a. The selection of criteria
b. The weighting of these criteria
c. The ratings of each system

As an example of the criteria-weight sensitivity, let us assume that the criteria weights are changed as follows:

Criteria	Weight Change	
	From	To
1	0.08	0.1
2	0.1	0.1 (no change)
3	0.13	0.1
4	0.09	0.1
5	0.12	0.1
6	0.07	0.1
7	0.11	0.2
8	0.1	0.2

With the same set of ratings, the total scores for each of the three alternatives now become

- Alternative A: 7.0
- Alternative B: 6.8
- Alternative C: 6.7

Thus, the modified criteria weights *completely reverse the order of preference.* This may have occurred, for example, if criteria 7 and 8 were related to cost and together comprise 40% of the total weight. If cost were considered that important, the preferred alternative would shift from C to A. This example leads to a simple but very important conclusion: No final system selection should be made without extensive sensitivity analyses!

9.8 MODELING AND SIMULATION

9.8.1 Performance Measurement and Software

Most of the time, modeling and simulation techniques are employed in order to calculate technical performance measures (TPMs) and to carry out trade-off studies between key measures and system parameters. The radar situation involving the detection of pulses in the presence of noise, discussed in Section 9.7.2, serves as a good example of how a model may be used to determine system performance.

The model of the detection process may be expanded to include the normal processing of pulses from the point at which they leave the transmitter to their reception at the threshold detector. Such a model may be constructed as a Parameter Dependency Diagram (PDD), a process developed by this author to model and analyze complex systems [9.10]. We start a PDD by identifying the key *output* parameters that we wish to compute. In the radar situation, these are the probability of detection, $P(d)$, and the false-alarm probability, $P(fd)$, as cited earlier. For each of these outputs, we then ask the question: What do these parameters depend on? As shown in Figure 9.6, $P(d)$ depends on three parameters: noise power (N), signal voltage (V), and detection threshold (T). In a similar vein, $P(fa)$ depends on N and T, but not V. All of these dependent parameters are also known as technical performance parameters (TPPs). If we continue to work backwards from these TPPs (N, V, and T), we can determine their dependent parameters until we come to the input signal, $5(\text{in})$, and the parameters on which it depends. This latter dependency is also

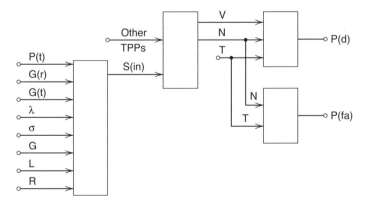

Figure 9.6. Illustrative parameter dependency diagram (PDD).

known as the radar range equation, with the following TPPs:

- Power transmitted, $P(t)$
- Gain of receiving antenna, $G(r)$
- Gain of transmitting antenna, $G(t)$
- Wavelength, λ.
- Target cross section strength, σ
- Receiver processing power gain, G
- Receiver power processing losses, L
- Range to target, R

The blocks in the FDD implicitly represent relationships or equations that relate the input parameters to the output parameters. If we are operating in the frequency domain, the boxes can be thought of as the transfer functions relating inputs to outputs. In all cases, the PDD is constructed initially without knowing the precise relationship between inputs and outputs. The structure of the PDD, however, makes it very clear as to what the key parameters are and the known or unknown relationships among these parameters. If a relationship is currently unknown, then a modeling effort is required to determine the necessary equations. The PDD is therefore a "roadmap" that explicitly shows the TPMs, the TPPs, and the known or unknown relationships between them. It is a performance "model" that represents how the key measures and parameters interrelate. Given the PDD and the necessary equations, the systems engineer is now in a position to carry out extensive trade-off studies and sensitivity analyses.

The preceding Parameter Dependency Diagramming (PDD) procedure is but one of many modeling techniques. If a model is particularly complex, the system engineer may wish to move to simulation, either by building a simulator to apply to the situation at hand or by using an existing, commercially available, simulation package. If workable in terms of the problem, the latter is highly recommended because there are numerous software packages available, at reasonable cost, to the systems engineering team. A list of such packages is provided in Exhibit 9.5 [9.15, 9.16, 9.17]. For the reader who is interested in the perspectives of one of the industry leaders in modeling and simulation, it is recommended that the words of A. Pritsker be taken very seriously [9.18].

Modeling and simulation is a specialized and rather complex subject, but it is essential that the systems engineering team master a variety of tools and techniques to be in a position to evaluate system performance on a quantitative basis. Until the system, or portions thereof, is actually built, there is really no other choice but to depend on modeling and simulation for performance assessments. The systems engineering team that faces this issue squarely will be competitive in the world of building large-scale systems. Those who do not are likely to be behind the power curve.

Exhibit 9.5: Selected Modeling and Simulation Software [9.15, 9.16, 9.17]

Name of Software Package	Builder/Publisher of Software
Achilles	In-Motion Technology
ADAS	Cadre Technologies Inc.
ALSSII	Productivity System
AutoMod	AutoSimulations Inc.
BATCHES	Batch Process Technologies Inc.
Best-Network	Best Consultants
CADmotion	SimSoft Inc.
Cinema Animation System	Systems Modeling Corp.
COMNET II.5	CACI Products Company
DISC++	Texas Tech University; Industrial Engineering Department
ESL	ISIM Simulation
Extend	Imagine That Inc.
FACTOR	Pritsker Corporation
FMS++	Texas Tech University; Industrial Engineering Department
G2	Gensym Corp.
GEMS-II	Lodestone II Inc.
General Simulation Sys.	Prediction Systems Inc.
GENETIK	Insight International Ltd.
GPSS/H	Wolverine Software Corp.
GPSS/PC	Minuteman Software
HOCUS	P-E International PLC
INMOD 1.8	Technical University—Sofia, Bulgaria
INSTRATA	Insight International Ltd.
InterSIM	OLM Holding Company
ISEE-SIMNON	Engineering Software Concepts Inc.
ISI-PC	Extech Ltd.
LANNET II.5	CACI Product Company
MAST Simulation Environ	CMS Research Inc.
Micro Saint & Animation	Micro Analysis/Design Simulation S/W
micro-GPSS	Stockholm School of Economics
MODSIM II	CACI Product Company
MOGUL	High Performance Software Inc.
MOR/DS	Holden-Day Publishing Company
NETWORK II.5	CACI Products Company
Packaging Lines Sim.Sys	Pritsker Corporation
Pascal Sim	University of Southhampton, UK
PASION	S. Raczynski, Mexico

Exhibit 9.5:   Selected Modeling and Simulation Software [9.15, 9.16, 9.17] (*Continued*)

Name of Software Package	Builder/Publisher of Software
PC Simula	Simula a.s., Norway
PCModel	SimSoft
PERCNET	Mitchell & Gauthier Associates
ProModel/PC	Production Modeling Corp. International
Proof Animation	Wolverine Software Inc.
Q+	AT&T Bell Laboratories
QASE RT	Advanced System Technologies Inc.
SES/workbench	Scientific and Engineering Software Inc.
SIGMA	The Scientific Press Inc.
SIMAN	Systems Modeling Corp.
SIMFACTORY II.5	CACI Products Company
SIMNET II	SimTec Inc.
SIMNON	SSPA Systems, Sweden
SIMNON	Engineering Software Concepts Inc.
SIMSCRIPT II.5	CACI Products Company
SIMSTARTER	Network Dynamics Inc.
SLAM/TESS	Pritsker Corporation
SLAMSYSTEM	Pritsker Corporation
Teamwork/SIM	Cadre Technologies Inc.
XCELL+	Pritsker Corporation

9.8.2   Modeling and Simulation in the Department of Defense

Some years ago, it might have been said that modeling and simulation (M&S) was "rediscovered" within the Department of Defense (DoD), based in part on the relative scarcity of money as well as the sheer value of these tools in designing and building large-scale systems. Terms like "synthetic environments" were used to support this rediscovery, which made a lot of sense then and makes even more sense today. Some aspects of measuring the performance of systems could be done *only* within the context of modeling and simulation. An example would be what is now called National Missile Defense. There is no way we could demonstrate the capabilities of that type of system, and its widespread advanced technologies, without a great deal of M&S.

If we look at the situation today, we see a lot of emphasis on understanding and refining our acquisition practices. One ingredient in that process is M&S, and therefore we also see the emergence of an Acquisition Modeling and Simulation Master Plan [9.19]. This plan is an attempt to "improve M&S

support to DoD acquisition for defining, developing, testing, producing, and sustaining capabilities" [9.20]. Some twenty-seven specific actions were set forth in order to support this intent. The basic purposes of such actions were cited as to:

- Support M&S activities beyond individual programs
- Support joint capabilities and systems-of-systems
- Remove M&S obstacles, and provide new approaches
- Support the positive interactions between DoD organizational components

Beyond these purposes, the master plan addresses five objectives. These are to:

Objective 1. Provide necessary policy and guidance.
Objective 2. Enhance the technical framework for M&S.
Objective 3. Improve model and simulation capabilities.
Objective 4. Improve model and simulation use.
Objective 5. Shape the workforce.

These activities are most welcome, especially as we have discovered that the world in which we now have to defend ourselves continues to change, at times rather drastically.

9.9 OTHER ARCHITECTURES AND TOOLS

9.9.1 Other Architectures

We have explored two architectural notions and constructs in some detail in this chapter. The first is this author's recommended approach, which is based on three well-defined steps and views:

- Synthesis
- Analysis
- Cost-effectiveness

A detailed illustration of this approach is presented in section 9.5.9, which describes the three key elements and the selection of a preferred architecture from among a set of alternatives. Seeking, defining, and evaluating alternatives represent an integral part of the author's seven aspects of the Systems Approach (see section 7.1).

A second approach, known as DoDAF (Department of Defense Architectural Framework) and discussed earlier, is built on three architectural views:

- An operational view
- A systems view
- A technical view

Each of these has subordinate views. Over time, we are going to continue to build on these well-accepted concepts. For an overview, see DoDAF documentation in the form of [9.6]:

Volume 1: definitions and guidelines
Volume 2: descriptions of each architecture product
Deskbook: additional information

In this section, we briefly note that there are indeed other notions that deal with architectures and architecting. One has been called the MoDAF approach, built and advanced by the United Kingdom Ministry of Defence. MoDAF attempts to extend DoDAF, principally by clarifying the precise nature of inputs and outputs, paying special attention to interoperability, and adding some new perspectives. The latter include:

- A strategic view
- An acquisition view
- A relationship to enterprise architecting

Since both DoDAF and MoDAF are dynamically changing, other new dimensions of MoDAF can be expected on a continuing basis.

Yet another notion in the architecting domain is that of developing enterprise architectures. An *enterprise architecture (EA)* can be thought of as an assemblage of software and hardware that supports the current and future business areas of the overall enterprise. An example of an approach to formulating an EA can be found at the highest levels of government, where, on behalf of the White House, the Office of Management and Budget (OMB) has been working on the Federal Enterprise Architecture (FEA). This business-driven notion applies across a set of functional areas, such as [9.21]:

a. Budget formulation and allocation
b. Information processing and sharing
c. Performance measurement
d. Collaboration between agencies
e. E-government
f. Component-based architectures

An EA perspective at the "state" level has been articulated as [9.22]:

> An Enterprise Architecture (EA) provides a strategic planning framework that relates and aligns information technology (IT) with the business functions that it supports.

Thus we see the inclusive scope of EA, dealing with all of the functions/business areas of the enterprise, and how they need to be supported both currently and into the future. This broad scope often implies that the EA becomes the province of the CIO (Chief Information Officer) of the enterprise. Equivalent notions apply to an EA in the commercial sector, with appropriately defined business areas.

Yet another architecture notion is the *service-oriented architecture* (*SOA*). This approach requires the ability to exchange information to and from disparate systems, creating services that can be deployed rapidly. In effect, SOA allows us to reliably "tap into" a set of existing information systems from remote locations using network connectivity. The Defense Information Systems Agency (DISA) has cited the SOA as an "architecture built primarily upon network available services" [9.23]. An SOA-based service thus needs a search capability, security, collaboration, interoperability, data accessibility, loose coupling, decentralization, and an appropriate implementation of standards. As of this writing, DISA was playing a key role within the DoD to make the SOA a widespread reality. If successful, it will make large contributions to the C^4I (command, control, communications, computers, and intelligence) world and provide for the rapid provision of system capabilities using smaller and more modular systems. It will also help to reduce duplications of applications and systems within and between enterprises and agencies. Communities of practice are also expected to help move this concept and its associated technologies and standards forward in a deliberate and productive manner.

9.9.2 Some Additional Tools

This author has been a strong advocate of the use of modeling and simulation tools to represent and analyze the behavior of complex systems. In Chapter 7, we saw a related topic—computer tool evaluation and utilization—as one of the thirty basic elements of systems engineering. In Chapter 8, we noted that one of the eight essential steps of requirements engineering is "automation of requirements analysis and allocation." Earlier in this chapter, we discussed Parameter Dependency Diagramming as a recommended approach. This chapter has also paid special attention directly to the overall topic of M&S. Another book by this author [9.24] contains a chapter on diagramming, namely, "Thinking through Pictures," as a way to help examine and manage complex systems. Continuing investments in M&S demonstrate

TABLE 9.12   Companies that Provide Modeling Language–Related Products

Artisan
I-Logix
Popkin
Sparx Systems
Telelogic
Vitech
Zachman

the need for such tools and confirm that they are making a substantial contribution.

Only two additional and specific approaches to modeling will be cited here, the Unified Modeling Language (UML) and the Systems Engineering Modeling Language (SysML). They are related to each another, and one might say they occupy a new and special place in the analysis of systems [9.25]. These languages, as expected, have some common elements, but they also have their own unique features. Indeed, it is conventional wisdom that SysML represents a subset of UML 2.0. Nonetheless, it appears that SysML occupies a more forward-looking position with considerable momentum and many supporters.

Diagrams are a central feature in SysML, with special emphasis on structure, behavior, requirements, and parametrics. Starting with these, we wish to "model" all significant aspects of systems, including their doctrine, organization, training, materiel, leadership, personnel, and facilities (DOTMLPF), as cited in the system acquisition standard known as 5000.2 [9.26].

A nonprofit organization named Object Management Group (OMG) is a key player in the field, especially as it, along with others, advances the state of the art with respect to SysML. The reader is urged to make beneficial use of the OMG Web site [9.27] and the information that it provides.

On the vendor side, many forward-looking software companies are offering excellent products that deal with UML, SysML, and related tools. A brief list is provided in Table 9.12. These companies can be reached via a conventional search leading to their Web sites:

A conventional search will reveal other companies that provide software that can be considered related to UML, SysML, and the various architectural approaches mentioned in this chapter.

9.10   SUMMARY

This chapter focused on the key systems engineering element of systems architecting. As suggested, this element is a cornerstone of the systems engineering process. Strong emphasis has been placed on the generation and evaluation of *alternatives,* and on avoiding the tendency to leap to judgment with respect to a preferred system architecture. The roles of evaluation criteria, an evaluation framework, technical performance measurement, trade-off and sensitivity analyses, and modeling and simulation were also discussed. An attempt has been made to focus on principles as well as on an illustrative demonstration of how these principles are converted into practice. The Appendix shows several additional examples of how these principles may be applied to a variety of systems architecting situations and problems.

QUESTIONS/EXERCISES

9.1 Contrast and discuss how an architect designs a house or building with the systems architecting approach of this chapter.

9.2 Contrast and discuss the process of architecting with that of carrying out all of the thirty elements of systems engineering.

9.3 Architect a personal computer system.

9.4 Use the architecting approach defined here to select a cost-effective
   a. automobile
   b. project management software system
   c. modeling and simulation software system

9.5 Write a three-page review of the essentials of E. Rechtin's book (see the References).

9.6 Cite a dozen evaluation criteria for
   a. an automobile
   b. a house
   c. a system of your choice

9.7 Architect and engineering (A&E) firms architect buildings of various types and then engineer them. In what ways does that relate to the building of systems, as described in this as well as earlier chapters?

9.8 The illustrative architecture in this chapter (Section 9.5.9) suggests that any of the three alternatives might be preferred, depending upon the scenario you envision. Define three scenarios that would likely lead to each of the three alternatives as the preferred architecture.

9.9 Confirm the detection and false alarm probability numbers in Tables 9.8 and 9.9.

9.10 Write a three-page paper comparing the current DoDAF and MoDAF approaches to architecting. Which approach do you find more satisfying? Why?

REFERENCES

9.1 Rechtin, E. (1991). *Systems Architecting*. Englewood Cliffs, NJ: Prentice Hall.

9.2 *The NASA Mission Design Process* (1992). Washington, DC: National Aeronautics and Space Administration, NASA Engineering Management Council.

9.3 *Phase C/D Requirements Specification for the Earth Observing System Data and Information System (EOSDIS)* (1990). Greenbelt, MD: Goddard Space Flight Center.

9.4 *Earth Observing System (EOS) Data and Operations System (EDOS) Statement of Work* (1992). Greenbelt, MD: Goddard Space Flight Center.

9.5 *Payload Data Services System (PDSS) Operations Concept Document and Development Specification* (1992). National Aeronautics and Space Administration, Marshall Space Flight Center.

9.6 *C4ISR Architecture Framework*, version 2.0 (1997). Washington, DC: Architectures Working Group, U.S. Department of Defense (DoD), December 18.

9.7 *Standard for Systems Engineering*, IEEE P1220 (1994). Piscataway, NJ: Institute of Electrical and Electronics Engineers (IEEE) Standards Department.

9.8 *Draft Recommended Practice for Architectural Description*, IEEE P1471/D5.2, (1999). New York: Institute of Electrical and Electronics Engineers (IEEE), December.

9.9 Eisner, H. (2004). "New Systems Architecture Views." Paper presented at the 25th National Conference of the American Society of Engineering Management (ASEM), Alexandria, VA, October 20–23.

9.10 Eisner, H. (1988). *Computer-Aided Systems Engineering*. Englewood Cliffs, NJ: Prentice-Hall.

9.11 The architecting of this particular system was carried out by Richard C. Anderson, a student at The George Washington University, using the procedure defined by this author. The original description of and architecture for this system is provided in Section A.5 of the Appendix.

9.12 Henley, E., and H. Kumamoto (1981). *Reliability Engineering and Risk Assessment*. Englewood Cliffs, NJ: Prentice Hall.

9.13 Charette, R. (1989). *Software Engineering Risk Analysis and Management*. New York: Multiscience Press.

9.14 *Systems Engineering*, Military Standard 499B (Draft) (1991). Washington, DC: U.S. Department of Defense.

9.15 Swain, J. (1991). "World of Choices: Simulation Software Survey," *OR/MS Today* (October): 81–102.

9.16 Swain, J. (1993). "Flexible Tools for Modeling," *OR/MS Today* (December): 62–78.

9.17 Swain, J. (2001). "Power Tools for Visualization and Decision-Making," *OR/MS Today* (February): 52–63.

9.18 Pritsker, A. (1990). *Papers, Experiences, Perspectives*. West Lafayette, IN: Systems Publishing.

9.19 Department of Defense, Office of the Under Secretary of Defense (Acquisition, Technology and Logistics) (2006). "Acquisition Modeling and Simulation Master Plan." Issued under the authority of the Systems Engineering Forum, April 17.

9.20 Schaeffer, Mark D. (2006). Foreword to Department of Defense, Office of the Under Secretary of Defense (Acquisition, Technology and Logistics), "Acquisition Modeling and Simulation Master Plan." Issued under the authority of the Systems Engineering Forum, April 17.

9.21 See www.whitehouse.gov/omb.

9.22 See www.vita.virginia.gov.

9.23 See www.disa.mil.

9.24 Eisner, H. (2005). *Managing Complex Systems—Thinking Outside the Box*. Hoboken, NJ: John Wiley & Sons.

9.25 Friedenthal, S. (2005). Systems Modeling Language (SysML) Overview, 21 slides, April 20; see sanford.friedenthal@lmco.com.

9.26 U.S. Department of Defense (2003). *Operation of the Defense Acquisition System*. Instruction 5000.2. Washington, DC: DoD, May 12.

9.27 See http://syseng.omg.org.

10

SOFTWARE ENGINEERING

10.1 INTRODUCTION

A headline in the business section of the *Washington Post* [10.1] declared "Fidelity Says It Reported Wrong Prices." The Fidelity was Fidelity Investments and it apparently provided out-of-date information on the value of its mutual funds. The reason? It was "because of a computer software glitch, Fidelity reported." The impact of this glitch was not divulged, but it very likely had many investors rather unhappy.

This type of problem—some variety of computer software glitch—is experienced every day across the country. Our systems contain increasing amounts of software and we have not yet been able to migrate software development from mostly art to mostly engineering or science. There are many reasons for this problem, ranging from the fast-moving pace of the state of the art in software creation to lack of corporate infrastructures that adequately support the processes and products inherent in software development. For these reasons, we treat software engineering, defined herein as a subset of systems engineering, in a special and separate chapter. This chapter provides only an overview of software engineering, which itself is an appropriate subject for an entire book [10.2, 10.3, 10.4]. This overview attempts to capture selected highlights of software engineering in terms of dominant issues that are faced by industry practitioners.

It is also recognized that software development and engineering, in the main, developed without an explicit connection to the methods and disciplines of systems engineering. Thus, we are at least partially concerned with the relationship between systems and software engineering. This matter is also recognized in a software standard (Mil-Std-498) discussed in the next section.

10.2 STANDARDS

Standards play an important part in both systems and software engineering. We explored Military Standard 499B in relation to systems engineering in previous chapters. With respect to software engineering, we briefly examine the following four standards: (1) Mil-Std-498; (2) DOD-Std-2168; (3) ISO Standards; and (4) IEEE Standards.

10.2.1 Mil-Std-498

This is a landmark standard established by the U.S. Department of Defense (DoD) [10.5], building on an earlier DoD Standard known as Mil-Std-2167A [10.6], which had a major influence on how companies developed software for the DoD as well as for other parts of the government, that is, government civil agencies.

The table of contents for the general requirements for software development is shown in Exhibit 10.1. This exhibit provides an overview of the scope of this standard. Under item 4.1, the detailed software development process is delineated, as shown in Exhibit 10.2.

Exhibit 10.1: General Requirements of DoD Software Standard

4.1 Software development process
4.2 General requirements for software development
    4.2.1 Software development methods
    4.2.2 Standards for software products
    4.2.3 Reusable software products
        • 4.2.3.1 Incorporating reusable software products
        • 4.2.3.2 Developing reusable software products
    4.2.4 Handling of critical requirements
        • 4.2.4.1 Safety assurance
        • 4.2.4.2 Security assurance
        • 4.2.4.3 Privacy assurance
        • 4.2.4.4 Assurance of other critical requirements
    4.2.5 Computer hardware resource utilization
    4.2.6 Recording rationale
    4.2.7 Access for acquirer review

Exhibit 10.2: The Software Development Process

a. Project planning and oversight
b. Establishing a software development environment

c. System requirements analysis
d. System design
e. Software requirements analysis
f. Software design
g. Software implementation and unit testing
h. Unit integration and testing
i. Computer Software Configuration Item (CSCI) qualification testing
j. CSCI/Hardware Configuration Item (HWCI) integration and testing
k. System qualification testing
l. Preparing for software use
m. Preparing for software transition
n. Integral processes:
    1. Software configuration management
    2. Software product evaluation
    3. Software quality assurance
    4. Corrective action
    5. Joint technical and management reviews
    6. Other activities

From these exhibits, we see the initial focus on the "system," which then moves to the specific software item considerations. The more limited view of the software development process in DoD-Std-2167A is also expanded. Finally, the last item listed (Exhibit 10.2, Other activities) includes:

- Risk management
- Software management indicators
- Security and privacy
- Subcontractor management
- Interface with software independent verification and validation (IV&V) agents
- Coordination with associate developers
- Improvement of project processes

In this standard, software is decomposed into

- Computer software configuration items (CSCIs)
- Computer software components (CSCs)
- Computer software units (CSUs)

which, to a large extent, parallels the decomposition notions previously described in systems engineering. That is, the systems engineer decomposes the system into functions and subfunctions. The view presented in this standard is that the systems (software) engineer breaks down the system into segments. These segments are then broken down into hardware and software configuration items. Once that breakdown is accomplished, the CSCIs are decomposed into CSCs and the CSCs into CSUs. This provides a challenge for the systems engineer, namely, to bring system functional decomposition in conformance with the preceding software breakdown structure. A way to meet that challenge is to associate one or more layers of functional decomposition with the "segment" element. Once the lowest level of functional decomposition is attained, then the HWCI and CSCI breakout is entirely appropriate. This issue, at times, represents a disconnect between how systems engineers, in distinction to software engineers, look at and analyze a system that is composed of both hardware and software.

This standard was intended to supersede DoD-Std-2167A by making certain improvements. Particular emphasis was placed on resolving issues in 2167A with respect to the following:

1. Improving compatibility with incremental and evolutionary development methods
2. Improving compatibility with nonhierarchical design methods, that is, object-oriented methods
3. Improving compatibility with computer-aided software engineering (CASE) tools
4. Providing alternatives to, and more flexibility in, the preparation of documentation
5. Providing clearer requirements for incorporating reusable software
6. Including the use of software management indicators
7. Putting more emphasis on software supportability
8. Improving links to systems engineering

Each of these suggests one or more key issues with respect to the overall subject of software development. As such, they are discussed further either later in this chapter or in Chapter 12, which deals with trends in software engineering. It should be noted, however, that this standard makes the explicit statement that it "is not intended to specify or discourage the use of any particular software development method." In addition, and in conjunction with other standards, it "provides the means for establishing, evaluating, and maintaining quality in software development products."

10.2.2   DoD-Std-2168

This standard [10.7], dealing with software quality evaluation, represented a companion standard to 2167A. Its basic function was to establish

requirements for evaluating the quality of software and associated documentation and activities for mission-critical computer systems. It also
may be applied to the independent verification and validation (IV&V) of
software.

The essence of this standard was to provide a framework for the preceding
evaluation. Evaluation matrices are defined for each of the products of the
eight steps identified as part of the software development process. The basic
evaluation criteria for this process are:

1. Adherence to required format and documentation standards
2. Compliance with contractual requirements
3. Internal consistency
4. Understandability
5. Technical adequacy
6. Appropriate degree of completeness
7. Traceability to indicated documents
8. Consistency with indicated documents
9. Feasibility
10. Appropriate requirements analysis, design, coding techniques used to
    prepare item
11. Appropriate level of detail
12. Appropriate allocation of sizing and timing resources
13. Adequate test coverage of requirements
14. Adequacy of planned tools, facilities, procedures, methods, and re-
    sources
15. Appropriate content for intended audience

The standard also defines software quality factors that may be included as
requirements in the software requirements specification, namely:

- Correctness
- Efficiency
- Flexibility
- Integrity
- Interoperability
- Maintainability
- Portability
- Reliability
- Reusability
- Testability
- Usability

We also note that this standard called for an evaluation of risk management, expressed in the following manner:

> The contractor shall evaluate the procedures employed and the results achieved by risk management throughout the software development cycle. This evaluation shall verify that risk factors are identified and assessed, resources are assigned to reducing risk factors, alternatives for reducing risk are identified and analyzed, and sound alternatives are selected, implemented and evaluated.

The subject of risk management arises explicitly in the context of a development process known as the Spiral Model, cited again later in this chapter. The matter of software quality is also closely related to reliability, which is also is discussed later in this chapter.

10.2.3 ISO Standards

The International Organization for Standardization (ISO) has been promulgating standards [10.8] that are becoming more widely recognized and accepted. ISO standards referred to in Mil-Std-498, discussed before, are

ISO/IEC 12207:   Software Life Cycle Processes
ISO/IEC 9126:   Quality Characteristics and Guidelines for Their Use
ISO 9001:   Quality System—Model for Quality Assurance in
                 Design/Development, Production, Installation and
                 Servicing
ISO 9000-3:   Guidelines for the Application of ISO 9001 to the
                 Development, Supply and Maintenance of Software

As implied by their titles, ISO 12207 and ISO 9000-3 apply directly to software development.

ISO 12207 is focused on three types of life-cycle processes, namely, primary, supporting, and organizational. Primary processes deal with such areas as acquisition, supply, development, operation, and maintenance. Supporting processes are concerned with documentation, configuration management, quality assurance, verification and validation, reviews, audits, and problem resolution. Organizational processes include management, infrastructure, improvement, and training.

ISO 9000-3 (Part 3 of ISO 9000) particularizes ISO 9001 to matters dealing with software. Key areas within this document are concerned with

- Management responsibility
- Quality systems
- Internal quality system audits

- Corrective action
- Quality system life-cycle activities
- Quality system supporting activities

In general, the ISO standards are broader than Mil-Std-498 with respect to software. They also revolve around the centerpiece of quality. Practitioners who wish to do business in the international community, however, should devote time to understand the main thrusts of the ISO standards to be successful.

10.2.4 IEEE Standards

Mil-Std-498 also refers to IEEE (Institute of Electrical and Electronics Engineers) standards that are related to software development. The IEEE standards referenced are cited in Exhibit 10.3. We note from the exhibit that there is a standard for software project management plans. The list of IEEE standards is ever-growing and represents a touchstone for both military and commercial practices with respect to software development as well as other endeavors.

Exhibit 10.3: Selected IEEE Standards

IEEE Std 730:	Standard for Software Quality Assurance Plans
IEEE Std 828:	Standard for Configuration Management Plans
IEEE Std 829:	Standard for Software Test Documentation
IEEE Std 830:	Recommended Practice for Software Requirements Specifications
IEEE Std 982.2:	Guide: Use of Standard Measures to Produce Reliable Software
IEEE Std 990:	Recommended Practice for Ada as a Program Design Language
IEEE STd 1008:	Standard for Software Unit Testing
IEEE Std 1012:	Standard for Software Verification and Validation Plans
IEEE Std 1016:	Recommended Practice for Software Design Descriptions
IEEE Std 1016.1:	Guide for Software Design Descriptions
IEEE Std 1028:	Standard for Software Reviews and Audits
IEEE Std 1042:	Guide to Software Configuration Management
IEEE Std 1044:	Standard Classification for Software Anomalies
IEEE Std 1045:	Standard for Software Productivity Metrics
IEEE Std 1058.1:	Standard for Software Project Management Plans
IEEE Std 1059:	Guide for Verification and Validation Plans

IEEE Std 1061:	Standard for Software Quality Metrics Methodology
IEEE Std 1063:	Standard for Software User Documentation
IEEE Std 1074:	Standard for Developing Software Life Cycle Processes
IEEE Std 1209:	Recommended Practice for the Evaluation and Selection of CASE Tools
IEEE Std 1219:	Standard for Software Maintenance
IEEE Std 1228:	Standard for Software Safety Plans
IEEE Std 1298:	Software Quality Management System

The reader with a further interest in IEEE standards may access all software (and other) standards through their standards catalog [10.9]. The *Software Engineering Standards Collection*, a compilation of software-related standards, contained twenty-seven IEEE standards and is obtainable at a modest price.

10.2.5 IEEE/EIA 12207 [10.10]

The 12207 standard, which may be thought of as a next step beyond Military Standard 498, focuses on life cycle processes that support software and information systems engineering. Since it is process-oriented, it falls in line with other endeavors (such as business process reengineering) that highlight process in such a way as to declare that if one gets the process right, most of the problem is solved. Although this point is debatable, one can make the solid point that the wrong process is likely to lead to no end of difficulty. This author's view is that a correct process is a necessary, but still insufficient, basis for success. The primary element that clears up the insufficiency is appropriate subject and domain knowledge that can be applied to the problem at hand.

This standard emphasizes three sets of life cycle processes that address the following concerns:

1. Primary areas
2. Supporting areas, and
3. Organizational areas

Life cycle processes that are considered primary deal with:

1. Acquisition
2. Supply
3. Development
4. Operation, and
5. Maintenance

Those processes that provide support to the primary areas include:

1. Documentation
2. Configuration management
3. Quality assurance
4. Verification

5. Validation
6. Joint review
7. Audit
8. Problem resolution

Finally, there are life-cycle processes with respect to significant organizational matters in the following areas:

1. Management
2. Infrastructure
3. Improvement
4. Training

The seventeen processes represented in the primary, support, and organizational areas are described in considerable detail in this standard. At the same time, Annexes in the standard suggest and provide guidance on how to tailor the standard to suit a particular software project. This standard is also a good example of the influence of process-thinking and the resultant emphasis on both defining all relevant processes and implementing them correctly.

10.3 SOFTWARE MANAGEMENT STRATEGIES

Exhibit 10.2 defined a version of the software development process in the form of some thirteen steps. However, there has been a considerable amount of dialogue and experimentation regarding the best sequencing of these types of steps. The earliest management approach was the so-called "waterfall model," which defined a linear sequence such that there was basically no revisitation or iteration of earlier steps. The basic elements of the waterfall model were:

1. System requirements analysis
2. System design
3. Software requirements analysis
4. Preliminary design
5. Detailed design
6. Coding and CSU test
7. CSC test and integration
8. CSCI testing

Indeed, DoD-Std-2167A [10.6] reinforced this linear perspective.

Moving beyond the waterfall model, Boehm suggested the notion of a "spiral model" [10.11]. This model showed several iterations of risk analysis, building prototypes as well as reviews and validations of requirements. Whereas the spiral model received a lot of attention as a new way of looking at a software process strategy, it was not formally accepted as part of a standard approach. Instead, Mil-Std-498 came on the scene with other "candidate program strategies," namely:

- A Grand Design strategy
- An Incremental strategy
- An Evolutionary strategy

Standard 498 defined these strategies as follows:

Grand Design. A "once through, do each step once" approach that determines user needs, defines requirements, designs the system, implements the system, tests, fixes, and then delivers

Incremental. Determines user needs and defines the system requirements and then performs the rest of the development in a sequence of builds; the first build incorporates part of the desired capabilities, the next build adds more capabilities, and so on, until the system is complete (also called a preplanned product improvement strategy)

Evolutionary. Develops a system in builds, but acknowledges that the user needs are not fully understood and all requirements cannot be completely defined up front; thus, user needs and requirements are refined in each successive build

The features of these three strategies may be articulated, in part, as follows [10.5]:

| | Program Strategy | | |
	Grand Design	Incremental	Evolutionary
Defines all requirements first?	Yes	Yes	No
Has multiple development cycles?	No	Yes	Yes
Fields interim software?	No	Maybe	Yes

The preferred approaches are the incremental and the evolutionary, depending on the circumstances surrounding the software project itself. A risk-opportunity analysis, however, is also suggested to arrive at a selection for any one particular project. An example of such a risk-opportunity evaluation is outlined in Table 10.1 [10.5].

TABLE 10.1 Sample Risk Analysis for Determining the Appropriate Program Strategy

Grand Design		Incremental		Evolutionary	
Risk Item (Reasons against this strategy)	Risk Level	Risk Item (Reasons against this strategy)	Risk Level	Risk Item (Reasons against this strategy)	Risk Level
Requirements are not well understood	H	Requirements are not well understood	H		
System too large to do all at once	M	User prefers all capabilities at first delivery	M	User prefers all capabilities at first delivery	M
Rapid changes in mission technology anticipated—may change the requirements	H	Rapid changes in mission technology are expected—may change the requirements	H		
Limited staff or budget available now	M				
Opportunity Item (Reasons to use this strategy)	Opp. Level	Opportunity Item (Reasons to use this strategy)	Opp. Level	Opportunity Item (Reasons to use this strategy)	Opp. Level
User prefers all capabilities at first delivery	M	Early capability is needed	H	Early capability is needed	H
User prefers to phase out old system all at once	L	System breaks naturally into increments	M	System breaks naturally into increments	M
		Funding /staffing will be incremental	H	Funding /staffing will be incremental	H
				User feedback and monitoring of technology changes are needed to understand full requirements	H
				Decision: Use this strategy	

Finally, it is important at this juncture to mention another development approach, known as "rapid prototyping." This approach, as the title suggests, involves the very quick building of at least one prototype in order to:

- Develop a better understanding of the key design elements
- Identify critical problems as early as possible
- Have a system mock-up to show to the customer
- Obtain explicit feedback from the customer

Rapid prototyping is highly recommended and can be utilized in conjunction with either the incremental or the evolutionary strategies described before. Project constraints may preclude its use but, whenever possible, it should be suggested as a viable approach to the customer. Under certain types of software system procurements within the general category of systems integration, a live test demonstration (LTD) is often a requirement prior to the award of a contract. An LTD is basically a form of rapid prototyping.

10.4   CAPABILITY MATURITY

One of the more interesting developments in software engineering has been that of the Capability Maturity Model (CMM). The CMM was formulated by the Software Engineering Institute (SEI) at Carnegie Mellon University, with the assistance of subcontractors. Work on the CMM was sponsored by the DoD and has had an enormous impact on the thinking, as well as the behavior, of a large number of industrial companies. It has also found its way into the programs of civil government agencies, for example, the Federal Aviation Administration (FAA).

Notions of the software CMM are generally attributed to Watts Humphrey [10.12] and his colleagues [10.13] at the SEI. The structure of the CMM is based on a five-level characterization of software maturity, namely:

1. An initial level
2. A repeatable level
3. A defined level
4. A managed level
5. An optimizing level

Initial Level. At this level, software development processes are mostly ad hoc, and therefore can be chaotic at times. Individual as well as group processes are generally not defined or well organized, and individual efforts and capabilities have a strong influence on the success, or lack of it, on a software development project.

Repeatable Level. Some processes are established at this level, mostly dealing with the tracking of schedule, cost, and software functionality. Processes established at the initial level can be repeated to the extent that they led to success. A repeatable level can be achieved by adopting similar earlier applications, but generally not with new applications.

Defined Level. At the defined level, the software processes for the team effort are documented, standardized, and integrated for the organization. That is, all projects within an organization follow standard software processes that are designed to be effective in the various domains of software engineering.

Managed Level. A key element of this level is measuring the software process as well as the results (products) of that process. Through such measurement, it is then possible to determine process and product efficacy and exercise control through changes when both are not yielding appropriate results.

Optimizing Level. At the optimizing level, continuous process improvement is achieved through a well-developed system of measurement, feedback, and change. New ideas and technologies for improvement are routinely explored with the objective of optimizing all software development processes and products.

Given these basic concepts, the next obvious question is: How is one to determine, or measure, the capability maturity level of an organization? Considerable attention has been focused on this matter, bringing the CMM from a conceptual framework to a real-world change mechanism and agent. Implementation is structured into two parts, namely:

1. A Software Process Assessment (SPA)
2. A Software Capability Evaluation (SCE)

For the SPA, the focus is an individual organization and how it can (a) identify the status of its software process, and (b) establish priorities with respect to improving that process. Thus, the SPA is an *internal* mechanism to help an organization determine where it is in terms of capability maturity and point the direction toward enhancing that level of maturity. The SCE, on the other hand, is used by agencies that are acquiring software systems to determine the CMM levels of various organizations. It is therefore an *external* view of an organization's capability for the main purpose of establishing which organizations are qualified to produce high-quality software and systems.

For both SPA and the SCE, *key process areas* (KPAs) are defined and profiled, and questionnaires are used as assessment and evaluation devices. Examples of key process areas for CMM levels 2 through 5 are shown in Exhibit 10.4.

Exhibit 10.4: Examples of Key Process Areas (KPAs)

CMM Level 2

- Requirements management
- Software project planning
- Software project tracking and oversight
- Software subcontract management
- Software quality assurance
- Software configuration management

CMM Level 3

- Organizational process focus
- Organizational process definition
- Training program
- Integrated software management
- Software product engineering
- Intergroup coordination
- Peer reviews

CMM Level 4

- Process measurement and analysis
- Quality management

CMM Level 5

- Defect prevention
- Technology innovation
- Process change management

Each key process area is rated as not satisfied (NS), partially satisfied (PS), or fully satisfied (FS).

With respect to the use of questionnaires as part of the CMM, there are some lessons to be learned. Often, project systems and software engineers tend to disregard processes that they consider to be less than fully rigorous and measurable. However, as the CMM construct demonstrates, it is indeed possible, through questionnaires, to obtain some type of measurement that can be used to make both qualitative and quantitative statements about organizations and the processes that they employ. This method of "measuring the unmeasurable," although perhaps not fully rigorous or even repeatable, can be put into place and can also have enormous real-world impacts. The project, systems, and software engineering managers must understand the role of such procedures and use them wherever and whenever they appear to be applicable. An unyielding stance that fails to acknowledge the utility of checklists and questionnaires normally leads to management difficulties and problems. Organizations are paying a great deal of attention to the CMM notions because they are having major impacts on what they do and how prepared they are to do business in the software development world.

Although essentially everyone appears to acknowledge the impacts of the CMM notions, not everyone is pleased by these impacts and their consequences. For example, in some circles, it is claimed that the "opposition (to the CMM) is loud and clear" [10.14]. However, the bottom line reported is that "the CMM, on balance, can be considered a very successful model, particularly when combined with TQM [Total Quality Management] principles." Examples of CMM implementations by Hughes and Raytheon supported this latter conclusion. For example, Hughes, in moving a division from level 2 to level 3, spent about $400,000 over a three-year period. The estimated return on that investment was $2 million annually. Raytheon invested almost $1 million annually in process improvements, achieving a 7.7:1 return on that investment and 2:1 productivity gains.

Based on extensive industry interest in adopting CMM and because some government agencies are indicating that competition will be limited to enterprises that achieve certain levels of capability, it is likely that the CMM, and extensions thereof, will continue to have a strong influence on the practice of software engineering. Thus, for all software-intensive systems, the Project Manager (PM), Chief Systems Engineer (CSE), and Chief Software Engineer are likely to have to pay considerable attention to the CMM.

Beyond the CMM for software lies a CMM concept and implementation for systems engineering (in distinction to software engineering). This issue is explored in Chapter 12.

10.5 METRICS

Various initiatives with respect to making vital measurements of the software development process and its products have been ongoing for many years. Such measurements are called software metrics, several of which are briefly described in this section.

10.5.1 Management Indicators

By way of introducing the subject of software metrics, perhaps the least controversial is a set of management indicators [10.5] that is clearly measurable in terms of factors in which management has a specific interest. Such a set is provided in Exhibit 10.5 and is a candidate for use on a software development project.

Exhibit 10.5: Candidate Management Indicators

1. *Requirements Volatility.* The total number of requirements and requirement changes over time.

2. *Software Size.* Planned and actual number of units, lines of code, or other size measurements over time.

3. *Software Staffing.* Planned and actual staffing levels over time.

4. *Software Complexity.* Complexity of each software unit.

5. *Software Progress.* Planned and actual number of software units designed, implemented, unit tested, and integrated over time.

6. *Problem/Change Report Status.* Total number, number closed, number opened in the current reporting period, age, priority.

7. *Build Release Content.* Planned and actual number of software units released in each build.

8. *Computer Hardware Resource Utilization.* Planned and actual use of computer hardware resources (such as processor capacity, memory capacity, etc.) over time.

9. *Milestone Performance.* Planned and actual dates of key project milestones.

10. *Scrap/Rework.* Amount of resources expended to replace or revise software products after they are placed under project-level or higher configuration control.

11. *Effect of Reuse.* A breakout of each of the preceding indicators for reused versus new software products.

Although this list is quantitative, more conventional metrics known in the industry are considered in the next sections.

10.5.2 COCOMO

COCOMO is an acronym meaning "Constructive COst MOdel." It was developed by Barry Boehm [10.15] and has been a cornerstone of the industry with respect to estimating the levels of effort and times required to develop software.

The basic equations for COCOMO, a formal cost-estimating relationship (CER), follow:

$$PM = C(KDSI)^X \tag{10.1}$$

$$TDEV = D(PM)^Y \tag{10.2}$$

$$PROD = \frac{DSI}{PM} \tag{10.3}$$

$$FTES = \frac{PM}{TDEV} \tag{10.4}$$

where   PM = person-months required to complete software
                        KDSI = thousands of delivered source instructions
                        TDEV = required development time
                        PROD = productivity
                        FTES = full-time equivalent staff needed
            C, D, X, Y = empirically derived constants

The environment within which the software development is being carried out
is defined as the *mode* of the effort. For Boehm's *organic mode*, characterized by a relatively small team, extensive experience, and a stable in-house
environment, Equations (10.1) and (10.2) become

$$PM = 2.4(KDSI)^{1.05} \qquad\qquad (10.5)$$

$$TDEV = 2.5(PM)^{0.38} \qquad\qquad (10.6)$$

To illustrate the results obtained by these COCOMO equations, assume that
the number of delivered source instructions is estimated to be 40,000. The
preceding two equations then yield

$$PM = 2.4(40)^{1.05} = (2.4)(48.1) = 115.4 \text{ person-months}$$
$$TDEV = 2.5(115.4)^{0.38} = (2.5)(6.076) = 15.2 \text{ months}$$

The productivity is thus

$$PROD = \frac{40{,}000}{115.4} 346.6 \text{ DSI/PM, or about } 16.5 \text{ DSI per day}$$

and the full-time equivalent staff required is

$$FTES = \frac{115.4}{15.2} = 7.6 \text{ people}$$

The COCOMO equations are nonlinear so that, for example, if the DSI
doubles to 80,000, the person-months and development time become 239
and 20 months, respectively, and the productivity drops slightly to 15.9 DSI
per day. Thus, the calculations are rather straightforward and yield important
planning estimates for the software developer.

    One of the more serious questions raised with respect to COCOMO (and
there are several issues that have been raised regarding COCOMO's use) is that
of obtaining good estimates of the delivered source instructions. Indeed, if they
are not valid or accurate, then the usual "garbage-in, garbage-out" admonition
applies. This issue has never really been solved, but various enterprises have
made attempts at trying to assure good input estimates. Among the best of
these approaches is to obtain multiple and independent estimates from key
personnel on the development team to see what the spread might be and

to ultimately converge on a workable consensus. This is a recommended approach when attempting to utilize COCOMO or its variants for which the fundamental input is the delivered source instructions or lines of code.

Other criticisms of COCOMO have focused on the fact that other variables, such as the language used, use or not of CASE tools, and so forth, are not explicitly accounted for. Some of this criticism is blunted by the extensions to COCOMO (e.g., REVIC, discussed in what follows) that take such variables into consideration in an explicit manner.

COCOMO II [10.16] is, of course, an update of COCOMO I, as described briefly above. The model is designed to accommodate three basic levels of granularity, as well as three stages in the life of software development. These levels refer to:

1. *Prototyping.* The input is sized in object points.
2. *Early Design.* The input is provided in source statements (lines of code) or function points, and there are seven effort multipliers.
3. *Post Architecture.* The input is stated in source statements or function points, and there are seventeen effort multipliers.

There are also five *scale factors* that replace the previous use of the organic, semidetached, and embedded modes.

The source statements form of the COCOMO II model basically takes the same form as that in COCOMO I. That is, person-months (PM) are calculated with the following formula:

$$PM\,(\text{effort}) = A\,(\text{size})^B$$

where PM is the effort in person-months, A is characterized by a set of effort multipliers (EMs), size is the number of source statements, and B represents a set of scale factors that model economies ($B < 1.0$) or diseconomies ($B > 1.0$) of scale. The value of B is itself a function of five scale factors, namely:

1. Precedentedness
2. Development flexibility
3. Risk resolution
4. Team cohesion
5. Process maturity

The value of A is also a function of seven or seventeen cost drivers, known as effort multipliers, depending upon whether one is using the early design or the post architecture version of the model. The reader is referred to the COCOMO II book [10.16] for complete lists of these effort multipliers, as well as how they may be used to determine the value of A in the basic COCOMO equation.

Thus, a central theme of COCOMO II is that it is a model that estimates person-months as a function of 12 (5 scale drivers + 7 effort multipliers) or of 22 (5 scale drivers + 17 effort multipliers) additional variables. These variables are factors that influence the effectiveness and efficiency of the software development effort. COCOMO II also spells out relationships for project scheduling, as was the case for COCOMO I. COCOMO II also expands its areas of consideration, including important topics such as reuse, reengineering, rapid application development, the use of commercial-off-the-shelf (COTS) software, software quality, productivity estimation and risk assessment. COCOMO II is considered to be another major step forward in understanding the quantitative relationships that surround the development of software.

10.5.3 REVIC

REVIC is an extension of COCOMO that has been developed in a software package by Raymond Kile [10.17]. It is structured around the basic COCOMO notions and extends them by considering variables that had not been previously incorporated into the COCOMO model. Such variables include programmer capability, applications experience, programmer language experience, requirements volatility, database size, use of software tools, management reserve for risk, and others. The software requests low, most probable, and high line of code input estimates for each code module. It is claimed that better answers are obtained with smaller modules. For each module, inputs are also requested with respect to modifications of the design and the code. Outputs are displayed regarding effort required (person-months), schedule (months), numbers of people, productivity (lines of code per person-month), and costs. By virtue of the three estimates for the input lines of code, a standard deviation is also computed and shown. REVIC is a very useful software package with a great deal of utility for the software developer.

The Air Force has made REVIC available to the software community and has encouraged additions and expanded applications. Software disks for REVIC can be obtained through the Air Force's Software Technology Support Center (STSC) [10.18]. The STSC also provides a variety of documents that describe the tools that can be used to support the software development process (see also Chapter 12).

10.5.4 Function Points

The use of function points is considered to be an alternative to the COCOMO model to estimate the effort required to develop software. Indeed, its leading proponent, Capers Jones, argues strongly that function point analysis (FPA) is much superior to COCOMO and its derivative methods [10.19].

A formulation of FPA claims that unadjusted function points (UFPs) are related to the numbers of inputs, outputs, master files, interfaces, and inquiries [10.20]. These components are defined in more detail in the literature [10.21].

In particular, the UFP count is a linear combination of these components, with weighting factors. The UFP thus derived is then multiplied by an adjustment factor that changes the UFP by at most ± 35% to develop the final function point count (FP). The adjustment factor is itself a function of some fourteen aspects of software complexity. One model for the estimation of effort, measured in work-hours, is then derived from the final function point count:

$$E = 585.7 + 15.12(FP) \tag{10.7}$$

Thus, we have a method whereby we can estimate effort in work-hours from the function points, which, in turn, are derivable from the user-related estimates of inputs, outputs, files, interfaces and inquiries.

The preceding FPA has some definite attractions, as pointed out in the literature [10.20]. First, it does not explicitly depend on the source instructions or lines of code estimates, which, as pointed out here, can be difficult to estimate accurately. It also does not depend on the language used or other variables such as the use of CASE tools, and so forth. Also, the FPA may be usable at the stage of development when only requirements or design specifications have been formulated. Finally, the FPA tends to be oriented to the user's view of the system, as reflected by the five input components.

FPA has been gaining support and is considered to be a more than viable alternative to COCOMO-type estimation procedures. It is also embodied in existing software tools (e.g., SPQR/20 and ESTIMACS™) that add credibility and ease of use. Here again, only time will tell which of these two methods, if either, becomes the preferred procedure in terms of developer usage and customer acceptance.

10.5.5 Reliability

The issue of reliability in the software arena is still in its formative stages and is likely to be in a state of flux as industry gains more experience with the various ways in which software reliability has been viewed. The procedure described here is based on the theoretical and empirical work done at Bell Labs [10.22]. Their recommended approach is embedded in what is known as the Basic Execution Time Model (BETM).

BETM can be explained by referring to Figure 10.1. The graph plots the software failure intensity (I) against the total number of failures experienced (N) and, in this model, it is described as a straight line. The implication is that the software in question starts out at a given failure intensity I_o and has a total number of defects N_o (the Y and X intercepts, respectively). BETM is a nonhomogeneous Poisson process, such that the failure intensity (I) varies and decreases linearly to zero as failures occur and are "repaired." Also, time is measured in computer program execution time, that is, the time actually spent by a processor in executing program instructions.

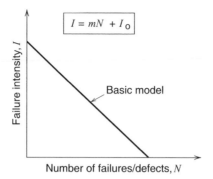

Figure 10.1. Basic execution time model of reliability.

The key issues associated with the use of the BETM are:

a. The initial failure intensity must be estimated.
b. The total number of failures, or defects, in the software must also be estimated.

For both issues, the software developer may not have much experience on which to base these initial estimates. As experience is gained, and empirical data derived on real programs, these estimates are likely to improve.

The equation describing the BETM, as represented in Figure 10.1, is simply

$$I = mN + L_o \qquad (10.8)$$

where
$I =$ failure intensity
$N =$ number of failures or defects
$m =$ slope of failure intensity/failure line
$I_0 =$ the initial estimate of failure intensity

To illustrate how this relationship leads to a probability estimate, assume that $I_0 = 60$ failures per CPU-hour, and the total number of estimated failures (N_0) is 300. These estimates fully describe the BETM line in Figure 10.1. Now let us further assume that we have experienced a total of 200 failures to date. The calculated slope of the line is $-60/300$, or -0.2, and Equation (10.8) becomes

$$I = -0.2N + 60$$

We then calculate the current failure intensity as

$$I = (-0.2)(200) + 60 = -40 + 60 = 20 \text{ failures per CPU-hour}$$

This so-called "instantaneous" failure intensity describes the current situation and can be used as the "failure rate" in a Poisson process, but only at the current time and situation. We can then ask the question: What is the probability that there will be k failures found in the next T hours of execution time? We can use the Poisson probability distribution to calculate this answer as

$$p(k) = \frac{(IT)^k \exp(-IT)}{k!} \tag{10.9}$$

If, for example, we take $k = 0$ and the number of hours $T = 0.1$ CPU-hour, the Poisson reduces to the exponential and becomes

$$P(0) = \exp[-(20)(0.1)] = \exp(-2) = 0.135$$

Thus, in this illustration, the probability that no defects will be found, that is, no failures experienced, in the next 0.1 CPU-hour is 0.135.

The BETM is, of course, not the only formulation of software reliability. Indeed, Bell Labs researchers argue that the logarithmic Poisson execution time model has high predictive validity and is only somewhat more complicated than is the BETM. Several other competitive models are presented and explored in their book [10.22], many of which have also been incorporated into the software package known as SMERFS, as discussed in the next section.

10.5.6 SMERFS

A particularly useful initiative on the part of the Navy led to the development of SMERFS (Statistical Modeling and Estimation of Reliability Functions for Software) [10.23]. SMERFS provides both the background mathematics as well as computer program disks that support the calculation of software reliability. As an example, one module in SMERFS deals with the execution of five software reliability models that yield reliability estimates for execution time data. These models are:

1. The Musa Basic Execution Time Model
2. The Musa Log Poisson Execution Time Model
3. The Littlewood and Verrall Bayesian Model
4. The Geometric Model
5. The NHPP Model for Time-Between-Errors

The first item on this list is the basis for the software reliability formulation in the previous section.

If the user is inclined to adopt an error-count approach to reliability estimation and prediction, the SMERFS formulation is able to support such an approach through the following five models:

1. The Generalized Poisson Model
2. The Nonhomogeneous Poisson Model
3. The Brooks and Motley Model
4. The Schneidewind Model
5. The S-Shaped Reliability Growth Model

Thus, developments such as SMERFS provide a good foundation for the rational consideration of how to deal with the difficult issue of software reliability, which itself is likely to be under debate for some time to come [10.24].

10.5.7 McCabe Metrics

One of the early developers of software metrics was T. McCabe, who wrote a seminal paper that identified a software complexity measure [10.25]. Since that time, the field has blossomed, but the metrics developed by McCabe's company have been plentiful and very well received by industry and government.

The desired characteristics of a complexity measure, according to McCabe, are as follows [10.26]:

- It should be closely related to the amount of work it takes to test or validate a program.
- It should conform to our intuitive notion of complexity.
- It should be straightforward for programmers to calculate the complexity measure.
- It would be useful if the measure could be automated.
- It should be objective; different people should get the same complexity for the same program.
- It should be language-independent.
- It should apply to structured and unstructured code.
- It should map into an operational step and actually drive the test effort.

The centerpiece of McCabe's original idea satisfying these characteristics is the notion of software cyclomatic complexity. Cyclomatic complexity is a measure of the logical complexity of a software module and of the minimum effort required to qualify that module. It is the number of linearly independent paths and, as such, represents the minimum number of paths that one should test.

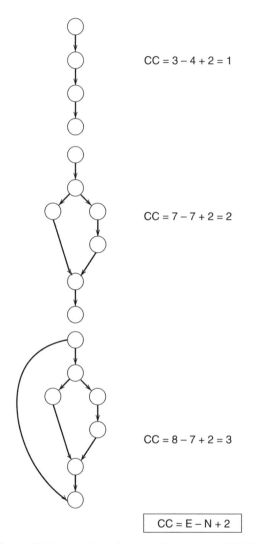

CC = 3 − 4 + 2 = 1

CC = 7 − 7 + 2 = 2

CC = 8 − 7 + 2 = 3

CC = E − N + 2

Figure 10.2. Illustration of cyclomatic complexity [10.26].

To illustrate the cyclomatic complexity metric, we refer to Figure 10.2, which shows three representations of the logical code, or flowgraph, of a module. Cyclomatic complexity can be calculated using the formula:

$$CC = E - N + 2 \qquad (10.10)$$

where CC = cyclomatic complexity
         E = number of edges or connections of the code
         N = number of nodes in the code

Thus, cyclomatic complexity is easy to compute and, in fact, has been automated by McCabe & Associates and is available as a software metrics tool. We note the significance of this metric in terms of identifying testing effort and paths for testing.

McCabe has gone considerably beyond the original complexity measure as described before, producing a variety of metrics as well as software packages that compute these metrics. As a further example, *essential complexity* measures the extent to which a module contains unstructured constructs. The procedure is that all structured constructs are removed, and then the cyclomatic complexity of the reduced flowgraph represents the essential complexity. Essential complexity has the advantages of

- Quantifying the degree of structuredness
- Revealing the quality of the code
- Predicting the maintenance effort
- Helping in the modularization process

In addition to the preceding, McCabe has produced a variety of software metrics that measure various aspects of software, including the following tools:

- McCabe IQ2^{TM}
- CodeBreaker^{TM}
- Design Complexity Tool^{TM}
- The Battlemap Analysis Tool^{TM}
- The McCabe Instrumentation Tool^{TM}
- The McCabe Slice Tool^{TM}

The reader with a further interest in McCabe metrics should contact McCabe & Associates directly (see [10.26] in the References). They are a leading purveyor, both nationally and internationally, of automated tools that compute software metrics.

10.6 THE SYSTEMS ENGINEER AND SOFTWARE ENGINEERING

This section might well be called "What the Systems Engineer Needs to Know about Software Engineering." A default answer to this question, one that takes very little new thought, might be the areas already defined as central to the CMMI (Integrated Capability Maturity Model). That is not a terrible answer since it addresses process areas that have been explored many times by many groups, especially the Software Engineering Institute (SEI) and the Department of Defense (DoD). However, the perspective here is somewhat different. This volume focuses on software engineering and its management, assuming that the reader already qualifies as a Systems Engineer. This implicitly means that the reader understands and practices key process areas within systems engineering. Such a Systems Engineer, or Chief Systems Engineer, has a special need to learn more about software engineering without becoming a software engineer.

In particular, here we define and briefly discuss what might be called a "top ten" list of what the Systems Engineer needs to know about software engineering:

1. Cost and schedule estimation
2. Other software metrics
3. Software architecting and modularity
4. Software development process
5. Satisfaction of requirements
6. Testing
7. Risk assessment and mitigation
8. Software RMA (reliability, maintainability-availability) and Warranties
9. Prototyping
10. Integration of COTS, GOTS, and reuse

Cost and Schedule Estimation. Cost and schedule estimates, which are critical, usually are based on COCOMO (the COnstructive COst MOdel) and/or function point analysis and are absolutely required to verify (or not) that there is sufficient funding and time to get the job done. Discrepancies between estimated and available values are grounds for great concern, especially since we have been missing the mark in so many new system developments.

Other Software Metrics. In addition to cost and schedule measurements, other software-related metrics have become virtually standard practice. These include numbers of defects, extent of reuse, number of delivered source instructions, complexity, and earned value. The reader also is referred to the software management indicators in section 10.5.

Software Architecting and Modularity. Decomposing software into its major functions and subfunctions remains a critical aspect of system and software architecting. Incorrect decomposition has been cited directly as a source of failure as well as of inappropriate work assignments. We remain in search of modules that can be worked on in parallel and with a minimum number of interactions and interfaces.

Software Development Process. Even though we have accepted evolutionary development and acquisition as the best overall process, many questions remain that have to do, for example, with preferred language, sequencing of builds, methods for verification and validation (V&V), and the correct levels of reuse.

Satisfaction of Requirements.   We continue to accept the notion of satisfaction of requirements as a touchstone for systems and software engineering. This is entirely appropriate. However, we also have said that requirements need to be reviewed and questioned in a systematic way so that we are able to make changes when some requirements are not correct.

Testing.   As with hardware, software systems are on track until real-world testing and the appearance of defects reveal that previously unknown failures have put our system distinctly off track. The time and effort needed to carry out unit and subsystem testing tend to be significantly underestimated. The reason: We are optimistic about test results, and such optimism is not sufficiently tempered by the degree of difficulty of the software task.

Risk Assessment and Mitigation.   The reader may recall that risk analysis was *not* one of the key process areas (KPAs) in the original software Capability Maturity Model (CMM) (see section 10.4). This omission was recognized, and the importance of risk measurement and mitigation has been acknowledged. In fact, it is so important that this author places it as one of the "top ten" on this list. Experienced systems and software engineers, starting off a new program, can accurately predict where the key risks are and what might be done to minimize their possible effects. Real-world risk mitigation is worth its weight in gold.

Software RMA and Warranties.   The RMA of software, each year, tends to be recognized as a difficult problem to address. Indeed, there are still many approaches to computing the reliability of software, each with its advantages and disadvantages. Reliability becomes especially important when and if a company is providing software warranties. RMA "models" give us the basis for looking realistically at the possible consequences, including costs, of offering alternative warranties and guarantees. Continuing to go back and "repair" software while it is in the field can be very costly, changing a successful program into an unsuccessful one over time.

Prototyping.   In general, we can tailor the evolutionary acquisition approach through a systematic and well-controlled prototyping program. Certain areas of software can be designated as critical and high risk, giving us the idea that we can develop early prototypes that will serve to reduce the risk. By tackling a difficult problem very early, we increase the likelihood of success, especially with respect to schedules and costs.

Integration of COTS, GOTS, and Reuse.   Many large-scale systems involve the software integration of COTS (commercial-off-the-shelf), GOTS (government-off-the-shelf), and reuse (software components from the contractor world). Each type of software usually originates in a different place and without due regard for the ultimate need for appropriate integration. Although the concepts are relatively simple, accomplishing anything resembling integration can have a degree of difficulty ranging from challenging to almost impossible. Just because software subsystems exist in demonstrable, testable packages, it does not follow that they are therefore easy to integrate.

10.7   SUMMARY

As suggested in this chapter, software development is in a state of flux, with continuing initiatives to try to bring it from largely an art to a well-understood and repeatable engineering process.

Researchers and observers have given us much to contemplate, and new technologies as well as preferred solutions are in abundance. By adding new perspectives, they also add a "confusion of plenty" as we struggle to find the right answers that will improve the way we develop software. As final examples, we now look at some suggestions made by selected observers of the scene, followed by the ten commandments of software engineering.

First, we have the analogy provided by N. Augustine, a leader in government, industry, and academia:

> Software is like entropy, it is difficult to grasp, weighs nothing, and obeys the second law of thermodynamics, i.e., it always increases [10.27].

Second, we have the observations of E. Rechtin, previously the President of the Aerospace Corporation:

> In architecting a new software program, all the serious mistakes are made on the first day.

> A team producing at the fastest rate humanly possible spends half of its time coordinating and interfacing [10.28].

Third, we cite points "Wirth" noting, taken from an article by Niklaus Wirth, the designer of the programming language Pascal:

> The way to streamline software lies in disciplined methodologies and a return to the essentials.

> The most difficult design task ... is the decomposition of the whole into a module hierarchy.

> The belief that complex systems require armies of designers and programmers is wrong [10.29].

Fourth, we are indebted to Frederick Brooks [10.30] for his many insightful contributions to software engineering, a few of which are cited here:

> From this process [i.e., Wirth's top-down methods] one identifies *modules* of solutions or of data whose further refinement can proceed independently of other work.

> A good top-down design avoids bugs in several ways

> Adding manpower to a late software project makes it later (Brooks' Law)

Finally, we complete this chapter with this author's ten commandments for software projects:

Commandment 1. Formulate a software development plan and process that seem to work and stick with them.

Commandment 2. Maintain continuity of personnel and high-performance software development teams (see also Chapter 6).

Commandment 3. Adopt and adapt the notions of the Capability Maturity Model, coupled with a commitment to continuous improvement.

Commandment 4. Always do a detailed risk-assessment and -mitigation analysis at the beginning of a project, and at selected points during the project.

Commandment 5. Begin and maintain a program of software measurement and metrics, to include keeping statistics on all projects.

Commandment 6. Develop a modest but effective software engineering environment (see Chapter 12), including a supportive infrastructure.

Commandment 7. Review the status of every serious software development project at least twice a month and at times weekly.

Commandment 8. Manage the software requirements, including challenging requirements that do not make sense.

Commandment 9. Do not allow documentation to overwhelm and undermine the development of the software or the time of your key software designers and developers.

Commandment 10. Integrate your software engineering efforts and team with your systems engineering methods and team.

QUESTIONS/EXERCISES

10.1 Which software management strategy do you favor? Why?

10.2 Define and discuss three reasons why software development projects run into problems. How would you avoid these problems?

10.3 What software standards appear to be most important today? Explain.

10.4 What is the highest capability-maturity level for software development in your organization? Cite and discuss three new initiatives that are necessary to achieve the next level in capability maturity.

10.5 Calculate the person-months, development time, productivity and full-time equivalent staff required for a software project involving 80,000 delivered source instructions. How does the productivity compare with the example in the text?

10.6 For a software project, it is estimated that the initial failure intensity is 60 failures per CPU-hour and the slope of the Basic Execution Time Model (BETM) line is -0.3. The total number of failures experienced to date is 150.

    a. What is a good estimate of the total number of failures or defects in the software?

    b. What is a good estimate of the current failure intensity?

    c. What is the probability that no defects will be found in the next 0.1 CPU-hour?

10.7 Show an example of how COCOMO II utilizes the five scaling drivers to determine a numeric value of the exponent in the COCOMO II equation.

10.8 For COCOMO II, define the seven and seventeen effort multipliers associated with the early design and post-architecture forms of the model.

10.9 Identify and discuss in three pages two approaches to software reliability that are different from the BETM model examined in some detail in this text.

10.10 Write a three-page discussion of software architecting, as described in the literature.

REFERENCES

10.1 Fromson, B. (1994). "Fidelity Says It Reported Wrong Prices," *Washington Post*, June 23.

10.2 Fox, J. (1982). *Software and Its Development*. Englewood Cliffs, NJ: Prentice Hall.

10.3 Marciniak, J., and J. Reifer (1990). *Software Acquisition Management*. New York: John Wiley.

10.4 Arthur, L. J. (1992). *Rapid Evolutionary Development*. New York: John Wiley.

10.5 *Military Standard—Software Development and Documentation*, Military Standard 498 (1994). Washington, DC: U.S. Department of Defense.

10.6 *Military Standard—Defense System Software Development*, DOD-STD-2167A (1988). Washington, DC: U.S. Department of Defense.

10.7 *Military Standard—Software Quality Evaluation*, DOD-STD-2168 (1979). Washington, DC: U.S. Department of Defense.

10.8 *ISO 9000,* Contact American National Standards Institute (ANSI). New York: International Organization for Standardization.

10.9 *IEEE Standards Products Catalog* (1995). Piscataway, NJ: Institute of Electrical and Electronics Engineers.

10.10 *Standard for Information Technology—Software Life Cycle Processes,* IEEE/EIA 12207.0-1996 (ISO/IEC 12207) (1998). New York: Institute of Electrical and Electronics Engineers, March.

10.11 Boehm, B., ed. (1989). *Software Risk Management,* New York: IEEE Computer Society Press.

10.12 Humphrey, W. (1987). *Characterizing the Software Process: A Maturity Framework,* CMU/SEI-87-TR-ll, DTIC Number ADA 182895. Pittsburgh: Software Engineering Institute.

10.13 Paulk, M. B. Curtis, and M. B. Chrissis (1991). *Capability Maturity Model for Software,* CMU/SEI-91-TR-24. Pittsburgh: Software Engineering Institute.

10.14 Saiedian, H., and R. Kuzara (1995). "SEI Capability Maturity Model's Impact on Contractors," *IEEE Computer Magazine* (January): 16–25.

10.15 Boehm, B. (1981). *Software Engineering Economics.* Englewood Cliffs, NJ: Prentice Hall.

10.16 Boehm, B., et al. (2000), *Software Cost Estimation with COCOMO II.* Upper Saddle River, NJ: Prentice Hall PTR.

10.17 Kile, Raymond L. (1992). REVIC, Version 9.1.1. Contact STSC, Hill AFB, Ogden, Utah.

10.18 Software Technology Support Center (STSC), Hill AFB, Ogden, UT: See also *CrossTalk: The Journal of Defense Software Engineering,* Ogden ALC/TISE, 7278 Fourth Street, Hill AFB, UT 84056-5205.

10.19 Jones, C. (1994). "Software Metrics: Good, Bad and Missing," *IEEE Computer Magazine* (September): 98–100.

10.20 Matson, J., B. Barrett, and J. Mellichamp (1994). "Software Development Cost Estimation Using Function Points," *IEEE Transactions on Software Engineering* **20**(4): 275–286.

10.21 Marciniak, J., ed. (1994). *Encyclopedia of Software Engineering.* New York: John Wiley.

10.22 Musa, J., A. Iannino, and K. Okumoto (1987). *Software Reliability.* New York: McGraw-Hill.

10.23 Farr, W., and O. Smith (1991). *Statistical Modeling and Estimation of Reliability Functions for Software (SMERFS) Users Guide,* NAVSWC TR 84-373, Revision 2. Dahlgren, VA: Naval Surface Warfare Center.

10.24 Lyu, M., F. Buckley, R. Tausworthe, T. Keller, and J. Musa (1995). "Software Reliability: To Use or Not To Use?," *CrossTalk* **8**(2): 20–26. Published by the Software Technology Support Center, Hill AFB, Ogden, UT.

10.25 McCabe, T. (1976). "A Complexity Measure," *IEEE Transactions on Software Engineering* **2**(4): 308–320.

10.26 McCabe & Associates Website: www.mccabe.com.

10.27 Augustine, N. (1982). *Augustine's Laws.* New York: American Institute of Aeronautics and Astronautics.

10.28 Rechtin, E. (1991). *Systems Architecting.* Englewood Cliffs, NJ: Prentice Hall.

10.29 Wirth, N. (1995). "A Plea for Lean Software," *IEEE Computer Magazine* (February): 64–68.

10.30 Brooks, F. P.,(1995). *The Mythical Man-Month.* Reading, MA: Addison Wesley Longman.

___11
SELECTED QUANTITATIVE RELATIONSHIPS

11.1 INTRODUCTION

This chapter has been reserved for a variety of important quantitative relationships so that they can be found in one location in this book. Both systems engineering and project management depend on the mastery of basic quantitative relationships, with an emphasis on the former. That is, it is really not possible to understand the details of systems engineering without mastering these relationships.

Several earlier chapters introduced key quantitative relationships, for example, the chapters containing discussions of requirements analysis, technical performance measurement, simulation, and modeling. Some of these relationships are reiterated here.

A central motivation for examining the relationships set forth in this chapter is to *predict* the performance of systems. In the early stages of system development and engineering, there is no real system that can be tested to determine its performance. Thus, we resort to "pencil and paper" studies that purport to tell us how the designed system will, we hope, perform in the real world. Today, and for the foreseeable future, the pencils and papers have turned into computers. In that sense, we must also master a variety of computer applications that contain the quantitative relationships appropriate to the task at hand.

This chapter is almost exclusively devoted to probability relationships, in distinction to other classes of mathematics (e.g., differential equations, transformation calculus, control systems theory, etc.). The basic reason for this is that many of the technical performance measures of systems are expressed in terms of probabilities. A few examples are:

- The detection and false-alarm probabilities for a radar system
- The response time probability for an on-line transaction-processing system (OLTP)
- The probability of a call going through in a telephone system
- The probability of having a system available to operate when called on to do so
- The probability of successful operation of a system over its lifetime
- The kill probability for a weapons system
- The probability of experiencing some type of disastrous failure that places peoples' lives in jeopardy (e.g., a nuclear plant incident)
- The circular error probability for a guidance system
- The distribution of trip times for a transportation system

As difficult as it might be to calculate these probabilities, we often have little choice because many of the analyses and trade-offs that we must carry out as systems engineers are based on these measures. This chapter contains but a small sampling of the theory available to us in this regard.

11.2 BASIC PROBABILITY RELATIONSHIPS

The centerpiece of probability theory is simply that we wish to express and then calculate the probability of the occurrence of a particular event. Such an event can be as simple as the toss of a die or as complex as the likelihood of a false signal in a city's subway system. Given this notion, we are then interested in considering several events (say, A and/or B), in the context of both events occurring, or possibly one or the other event occurring, as discussed in what follows.

We summarize the elementary probability relationships here:

$$P(A) \geq 0 \text{ for every event} \tag{11.1}$$

$$P(C) = 1 \text{ for the certain event C} \tag{11.2}$$

$$P(A \text{ or } B) = P(A + B) = P(A) + P(B), \text{ if } A \text{ and } B$$
$$\text{are mutually exclusive} \tag{11.3}$$

$$P(A \text{ and } B) = P(AB) = P(A|B)P(B) = P(B|A)P(A) \tag{11.4}$$

$$P(A \text{ and } B) = P(AB) = P(A)P(B)$$
$$\text{for independence between } A \text{ and } B \tag{11.5}$$

$$P(A) = P(A|E)P(B) + P(A|\bar{B})P(\bar{B}) \text{ where } (B + \bar{B})$$
$$\text{constitutes the certain event} \tag{11.6}$$

Example. The probability of throwing a 3 with a single die is simply 1/6.

Example. The probability of throwing a 3 or a 4 with a single die toss is
$P(3 + 4) = P(3) + P(4) = 2/6 = 1/3$.

Example. The probability of throwing a 3 on one toss and a 4 on the next
toss is $P(3 \text{ and } 4) = P(3)P(4) = (1/6)(1/6) = 1/36$.

Although these relationships and examples are simple and highly intuitive, in
systems engineering and project management we may be faced with simple
notions for which it is rather complex to develop an answer. For example, in
the context of project risk, we may ask the question: What is the likelihood
(probability) that instead of completing our project in three months (the
scheduled time), it is completed in four months (a one-month overrun)? If
the answer is close to "1," the project schedule is in great jeopardy and the
penalties for missing the due date may be quite high. It would be therefore
rather important to find another schedule solution that does not lead to this
problem. Not only is an issue of this type very important, developing the
necessary probability estimates is anything but simple, especially for a large
project.

In terms of a system performance requirement, we may ask this question:
What is the probability that our design for an OLTP (on-line transaction-
processing) system will meet the response time requirement of less than or
equal to seven seconds? There are only two events in this question, either
meeting the requirement or not. Here again, the question can be posed in
simple terms, but it is a major task to develop an answer in which we have some
confidence. Indeed, the simulation of our overall computer system design may
be necessary to do so.

11.2.1 Discrete Distributions

The notion of probability distributions is that of describing the entire probability scheme for the question at hand. For the die toss, the overall scheme
is simple: For each toss, there are six possible events (outcomes) and the
probability of occurrence for each of these outcomes is the same, namely,
1/6. Because each event is discrete, the description is called a discrete probability distribution, and may be expressed as $P(X) = 1/6$ for all six possible
events. This is called a uniform distribution because all probabilities are equal.
Examples of other types of discrete distributions are:

- The triangular distribution
- The binomial distribution
- The Poisson distribution

Each distribution is briefly discussed later in this chapter.

11.2.2   Continuous Distributions

For continuous distributions, the variable of interest is considered to be a continuous variable. For example, if we spin a pointer that is fixed at the center, we can assume that the resultant angle from some reference angle is more or less continuous. If we are attempting to hit the center of a bull's-eye, the distance from the center is also a continuous variable. This might apply to a situation in which we are trying to point an instrument aboard a spacecraft toward a certain point in space, for example, a star. Continuous probability distributions are normally expressed by formula in terms *of probability density* functions [$p(x)$] and *cumulative distribution* functions [$F(x)$], the latter being the integral of the former, and yielding direct estimates of probability values. Later in this chapter, we briefly explore the following types of continuous distributions:

- The normal (Gaussian) distribution
- The uniform distribution
- The exponential distribution
- The Rayleigh distribution

In general, we will use the following notations:

- $P(x)$ for a discrete variable and distribution
- $p(x)$ for a continuous *density function* distribution, and
- $F(x)$ for a continuous *cumulative distribution function* (CDF)

11.2.3   Means and Variances

The *mean value* of a distribution may be viewed as the average or expected value of that distribution. It is found by weighting (multiplying) all values of the variable with the values of the probabilities or probability densities, and summing the results. The relationships for the discrete and continuous cases, respectively, can be defined as:

$$\text{Mean value of } X = m(X) = \Sigma X P(X), \text{ for all } X \qquad (11.7)$$
$$\text{Mean value of } x = m(x) = \int x p(x)\, dx \qquad (11.8)$$

The mean value of x is often expressed as the expected value of x, or $E(x)$.

> *Example.* The mean value of the single die tossing situation is $(1)(1/6) + (2)(1/6) + (3)(1/6) + (4)(1/6) + (5)(1/6) + (6)(1/6) = 3.5$. We note that this result leads to a mean value that is different from all possible values of the variable.

Example. The mean value for the distribution generated by tossing two dice is 7, as all of us know intuitively. As an exercise, the reader may wish to prove this by developing the overall (triangular) distribution, where values range from 2 to 12.

The *variance* of a distribution can be found by weighting all values of the probabilities or probability density with the expression $(x - m)^2$. That is, the mean value is subtracted from each value of the variable, the result squared (yielding a positive number), weighted with the probability values, and then summed. The variance is always a positive number, whereas the mean value may be negative. The variance is a measure of the "spread" of the distribution and is extremely important in, for example, carrying out error analyses of systems. In formal terms, the variance, for the discrete and continuous cases, respectively, may be found as:

$$\text{Variance} = \sigma^2 = \Sigma(X - m)^2 P(X), \text{ for all values of } X \quad (11.9)$$
$$\text{Variance} = \sigma^2 = \int (x - m)^2 p(x)\, dx \quad (11.10)$$

Example. The variance for the single die toss is calculated formally as:
$(1 - 3.5)^2(1/6) + (2 - 3.5)^2(1/6) + (3 - 3.5)^2(1/6) + (4 - 3.5)^2\ (1/6)$
$+ (5 - 3.5)^2(1/6) + (6 - 3.5)^2(1/6) = 2.9$

By definition, the square root of the variance is the *standard deviation* or root-mean-square value (σ) of the distribution.

On occasion, we are interested in the ratio of the standard deviation to the mean value, or its reciprocal. This tells us something about the significance of the standard deviation, or variance, in the problem at hand. If the deviation is small compared to the mean value, we may be less concerned about its influence or effect. For example, if we were to calculate the standard deviation in our time schedule for a project as two weeks in a hundred-week project, we are perhaps less concerned than if the overall project were scheduled for eight weeks.

Another situation in which the preceding ratio shows itself is in a communications system. A critical performance measure in such a system is the signal-to-noise ratio *(S/N)*, where S is the signal power and N is the noise power. This is, more or less, equivalent to a ratio of the square of the signal value to the variance of the noise distribution. Thus, we see an immediate application of the notions of mean value and variance to the theory and practice of communications systems.

11.2.4 Sums of Variables

The notion of a sum of random variables has already been introduced here in dealing with the expression:

$$P(A \text{ or } B) = P(A + B) = P(A) + P(B) \text{ if } A \text{ and } B \text{ are}$$
$$\text{mutually exclusive} \quad (11.11)$$

In this same context, if X and Y are random variables, then the distribution of their sum, Z, is often of immediate interest. This may be explored by means of a further look at the matter of tossing a pair of dice, where one variable, X, is the result of tossing one die and the other variable, Y, is the result of tossing the second die. We are thus interested in the distribution of the sum, that is, the results obtained when the faces shown on each die are added.

Clearly, and intuitively, there are eleven possibilities for the sum when two dice are thrown, namely, the discrete numbers 2 through 12. However, as we know, not all of these have the same or equal probabilities, as was the case with a single die. The probability distribution of Z, the sum of X and Y, can be determined to be the following:

$$
\begin{array}{ll}
P(Z = 2) = 1/36 & P(Z = 8) \;\; = 5/36 \\
P(Z = 3) = 1/18 & P(Z = 9) \;\; = 1/9 \\
P(Z = 4) = 1/12 & P(Z = 10) = 1/12 \\
P(Z = 5) = 1/9 & P(Z = 11) = 1/18 \\
P(Z = 6) = 5/36 & P(Z = 12) = 1/36 \\
P(Z = 7) = 1/6 &
\end{array}
$$

If this distribution is plotted, it will form a "triangular" distribution. In formal terms, the distribution of a sum is the convolution of the individual distributions. Convolving two uniform distributions leads to the triangular distribution. In the continuous case, the convolution integral is expressed as

$$
p(z) = p(x + y) = \int g(x)h(z - x)\,dx \tag{11.12}
$$

The next areas of interest with respect to sums of random variables are the values of the means and variances of the sum.

The mean value of the sum of two random variables is

$$
\mathrm{Mean}(Z) = E(Z) = \mathrm{mean}(X + Y) = \mathrm{mean}(X) + \mathrm{mean}(Y) \tag{11.13}
$$

that is, the mean value of a sum is the sum of the mean values.

> *Example.* The mean value of the sum of throwing two dice is $3.5 + 3.5 = 7$. It also can be verified that the same result is obtained by finding the mean value of the sum using the definition of a mean value and the previously cited full distribution for the sum.

In the case of the variance of a sum, it is the sum of the variances only when the individual distributions are independent, that is, when $P(XY) = G(X)H(Y)$. This can be expressed as

$$
\sigma^2(Z) = \sigma^2(X + Y) = \sigma^2(X) + \sigma^2(Y) \text{ when } X \text{ and } Y \text{ are independent} \tag{11.14}
$$

The situation in which the individual variables are not independent is dealt with later in this chapter.

Adding variances for independent random variables is critically important in error analyses of systems, as alluded to in Chapter 8 as well as later in this chapter. In these analyses, it is often assumed that one has a set of independent and additive random variables and that the overall error is therefore the sum of the individual errors, with all errors expressed as variances. If there are coefficients that relate these additive variables together, then the appropriate relationships are as follows: If $z = c_1 x + c_2 y$, where c_1 and c_2 are coefficients, then for additive independent variables,

$$\sigma^2(z) = c_1^2 \sigma^2(x) + c_2^2 \sigma^2(y) \tag{11.15}$$

and similarly for a set of coefficients c_1 through c_n.

The systems engineer also must be aware of the association between errors and variances as, for example, one or two sigma values. At times, errors are taken at the one-sigma value, and at other times, the two-sigma value. This choice depends on the degree to which it is important to keep errors within certain bounds. This issue is reexamined in the sections dealing with the normal (Gaussian) distribution and error analyses.

11.2.5 Functions of Random Variables

Functions of random variables arise when we make a functional transformation and then wish to examine the resultant distributions. As an example, if we transform a random variable X into a random variable Y by means of the transformation $Y = g(X)$, where $g(\ )$ is some arbitrary function, and we know the distribution of X, we may then inquire into the distribution of Y.

To make this notion more concrete, consider the transformation of random variables expressed as

$$Y = mX + b \tag{11.16}$$

where X and Y are random variables and m and b are constants. This simple transformation converts the distribution of X into some other distribution of Y. Because this particular function is linear, we would expect a preservation in the form of the distribution. Another example is the so-called square-law detector device that modifies the input variable X into an output variable Y by means of the relationship

$$Y = X^2 \tag{11.17}$$

Just as X is processed, so is the distribution of X through this particular square-law transformation. In this case, we would expect the form of the distribution to be modified because this is a nonlinear relationship.

From a systems engineering perspective, one of the values of understanding the transformation of random variables is to track how such a variable behaves as it is processed through some type of system. It has special relevance, as well, to transformations from one coordinate system to another, for example, going from rectangular to polar coordinates. This issue is discussed in somewhat greater detail in a later section.

11.2.6 Two Variables

Often, we are concerned with the *joint* behavior of two random variables. We may describe such behavior in terms of a *joint distribution* (density function) such as $p(x, y)$. As suggested before, if such a distribution is partitionable, as

$$p(x, y) = g(x)h(y) \tag{11.18}$$

then the two variables x and y are independent.

> *Example.* We may develop a simple joint distribution when tossing two dice, where each die represents a random-variable generator. The probability of obtaining any given pair of numbers, say, 3 and 5 [a 3 on the "X" die and a 5 on the "Y" die is $P(X, Y) = P(3, 5)$]. For any given pair, of which there are 36 possibilities, the probability is clearly 1/36. Therefore, this is a uniform discrete joint distribution that has only one value, namely, $P(X, Y) = 1/36$.

There is also a mean or expected-value concept when dealing with a joint distribution. This may be expressed as

$$E(xy) = \int \int xyp(x, y)\, dx\, dy \tag{11.19}$$

As might be expected, if x and y are independent, the preceding yields the product of the expected, or mean, values of the individual distributions.

11.2.7 Correlation

If two random variables bear some relationship to one another, they may be correlated. This notion has a specific meaning in probability theory. First, we define a term known as the *covariance* of two distributions, given as

$$\text{Cov}(xy) = E\{[x - E(x)][y - E(y)]\} = E(xy) - E(x)E(y) \tag{11.20}$$

Clearly, if the two variables are independent, then $E(xy) = E(x)E(y)$ and the covariance reduces to zero.

For correlated random variables, we may also inquire into the effect on the variance of the sum of such variables, that is, var$(x + y)$. In such a case, the variance has a covariance term, as

$$\text{var}(x + y) = \text{var}(x) + \text{var}(y) + 2\,\text{Cov}(xy) \tag{11.21}$$

A formal correlation coefficient may also be calculated from the definition:

$$\text{Correlation coefficient} = \frac{\text{Cov}(xy)}{\sigma(x)\sigma(y)} \tag{11.22}$$

By dividing by the product of the individual standard deviations, the correlation coefficient is normalized to values between -1 and $+1$.

11.3 THE BINOMIAL DISTRIBUTION

The specific discrete distribution known as the binomial distribution arises when there are repeated independent trials with only two resultant possibilities. If we call the probability of success p and the probability of failure q, where $(p + q = 1)$, then the distribution becomes

$$\begin{aligned} P(x) &= \binom{n}{x} p^x q^{n-x}, \quad \text{for } x = 0, 1, 2, 3, \ldots, n \\ &= 0, \qquad\qquad\qquad \text{otherwise} \end{aligned} \tag{11.23}$$

This distribution defines the probability of exactly x successes in n independent trials.

Example. We often assume that the bits of data that are transmitted in a digital communication system are independent from one another and that there is a bit error rate (BER) of some value, for example, 10^{-8}. For purposes of illustrating the binomial, we assume a character that is 8 bits long and that the BER is 10^{-3}. If receiving an individual bit without error is defined as success, then the probability of "success" is 0.999 and the probability of "failure" is 0.001. The probability that we will have eight successes (no errors in the transmission of a character) is then

$$P(8) = \binom{8}{8} (0.999)^8 (0.001)^0 = (0.999)^8 = 0.992$$

In a similar fashion, one can then calculate the probabilities of no error in a set of bits (a byte), characters (a word), a series of words (a message), and so forth.

Example. If, when throwing a die, an odd number is success and an even number failure, the probability of exactly four successes in ten trials is

$$P(4) = \binom{10}{4}(0.5)^4(0.5)^6 = (210)(0.0625)(0.015625) = 0.205$$

The mean value of the binomial is equal to np, which is the expected number of successes in n trials. This is in consonance with our intuition as we, for example, would expect to have 40 successes in 100 trials if the probability of success on each trial were 0.4.

11.4 THE POISSON DISTRIBUTION

The Poisson distribution was introduced briefly in Chapter 10, Section 10.5.5, which dealt with the issue of software reliability. The Poisson is a discrete distribution given by the following formula:

$$P(k) = \frac{(\lambda t)^k \exp(-\lambda t)}{k!} \quad \text{for } k = 0, 1, 2, \ldots, n \qquad (11.24)$$

where $P(k)$ is the probability of exactly k events of interest, λ is the rate at which such events are occurring, and t is the time (or space) over which the events are occurring. This distribution may be used in situations for which events happen at some rate and we wish to ascertain the probability of some number of events occurring in a total period of time, or in a certain space.

Example. Cars are passing a toll booth at an overall rate of about 120 cars per hour. The probability that exactly three cars will pass through the toll booth in a period of 1 minute would be

$$P(3) = \frac{[(2)(1)]^3 \exp[-(2)(1)]}{3!} = 0.18$$

The Poisson distribution, when $k = 0$, reduces to $P(0) = \exp(-\lambda t)$, which is the exponential distribution. For example, if failures were occurring in a system at a rate of λ, and we were concerned with the likelihood of having no failures in some period of time, t, we would use the exponential to make this calculation. The subject of reliability is examined again later in this chapter.

11.5 THE NORMAL (GAUSSIAN) DISTRIBUTION

The normal distribution, sometimes also called the Gaussian distribution, is a continuous distribution that is very commonly used. Its shape is as shown in Figure 11.1(a), the familiar bell-shaped curve.

One common formula for the density function for the normal distribution, shown in Figure 11.1, is

$$P(x) = \frac{1}{\sigma\sqrt{2\pi}} \exp\left(\frac{-x^2}{2\sigma^2}\right) \qquad (11.25)$$

The normal distribution (in this case) is symmetric about $x = 0$ and has the standard deviation σ as a parameter.

If we wish to calculate probabilities, however, it is necessary to integrate a continuous density function. Formally, the cumulative distribution function (CDF) is found, for the continuous case, as

$$F(x) = \int_{-\infty}^{x} p(x)\,dx \qquad (11.26)$$

$F(x)$ calculates the probability that random variable x is equal to *or less than* some particular value. For example, in Figure 11.1, we would calculate the probability that the random variable is less than or equal to zero as 0.5 by integrating from minus infinity to zero. The CDF for the normal distribution with zero mean value is also plotted in Figure 11.1(b), and its ordinate values correspond directly to probabilities.

The normal distribution is not readily integrable, so that we resort to table lookups in order to calculate other than the simplest probability cases. Further, these tables have been developed for the "standard normal," which has a mean value of zero and a standard deviation of unity.

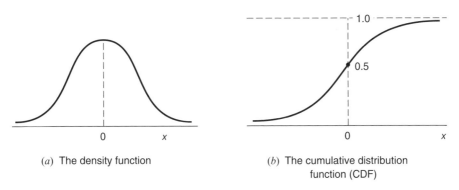

(a) The density function

(b) The cumulative distribution function (CDF)

Figure 11.1. The normal (Gaussian) distributing.

Example. For the normal distribution shown in Figure 11.1, the probabilities that the random variable is equal to or less than one and two sigma (σ) to the right of the mean are:

For one sigma, the value is $0.5 + 0.3413 = 0.8413$

For two sigma, the value is $0.5 + 0.4772 = 0.9772$

The values 0.3413 and 0.4772 were obtained by referring to a standard normal table for the argument equal to one and two, respectively. The probabilities that the variable lies within plus-and-minus one and two sigma from the mean are simply 0.6827 and 0.9545, respectively. Thus, about 95% of the distribution lies within plus-and-minus two sigma from the mean.

Example. We assume that the critical path of a simple PERT network consists of five independent sequential activities, with the following estimates of optimistic, most likely, and pessimistic times in weeks:

Activity	Optimistic Time	Most Likely Time	Pessimistic Time
A	1	2	3
B	2	4	6
C	4	6	8
D	1	3	5
E	4	6	8

The questions regarding this example are as follows:

1. What is the expected time for the project?
2. What is the standard deviation in the project's end date?
3. What is the likelihood that the project will exceed 23 weeks?

Using the formulas in Section 4.2 of Chapter 4, we calculate the expected times and variances for each of the preceding activities as follows:

Activity	Expected Time	Variance
A	2	1/9
B	4	4/9
C	6	4/9
D	3	4/9
E	6	4/9

The expected time for the project is the sum of the preceding expected times because the path described is the critical path. Its value is 21 weeks. The variance of the end date is the sum of the variances of each activity, as shown before, and is equal to 17/9, or 1.89. The standard deviation is its square root, or 1.37 weeks. If we now *assume* that the end-date distribution is normal, we can calculate the probability of the project exceeding 23 weeks, using the normal table and the fact that the mean (expected) value is 21 weeks. The time period between 23 and 21 weeks is 2 weeks, representing 1.46 sigma from the mean. From a table lookup at the 1.46 point, we obtain the area from the mean as 0.4279. The area to the right of that is therefore $0.5 - 0.4279$, or 0.0721. This corresponds to the probability that the project end date will exceed 23 weeks.

In the preceding example, it is demonstrated that the normal need not have a mean value of zero. If the distribution is shifted to the right by the value m, then it has a mean value equal to m and m is subtracted from the value of x in Equation (11.25). The mean value m and the standard deviation σ can be independently selected because it is a two-parameter distribution. A small standard deviation "narrows" the distribution and a large standard deviation broadens it.

The normal distribution is often used in analyzing communications systems to represent the noise distribution. As alluded to earlier, the noise power (N) is equivalent to the variance of the normal distribution, that is, $N = \sigma^2$. The normal distribution will be examined again in this chapter in the discussion of detecting a signal in noise.

11.6 THE UNIFORM DISTRIBUTION

As the name implies, the uniform distribution is "flat" over its entire range, as shown in Figure 11.2. The discrete case of the uniform distribution was discussed earlier in this chapter with respect to the toss of a die where all the probabilities are equal.

The uniform distribution shows up in the phase relationship when we are converting two independent normal distributions with the same variance from

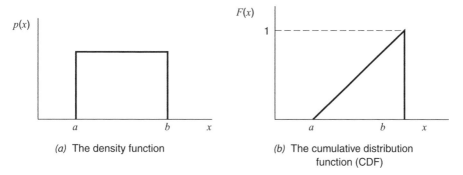

(a) The density function   (b) The cumulative distribution
                                      function (CDF)

Figure 11.2. The uniform distribution.

rectangular to polar coordinates. Such a case, applied, for example, to shooting at a bull's-eye, means that the shot is equally likely to be at any angle over the range zero to 2π radians.

The mean value for the uniform distribution that has a range from a to b is $(a + b)/2$ and its variance is $(b - a)^2/12$. We also have seen that adding two uniform random variables leads to a triangular distribution.

11.7   THE EXPONENTIAL DISTRIBUTION

The exponential distribution is illustrated in Figure 11.3 and has the following density function:

$$
\begin{aligned}
p(x) &= \lambda \, \exp(-\lambda x), &&\text{for } x \geq 0 \\
&= 0, &&\text{for } x < 0
\end{aligned}
\tag{11.27}
$$

Also shown in the figure is the CDF, which starts at zero and approaches the value of unity asymptotically.

We have seen the exponential distribution as derivable from the Poisson when $k = 0$. This distribution is widely used in reliability theory wherein the value of λ is taken to be a constant failure rate for a part of a system (e.g., a component). In such a case, variable x is converted into a time variable, t, and the CDF is found by integrating the preceding relationship, yielding

$$
F(t) = 1 - \exp(-\lambda t), \quad \text{for } t \geq 0
\tag{11.28}
$$

Because this is a failure distribution (i.e., represents the failure behavior), it can be converted into a reliability formula as

$$
R(t) = 1 - F(t) = \exp(-\lambda t), \quad \text{for } t \geq 0
\tag{11.29}
$$

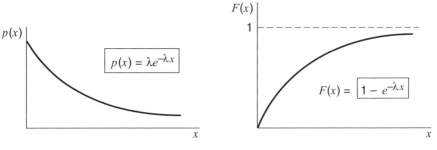

(a) The density function   (b) The cumulative distribution function

Figure 11.3. The exponential distribution.

where $R(t)$ is the probability of successful (failure-free) operation to time t. This is the very familiar expression of the reliability of a system, or component, with a constant failure rate. As indicated in Section 8.7.2 of Chapter 8, the failure rate and the mean time between failures (MTBF) are reciprocals of one another. The MTBF, or $1/\lambda$, is the mean value of this distribution because it represents the mean time to failure.

11.8 THE RAYLEIGH DISTRIBUTION

The Rayleigh distribution density function has the following form:

$$p(x) = \frac{x}{\sigma^2} \exp\left(\frac{-x^2}{2\sigma^2}\right), \quad \text{for } x \geq 0 \qquad (11.30)$$
$$= 0, \quad \text{for } x < 0$$

This distribution shows up in the aforementioned case when converting two independent normal distributions with the same variance from rectangular to polar coordinates. This a very common situation because it is the case of looking for the radius distance from the origin in a polar coordinate system. Just as the angle distribution in polar coordinates are uniformly distributed (see Section 11.6), the radial distance from the origin is Rayleigh-distributed. Thus, for the bull's-eye example, the vector from the origin obeys the Rayleigh law under the circumstances described before.

Unlike the normal distribution, the Rayleigh can be integrated in a straightforward manner. Thus, the CDF for the Rayleigh becomes

$$F(x) = 1 - \exp\left(\frac{-x^2}{2\sigma^2}\right), \quad \text{for } x \geq 0 \qquad (11.31)$$

and represents, as usual, the probability that the variable is less than or equal to a particular value of x.

We can look at some of the numbers associated with the Rayleigh distribution. At the value of the radius equal to sigma, equation 11.31 can be reduced to:

$$F(x) = 1 - \exp(-\sigma^2/2\sigma^2) = 1 - \exp(-.5) = 1 - .607 = .393$$

This means that the probability that the random variable corresponds to one sigma or less is equal to .393. If we are interested in two or three sigma, this value increases to .865 and .989, respectively. Recall that for the normal distribution, the plus and minus one-, two-, and three-sigma values are .683, .954, and .997.

The angle is uniform over the interval from zero to 360 degrees. For example, the probability that the angle lies in the interval from 270 to 300 degrees can be found as:

$$(300 - 270)/360 = 30/360 = 1/12 = 0.833$$

Given these conditions, using the Rayleigh and the uniform distributions, we are able to calculate the approximate vector and its angle corresponding to a particular latitude and longitude position on a surface. This procedure is obviously helpful for position determination and location, as in a search and rescue operation.

11.9 ERROR ANALYSES

The purpose of error analyses is to identify all critical sources of error and be in a position to control the magnitudes of such errors in all cases for which they may significantly detract from system performance. An error analysis example was set forth in Chapter 8 with respect to pointing at a target in a shipboard environment.

The sequence of steps in an error analysis is as follows:

1. Identify all significant error sources.
2. Develop a computational "model" that relates the errors to one another.
3. Estimate the magnitudes of the significant errors.
4. Allocate error budgets, where necessary.
5. Continue to estimate, predict, and control errors throughout the project.

The ultimate benefit of a formal error analysis is to assure that the system meets all requirements. The greatest leverage is obtained when this is accomplished prior to the actual building of the system. In that way, backtracking and reengineering are avoided, together with the penalties in cost and schedule that are usually involved.

Errors are often associated with the standard deviation of some error distribution. Unless otherwise specified, the systems engineer must decide how to relate the error requirement (e.g., pointing error) to the error distribution. A "5% solution" is often adopted, which means that the error corresponds to the two-sigma value. In more concrete terms, if the pointing error requirement was stated in terms of, for example, 1 degree, then this requirement would be associated with the two-sigma value of the error distribution. The standard deviation of that distribution would then be limited to at most 0.5 degree. If the design were to be even more rigorous, three- and four-sigma values might be used, but these choices impact the design and might be difficult to achieve.

A brief example illustrates some of the issues involved in the preceding. Let us assume that an on-line transaction processor (OLTP) requirement is stated as "99% of the time, the system must respond to a request for service in less than or equal to seven seconds." After some analysis, it is concluded that:

- The error distribution in response time may be well described by the normal distribution.
- Analysis shows that the average response time has been calculated to be four seconds.

From a normal probability table, the value 0.4900 (yielding a 0.99 probability) corresponds to 2.33 sigma from the mean. This constrains the "distance" from seven to four seconds to be equal to 2.33 sigma, that is, 2.33 sigma is set equal to $7 - 4 = 3$ seconds. From this data, we conclude that the value of sigma should be no greater than $3/2.33 = 1.29$ seconds. This establishes an "error budget" for the further design of the system.

The same problem can be viewed somewhat differently if we keep the seven-second requirement but the 99% is not part of the stated requirement. In such a situation, we may conclude that we wish the range from the mean of four seconds to the constraint of seven seconds to correspond to the two-sigma value of the distribution. That implies that $2\sigma = 3$ seconds and $\sigma = 1.5$ seconds. Two sigma from the mean, from the normal table, yields the value 0.4772. When this is added to the left half of the distribution, we obtain the probability of $0.5 + 0.4772 = 0.9772$. Our conclusion is that having a sigma of 1.5 leads to a probability of 0.9772, somewhat smaller than the previous case described earlier. In other words, if the error budget were not exceeded, we would satisfy the required system response time 97.7% of the time.

> *Example.* If $Z = 2X + 3F$, where X and Y are independent error variables, and $m(X) = 6$, $\sigma(X) = 4$, $m(Y) = 5$, and $\sigma(Y) = 7$, find the mean value of the random variable Z and the allowable error variance of Z. The mean value of Z is found simply as $(2)(6) + (3)(5) = 27$. The allowable variance of Z is equal to $(2^2)(16) + (3^2)(49) = 505$.

11.10 RADAR SIGNAL DETECTION

11.10.1 Detection and False Alarm Probabilities

In Chapter 9, Section 9.7.2, a radar-detection trade-off example was introduced. In this example, we had a signal plus additive Gaussian noise. The signal voltage was equal to V and the noise power (variance) equal to N, with threshold detection at a voltage value of T. Of interest were the calculations of the detection probability, $P(d)$, and the false-alarm probability, $P(fa)$. With the information in this chapter, we are now in a position to carry out a simple quantitative analysis.

Specifically, we calculate the detection and false-alarm probabilities when the noise and threshold are kept constant and the signal level is increased. Specific assumed values are:

Case 1: $V = 4$ volts, $N = 4$ watts, $T = 1$ volt
Case 2: $V = 8$ volts, $N = 4$ watts, $T = 1$ volt
Case 3: $V = 10$ volts, $N = 4$ watts, $T = 1$ volt

In all cases, the noise power remains the same, so that because $\sigma^2 = N$, the noise standard deviation is equal to 2 volts.

For Case 1, the range from the mean value of 4 volts to the threshold of 1 volt is 3 volts. Therefore, the threshold is at 1.5 sigma from the mean. At a value equal to 1.5, from the normal probability table, the corresponding area under the right-hand portion of the normal is 0.4332. When added to 0.5, a detection probability of $P(d) = 0.9332$ is obtained. For noise alone (no signal present), we may calculate the false-alarm probability by recognizing that the mean value of zero is only 1 volt from the threshold. That is, the threshold is only $1/2 = 0.5$ sigma from the mean. The corresponding table lookup value is 0.1915. However, in this situation, this must be subtracted from 0.5 to yield the correct answer of 0.3085.

In Case 2, the signal is increased to 8 volts with the other parameters remaining the same. This means that the distance from the threshold is $8 - 1 = 7$ volts, which represents 3.5 sigmas from the mean. At a value of 3.5, the normal table lookup gives us a value of approximately 0.49975. When added to 0.5, the detection probability for this case becomes 0.99975. Because neither the threshold nor the noise variance was changed, the false-alarm probability remains the same as in Case 1.

For Case 3, the signal is further increased to 10 volts. The distance from the threshold is now $10 - 1 = 9$ volts, which in this example is 4.5 sigmas. The table lookup results in a value of 0.499997, which when added to 0.5 yields a detection probability of 0.999997. As with Case 2, the false-alarm probability remains the same as in Case 1.

Although these values have been chosen so as to be amenable to lookup in the normal probability integral table, and are not necessarily realistic for a radar system, they illustrate the way in which one might make detection and false-alarm probability calculations, using the material presented in this chapter.

Example. In a pulse-signal-detection situation, (a) if the noise power is 9 watts, and the false-alarm probability is 0.0099, where is the threshold? In this case, $\sigma = 3$ and the area under the normal distribution is $0.5 - 0.0099 = 0.4901$. This yields a value from the normal table of 2.33. The threshold is therefore at 2.33 $\sigma = (2.33)(3) = 6.99$ volts. For part (b), what is the root-mean-square (rms) signal-to-noise ratio to achieve a detection probability of 0.9772? For $0.9772 - 0.5 = 0.4772$, the normal table yields a value of 2, in which case $2\sigma = (2)(3) = 6$. This value, when added to 6.99, results in an rms signal of approximately 13 volts. Thus, the rms signal-to-noise ratio is approximately $13/3 = 4.33$.

These examples illustrate the way in which signals might be detected in the presence of noise. The signal pulse is corrupted by additive Gaussian noise during transmission, so that at the receiver, both signal and noise are present. The threshold detection scheme simply compares the signal-plus-noise value to the threshold and decides that a signal is present when the threshold is exceeded.

In the section in this chapter dealing with the binomial distribution, we calculated the probability of no errors in a sequence of eight pulses that represented an alphanumeric character. We now can see how this section and that previous section fit together in that we are now in a position to calculate the error probability for a single pulse. Although there is much more to be explored in this regard, such as errors of both types, we hope that the reader can see more concretely the value in the preceding Gaussian error model and, as well, the binomial computation from the previous section.

11.10.2 Another Threshold Concept

Radar detection trade-off examples in chapter nine show how a threshold can be determined, under the given decision rule, if both the noise and the false alarm probability, $P(fa)$, are specified. A somewhat more theoretical approach to establishing a threshold is set forth here, based on the notion of a *likelihood ratio*. This ratio is used in hypothesis testing and also can be developed from a general *risk* formulation, as below.

We start with the risks of making decision selections S_1 and S_2 which will be defined as [11.4]:

$$\text{Risk}(S_1) = C_{11}p(x_1/y_j) + C_{21}p(x_2/y_j) \qquad (11.32)$$
$$\text{Risk}(S_2) = C_{12}p(x_1/y_j) + C_{22}p(x_2/y_j) \qquad (11.33)$$

where $\quad C_{ij} = \text{costs}$
$p(x_i/y_j) = \text{probability that } x_i \text{ was transmitted, given that } y_j \text{ was received (in a generalized channel)}$

We then make Selection S_1 if the:

$$\text{Likelihood Ratio} \quad L = \frac{p(y_j/x_1)}{P(y_j/x_2)} > \frac{p(x_2)}{p(x_1)} \frac{C_{21} - C_{22}}{C_{12} - C_{11}} \qquad (11.34)$$

We can interpret the two values of x_i as:

$x_1 = $ signal plus noise
$x_2 = $ noise alone

and also interpret the costs as:

$C_{11} = C_{22} = 0$
$C_{12} = C_n = $ cost of a failure to detect a signal when it is present
$C_{21} = C_f = $ cost of a false alarm

So, in distinction to the previous formulation, this risk analysis brings costs into the consideration when decisions are made.

The likelihood ratio, L, becomes:

$$L = \frac{p(y_j/x_1)}{p(y_j/x_2)} > \frac{p(x_2)C_f}{p(x_1)C_n} = B, \quad \text{for selection } S_1 \tag{11.35}$$

We continue to assume that the signal plus noise distribution is additive Gaussian noise of mean m and standard deviation σ, whereas the noise only distribution is the same, except for the mean value, which is zero. The likelihood ratio then becomes:

$$L = \exp[(2y - m)/2\sigma^2] \tag{11.36}$$

And the decision boundary can be found as:

$$\ln L = \frac{(2y - m)}{2\sigma^2} \tag{11.37}$$

We can then solve for values of y as

$$y = \frac{m}{2} + \sigma^2 \ln B \tag{11.38}$$

which basically represents the threshold that will be selected.

This can be illustrated by numerical values, assuming the following:

$p(x_1) = 0.6$
$C_n = 3$
$C_f = 5$
$m = 14$ volts
$\sigma = 3$

The value of B then becomes, from equation 11.35:

$$B = \frac{p(x_2)C_f}{p(x_1)C_n} = \frac{0.4\ 5}{0.6\ 3} = 2/1.8 = 1.11$$

From this value of B we are able to find the threshold as:

$$y = \frac{14}{2} + (3)^2 \ln 1.11$$
$$y = 7 + .948 = 7.948 \text{ volts}$$

This equation may be interpreted as:

Decide that a signal was sent if the received signal is greater than approximately 7.95. If less than that value, decide that there was no signal, only noise.

This formulation shows that at least one alternative approach is available from the rich field of decision theory applied to detecting signals in the presence of noise.

11.11 SYSTEM RELIABILITY

The discussion in Section 11.7 developed the exponential distribution as most directly applicable to the matter of component and system reliability. In this section, we further elaborate on a few reliability calculations.

We define the hazard function for a system or component as

$$h(t) = \frac{f(t)}{1 - F(t)} \tag{11.39}$$

If the distribution in question is exponential, then

$$h(t) = \frac{\lambda \exp(-\lambda t)}{\exp(-\lambda t)} = \lambda \tag{11.40}$$

which is therefore a constant hazard or failure rate. Under these conditions, the system or component has no "memory" and, technically, does not wear out. For most types of purely electronic equipment, this is a good approximation. For mechanical equipment, this is usually a poor model and should not be used. Instead, the Weibull distribution is a better choice.

For the exponential, then, the probability of failure-free operation to time t is

$$R(t) = \exp(-\lambda t) \tag{11.41}$$

where $R(t)$ is the reliability and λ is the constant failure rate, which, in turn, is equal to 1/MTBF.

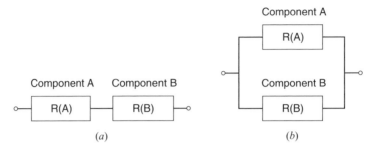

Figure 11.4. Two reliability configurations.

11.11.1 Components in Series

If two components (that comprise a system) are placed in a "series" reliability configuration, as in Figure 11.4(a), it means that *both* must be operative for the system to be working properly. The reliability of the system is therefore

$$R_s = R(A)R(B) = \exp(-\lambda_a t)\exp(-\lambda_b t)$$
$$= \exp[-(\lambda_a + \lambda_b)t] \tag{11.42}$$

This is the basis for the simple addition of failure rates for components when considering the reliability of a system.

Because the MTBFs and failure rates are reciprocals of one another,

$$\lambda_s = \lambda_a + \lambda_b \tag{11.43}$$

$$\frac{1}{\text{MTBF}_s} = \frac{1}{\text{MTBF}_a} + \frac{1}{\text{MTBF}_b} \tag{11.44}$$

as described in Chapter 8, Section 8.7.2. Extension to many components in a series reliability configuration is immediate.

> *Example.* The probability that the system described in Figure 8.3 will survive without failure for 500 hours can be computed by adding the given failure rates of 0.0004, 0.0005, 0.0006, and 0.0005, yielding a system failure rate of 0.002. The reliability of the system is therefore exp $(-.002t)$, where, in this case, $t = 500$ hours. This then reduces simply to $R = \exp(-1) = 0.368$. We note that this is the result for any simple exponential system in terms of failure-free operation to its MTBF.

11.11.2 Components in Parallel

A parallel reliability configuration, as in Figure 11.4(b), means that *at least one* of the components must be operative in order for the system to be working. For two components in parallel, the system reliability therefore can be expressed as

$$R_s = 1 - [1 - R(A)][1 - R(B)] \tag{11.45}$$
$$= 1 - [1 - \exp(-\lambda_a t)][1 - \exp(-\lambda_b t)] \tag{11.46}$$

and the failure rates are not additive for such a system. The parallel configuration introduces redundancy, and thus improves the reliability of the system, with the penalty being the addition of the redundant component. This is necessary when it is extremely important to keep a system on the air, such as with a manned spacecraft or an air traffic control system.

Example. If we take the system in the previous example and place it in a redundant configuration, the reliability then becomes

$$R_s = 1 - (1 - 0.368)(1 - 0.368) = 1 - (0.632)^2 = 0.6$$

Thus, the reliability has improved from 0.368 to 0.6 by adding simple redundancy, for which one pays the price of duplicating this piece of equipment.

Example. What is the probability of successful operation for 100 hours for a system with two subsystems with MTBFs of 200 and 300 hours when (a) the two subsystems are in "series," and (b) the two subsystems are in "parallel"? For part (a), $R_s = \exp(-100/200) = 0.6065$ and $R_2 = \exp(-100/300) = 0.7168$. The product of these represents the series case, which yields the result 0.434. In part (b), we have an overall reliability of $1 - (1 - 0.6065)(1 - 0.7168) = 1 - (0.3935)(0.2852) = 0.8885$.

11.11.3 Non-Constant Failure Rates

Earlier we saw that the hazard function for the exponential distribution turned out to be a constant equal to lambda (λ). This constant failure rate is the basis for a considerable part of the theory of reliability, but not all of it. We must take account of the fact that the hazard is not a constant in many situations. Indeed, when wear-out is experienced, we need to look at a more complex hazard.

An example is one in which the hazard function is:

$$h(t) = \alpha \lambda t^{\alpha - 1} \tag{11.47}$$

where both α and λ are greater than zero.

For this situation, the density function turns out to be the Weibull distribution, which can be expressed as:

$$f(t) = \alpha \lambda t^{\alpha - 1} \exp(-\lambda t^{\alpha}) \tag{11.48}$$

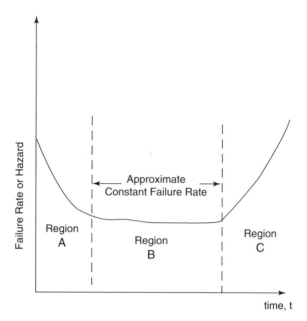

Figure 11.5. Changing Failure Rates or Hazards

Lambda and alpha are the two parameters of this distribution that makes it considerably more complex than the exponential. However, when alpha becomes unity, it reduces to the exponential case. The mean value and variance for the exponential are $1/\lambda$ and $1/\lambda^2$, respectively, but the corresponding values for the Weibull are more complicated and relate to the Gamma function. Nevertheless, the Weibull distribution function is used extensively when the failure rate is not a constant.

The situation is shown, in part, by the sketch in Figure 11.5. This figure shows a nonconstant hazard function, or failure rate, with three distinct regions. One region (B) shows the failure rate as approximately constant. To the left of that we see a region (A) in which the failure rate is decreasing from some initial value. This may be considered the burn-in or bake-out period, also at times called the infant mortality period. During this period, components are considered to not be "stabilized," and we try to screen them out of populations of components. Success in doing so moves these components from the nonconstant region (A) to the constant region (B). After some time, we enter region C in which the failure rate (or hazard function) begins to increase. The component is ultimately wearing out, although this may take quite a long period of time. The overall shape of this curve suggests its name, well known as the "bathtub" curve.

The systems engineer must be aware of the possibility that various parts of the system may be subject to wear-out. Analyses of these parts from a reliability point of view may recognize the wear-out phenomenon by using the Weibull distribution. Although this distribution has its complications, we need to make sure that we are not making invalid assumptions about failure characteristics of the systems we are analyzing and building. The tools are available; we just need to use them. Many books on reliability expand on and explain the way in which the Weibull distribution is used.

11.12 SOFTWARE RELIABILITY

Another application of the probability relationships discussed in this chapter deals with software reliability. As indicated in Chapter 10, there are several models for calculating the reliability of software. The one selected for explanation in the last chapter was the so-called basic execution time model (BETM). In that model, the failure intensity (I) was decreasing linearly with the increasing number of failures/defects. Because failure intensity and failure rate are basically the same notion, we did not have the simple constant failure-rate situation. A new failure intensity had to be found as the number of failures/defects increased and were discovered. Knowing the new failure intensity then allowed us to utilize the Poisson distribution to calculate the reliability. For the case in which we were inquiring into the probability of exactly zero failures in time t, the Poisson reduced to the exponential case, as shown in the discussion in Section 10.5.5.

As suggested in the previous chapter, many other software reliability approaches depend heavily on probability relationships. For that essential reason, it is important, especially for the software systems engineer, to master the elements of probability theory.

11.13 AVAILABILITY

The availability of a system, as previously discussed in Sections 7.3.17 and 8.6.4, is the probability that a system will operate when called on at random to do so. A mean-value approach to availability (see Chapter 8) defines it as

$$A = \frac{\text{MTBF}}{\text{MTBF} + \text{MDT}} \tag{11.49}$$

where $A =$ availability
$MTBF =$ mean time between failures
$MDT =$ mean down time

Availability can be viewed as the percentage of time that the system is operative, on the average, in relation to the total time.

> *Example.* If the failure rate for a system is 0.01 failure per hour and the mean-time-to-repair distribution is uniform in the range 2 to 8 hours, what is the system availability? The MTBF, from the failure rate, is 1/0.01, or 100 hours. The mean down time is taken to be the average value of the repair-time uniform distribution, which is calculated as $(2 + 8)/2 = 5$ hours. From the preceding formula, the availability is $100/(100 + 5) = 0.952$.

11.14 A LEAST SQUARES FIT

We often run into systems engineering problems in which we are trying to fit a line or a curve to a set of data points. Thus, we are converting from these data points into a formula that can be used to represent the empirically derived data set. This notion was discussed under the topic of *cost estimating relationships* (CERs), which allowed us to estimate an element of system costs as a function of a limited set of variables. The COCOMO discussion in Chapter 10 examined a good example. In this section, we look more closely at the mathematics of perhaps the simplest of such formulations: the case in which we have a set of data points and wish to obtain the best fit of a *line* to these points. Refer to Figure 11.6, which shows the resultant line for a set of four data points.

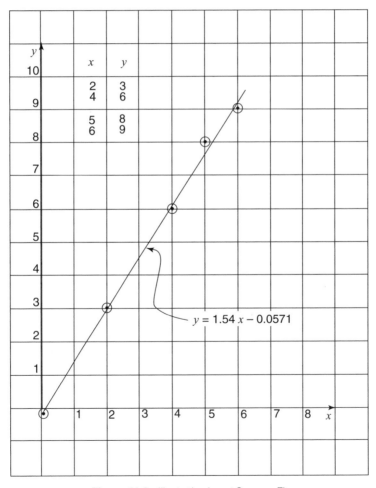

x	y
2	3
4	6
5	8
6	9

$y = 1.54\,x - 0.0571$

Figure 11.6. Illustrative Least Squares Fit

In general, when we are curve-fitting, we use the criterion of minimizing the resultant *mean-square* value between data points and the curve in question. This is an old idea that is directly related to mean-square errors and the notion of variance. As such, it has become a convention that we have agreed on as a preferred procedure. If we have limited ourselves to a line, then the least squares fit of that line may be found by using these relationships for the slope (m) and the y-intercept (b) of that line:

$$m = \frac{n\Sigma xy - \Sigma x \Sigma y}{n\Sigma x^2 - (\Sigma x)^2} \tag{11.50}$$

$$b = \frac{\Sigma y \Sigma x^2 - (\Sigma x \Sigma xy)}{n\Sigma x^2 - (\Sigma x)^2} \tag{11.51}$$

We will work out an example in which the four data points are as shown next, along with calculations that are needed for the preceding equations.

x	y	x^2	y^2	xy
2	3	4	9	6
4	6	16	36	24
5	8	25	64	40
6	9	36	81	54
17	26	81	190	124 = sums

number of data points $= n = 4$
mean value of $x = m(x) = (\Sigma x)/n = (17)/4 = 4.25$
mean value of $y = m(y) = (\Sigma y)/n = (26)/4 = 6.5$

Proceeding with the equations for the line's slope (m) and y-intercept (b), we have:

$$m = \frac{(4)(124) - (17)(26)}{(4)(81) - (17)^2} = \frac{54}{35} = 1.54$$

$$b = \frac{(26)(81) - (17)(124)}{35} = \frac{-2}{35} = -0.057$$

Therefore, the line may be described as:

$$y = 1.54x - 0.0571$$

This is the least squares fit to the four given data points. Note that the y-intercept is close to zero and almost passes through the origin. The reader is urged to double-check the calculation and continue with the related example at the end of this chapter.

11.15 SUMMARY

In this chapter, we presented selected quantitative relationships that support systems engineering and project management. Most of these relationships were drawn from the field of probability theory. Other domain-specific relationships are too numerous to be considered here, but may have to be addressed, depending on the domain knowledge necessary with respect to a given project (e.g., guidance, control, or aerodynamics).

The essential purpose of mastering these and other relationships is to measure the performance of the system that is being designed and built. Other purposes include the effective management of the overall systems engineering effort and the support of the quantitative aspects of all of the thirty elements of systems engineering.

A brief summary of the most significant relationships covered in this chapter is provided in Exhibit 11.1.

Exhibit 11.1: Summary of Quantitative Relationships

General

$$P(A + B) = P(A) + P(B), \quad \text{if } AB = 0$$
$$P(AB) = P(A|B)P(B) = P(B|A)P(A)$$
$$= P(A)P(B), \quad \text{if } A \text{ and } B \text{ are independent}$$

Mean value of $x = m(x) = E(x) = \int xp(x)\,dx$

Variance $= \sigma^2 = \int (x - m)^2\, p(x)\,dx$

$\text{Mean}(X + Y) = \text{mean}(X) + \text{mean}(Y)$

$\sigma^2(X + Y) = \sigma^2(X) + \sigma^2(Y)$, when X and Y are independent

$E(xy) = \int\int xyp(x, y)\,dx\,dy$

$\text{Cov}(xy) = E(xy) - E(x)E(y)$

Correlation coefficient $\dfrac{\text{Cov}(xy)}{\sigma(x)\sigma(y)}$

Cumulative distribution function (CDF) $= \int p(x)\,dx$

Specific Distributions and Applications

Binomial: $P(x) = \begin{pmatrix} n \\ x \end{pmatrix} p^x q^{n-x}$

Poisson: $P(k) = \dfrac{(\lambda t)^k \exp(-\lambda t)}{k!}$

Normal: $p(x) = \dfrac{1}{\sigma\sqrt{2\pi}} \exp\left[\dfrac{-(x - m)^2}{2\sigma^2}\right]$

Exponential: $p(x) = \lambda \exp(-\lambda x)$

Rayleigh: $p(x) = \dfrac{x}{\sigma^2} \exp\left(\dfrac{-x^2}{2\sigma^2}\right)$

Reliability: $R(t) = e^{-M} = \exp(-t/\text{MTBF})$

Series reliability $= R(A)R(B) = \exp[-(\lambda_a + \lambda_b)t]$

Parallel reliability $= 1 - [1 - \exp(-\lambda_a t)][1 - \exp(-\lambda_b t)]$

Availability $= \dfrac{\text{MTBF}}{\text{MTBF} + \text{MDT}}$

QUESTIONS/EXERCISES

11.1 For the binomial example in Section 11.3, calculate the probability of *either* one or zero errors.

11.2 For a roulette wheel with 18 red, 18 black, a zero, and a double zero, what is the probability of:
  a. winning when you bet on black?
  b. losing when you bet on red?

11.3 For a normal distribution with a mean value of 6 and a variance of 9, what is the probability that the random variable will exceed:
  a. 10
  b. 12
  c. 14

11.4 What is the probability of successful operation for 200 hours for a system with three subsystems with MTBFs of 100, 200, and 300 hours when the subsystems are:
  a. in a "series" reliability configuration?
  b. placed in a redundant configuration?

11.5 If the failure rate is 0.02 failure/hour and the mean-time-to-repair distribution is uniform in the range 2 to 10 hours, what is the system availability?

11.6 For threshold detection of a radar pulse in Gaussian additive noise, the pulse voltage is 14 volts, the threshold is set at 5 volts, and the noise power is 9 watts.
  a. Find the detection and false-alarm probabilities.
  b. Where would you put the threshold to obtain a false-alarm probability of approximately 0.02? What is the resultant detection probability?

11.7 The three one-sigma errors of a system, where the errors are additive and independent, are in the ratio 3:4:5, and the total allowable error variance is 0.5. What are the maximum values of the three independent errors?

11.8 What is the probability that a system will operate without failure:
  a. to its mean-time-between-failure (MTBF), and
  b. twice its mean-time-between failure?

11.9 For the illustrative least squares fit in the text, estimate the covariance of x and y and also the correlation coefficient.

11.10 For the Rayleigh distribution, how many "sigma" will correspond to a probability value of 0.5? Try also with 0.95.

REFERENCES

11.1 Feller, W. (1950). *An Introduction to Probability Theory and Its Applications*, Vol. 1. New York: John Wiley.

11.2 Parzen, E. (1960). *Modern Probability Theory and Its Applications*. New York: John Wiley.

11.3 Lloyd, D., and M. Lipow (1962). *Reliability: Management, Methods and Mathematics*. Englewood Cliffs, NJ: Prentice Hall.

11.4 Eisner, H. (1988). *Computer-Aided Systems Engineering*. Englewood Cliffs, NJ: Prentice Hall.

PART IV
TRENDS, PERSPECTIVES, AND INTEGRATIVE MANAGEMENT

_____12

SYSTEMS/SOFTWARE ENGINEERING AND PROJECT MANAGEMENT TRENDS

12.1 INTRODUCTION

Trends in systems engineering and project management, to a great extent, are based on two primary factors:

1. The demands placed on the project manager and the engineering team, particularly as systems become larger and more complex
2. The advances in methods and technology that can be utilized to respond to the preceding

The problems that we face as project managers and systems engineers are plentiful and we have little choice but to be aware of the new options that are continually becoming available. This can be done, at least in part, by tracking the trends and adopting new solutions where they have proven to be useful. This chapter provides an overview of trends in systems engineering, software engineering, and project management. Over the years, as new issues arise, new trends will develop. Many of the trends examined here may be expected, however, to be with us for quite some time.

12.2 SYSTEMS ENGINEERING TRENDS

In broad terms, most system developers are attempting to make continuous improvements in their systems engineering capabilities and processes. When successful, they lead to building systems within the specified and required constraints of time (schedule), cost (budget), and technical performance.

12.2.1 International Council on Systems Engineering (INCOSE)

In the early 1990s, a new organization, the National Council on Systems Engineering (NCOSE), was formed to recognize the disciplines of systems engineering and improve its practice. Changing its name and scope, the current INCOSE has made important contributions to the field of systems engineering.

As an example, the inaugural issue [12.1] of the Journal of INCOSE dealt with the following selected issues:

- The basics of systems engineering
- Systems architecting
- Relationships between systems and software engineering
- Model-based systems engineering
- Systems thinking
- Case studies

Moving forward in time to the 5th Annual International Symposium of INCOSE [12.2], we see the following selected topics listed in the technical program:

- Systems engineering management
- Systems engineering tools
- Systems engineering processes and methods
- Requirements management
- Measurement
- Emerging applications
- Education and training

The comprehensive scope of INCOSE considerations is likely to make it an important force in the development and application of the discipline of systems engineering for many years to come. The professional systems engineer is thus encouraged to join INCOSE and participate in its activities.

12.2.2 System of Systems (S2) Engineering

Many practitioners of systems engineering have, in effect, been working on a "system of systems." Very large systems tend to be systems of systems, such as our national communications system and our national air transportation system. But even in a much narrower context, we also find systems of systems.

An example is the national air defense system with its many elements that deal with threat warning, attack assessment, and system response.

As we move more into the "information age," we are finding large numbers of information "systems of systems." Examples include the National Information Infrastructure or Highway, the Internet, and other customized networks, and the merging of computers and communications technologies and systems.

With the emergence and prevalence of systems of systems, the matter of how the systems engineer and project manager should deal with this situation has arisen. A partial response is the notion of system of systems (S2) engineering.

A system of systems perspective is applicable when one or more of the following circumstances is operative [12.3]:

- A variety of related systems are being acquired independently, with each such system subject to the usual systems engineering disciplines
- The schedule relationship between the related systems is arbitrary and asynchronous
- There is an overriding system of systems mission, that is, each system must interoperate with the others so as to provide and integrate capability and response

For the previous situations, the recommended structure [12.3] for system of systems engineering contains the elements shown in Exhibit 12.1. These elements are to be performed by a systems engineering team that has the charter to attempt to optimize the performance, schedule, and cost of the system of systems. In addition, it is suggested that these elements be supported by a set of systems engineering tools (see Section 12.2.7).

Exhibit 12.1: Elements of Systems of Systems (S2) Engineering

1. Integration engineering
   1.1 Requirements
   1.2 Interfaces
   1.3 Interoperability
   1.4 Impacts
   1.5 Testing
   1.6 Software verification and validation
   1.7 Architecture development
2. Integration management
   2.1 Scheduling
   2.2 Budgeting/costing
   2.3 Configuration management
   2.4 Documentation

3. Transition management
   3.1 Transition planning
   3.2 Operations assurance
   3.3 Logistics planning
   3.4 Preplanned product improvement (P3I)

Building upon these notions of S2 engineering is that of Rapid Computer-Aided System of Systems Engineering (RCASSE). This construction recognizes that system development schedules for large-scale systems are rarely satisfied. This leads to situations in which by the time a system is fielded, it is almost obsolete, especially in the domain of information technologies.

RCASSE emphasizes a rapid, disciplined, and computer-supported design process, with the following ten steps executed over a nominally specified six-month period [12.4]:

1. Mission engineering
2. Baseline architecting
3. Performance assessment
4. Specialty engineering
5. Interfaces/compatibility evaluation
6. Software issues/sizing
7. Risk definition/mitigation
8. Scheduling
9. Preplanned product improvement (P3I)
10. Life-cycle cost-issue assessment

This reengineering of the more conventional systems engineering process is designed to be used when there exist:

- A system of systems situation
- Extreme schedule pressure
- A sophisticated development team with access to and experience with a variety of systems engineering tools

It focuses on only the essential elements in an attempt to converge quickly on a baseline system that satisfies the requirements.

It may be expected that further efforts to adapt the elements of systems engineering to a system of systems environment will be forthcoming as we continue to be faced with situations of this nature and the problems that they entail.

12.2.3 Capability Maturity Model (CMM) for Systems Engineering

Chapter 11 briefly described the capability maturity model (CMM) for software. A current trend is to develop a CMM for systems engineering. It is likely that this effort will be successful and have a significant impact on the systems engineering community.

The Software Engineering Institute (SEI), primary developers of the software CMM, has taken a key position in formulating a CMM for systems engineering (SE-CMM). An important ingredient in its structure is the identification of process areas (PAs), which it considers to be key elements of systems engineering [12.5, 12.6]. PAs have been broken down into three main categories, namely, (1) project, (2) organizational, and (3) engineering. Approximately seventeen preliminary PAs have been established under these three categories. The structure also breaks each PA into base practices (BPs).

The approach to the SE-CMM appears to have strong similarities to that of the CMM for software. It is a maturity model for systems engineering and a related method for assessment. As the software CMM had five levels of maturity (see Chapter 11), this SE-CMM has six levels:

Level 0: Initial
Level 1: Performed
Level 2: Managed
Level 3: Defined
Level 4: Measured
Level 5: Optimizing

These levels are intended to differentiate the process capability of the organization.

INCOSE has also been investigating a CMM for systems engineering. In this conception, some fifteen key focus areas have been defined under three categories [12.7]:

* Engineering Process
  — System requirements
  — System design
  — System integration and verification
  — Integrated engineering analysis
* System Management
  — Planning
  — Tracking and oversight
  — Subcontract management
  — Intergroup coordination

- — Configuration management
- — Quality assurance
- — Risk management
- Organizational
  - — Process management
  - — Training
  - — Technology management
  - — Environment and tool support

These key focus areas, at this time, are not precisely the same as the process areas (PAs) of the SE-CMM, but they have the same basic thrust. There are also other differences, and similarities, between the INCOSE approach and the SEI approach to a systems engineering CMM.

With the above SEI and INCOSE approaches as background, three organizations moved forward in order to synthesize these approaches. These organizations were the Electronics Industry Association (EIA), INCOSE, and the Enterprise Process Improvement Collaboration (EPIC). The net result was SECM, the Systems Engineering Capability Model, as embodied in EIA/IS-731 [12.8] (see also Chapter 2). This standard has two parts—the SECM model and the SECM appraisal method. It also has a total of nineteen focus areas, distributed under technical, management, and environment categories.

Another notable piece of work was the formulation of the iCMM (the Integrated Capability Maturity Model) by the Federal Aviation Administration (FAA) [12.9]. This model was focused upon the acquisition of software intensive systems and contains the process areas (PAs) listed below:

- Life Cycle or Engineering Processes
  - PA 01—Needs
  - PA 02—Requirements
  - PA 03—Architecture
  - PA 04—Alternatives
  - PA 05—Outsourcing
  - PA 06—Software development and maintenance
  - PA 07—Integration
  - PA 08—System test and evaluation
  - PA 09—Transition
  - PA 10—Product evaluation
- Management or Project Processes
  - PA 11—Project management
  - PA 12—Contract management

PA 13—Risk management

PA 14—Coordination

- Supporting Processes

    PA 15—Quality assurance and management

    PA 16—Configuration management

    PA 17—Peer review

    PA 18—Measurement

    PA 19—Prevention

- Organizational Processes

    PA 20—Organization process definition

    PA 21—Organization process improvement

    PA 22—Training

    PA 23—Innovation

Finally, with respect to this overall topic, the SEI set forth the CMMI, also an integrated version of the Capability Maturity Model [12.10]. This form of the CMM concept was intended to provide "guidance for improving your organization's processes and ability to manage the development, acquisition, and maintenance of products and services."

The significance of the preceding, however, is not that there are different approaches to formulating a systems engineering or integrated CMM. Rather, this body of work represents a trend toward better understanding of the elements of systems engineering and the improvement of the internal processes of systems engineering and its management. This can only have a beneficial effect on both the theory and practice of systems engineering.

12.2.4 Systems Architecting

All of Chapter 9 is devoted to a key element of systems engineering, namely, the architecting of the system. As important as this element is, one can expect that efforts will continue to attempt to define and develop how architecting is to be accomplished. Because it is fundamentally a design or synthesis process, it differs from analysis in that one is attempting to "invent" a new configuration that may not have existed before. Fascination with the creative process of architecting a new system is not misplaced.

As stated in Chapter 9, E. Rechtin has played a key role in studying the sometimes mysterious process of system architecting. Beyond his seminal books in this area [12.11, 12.12], Dr. Rechtin has continued to explore the subject of systems architecting. For example, in examining the foundations of systems architecting [12.13], he indicates that systems architecting "is focused and scoped by six core concepts or ideas: the systems approach, purpose orientation, ultraquality, modeling, experience-based heuristics, and certification." At about the same time, he also commented on the role and responsibility of the systems architect [12.14]. Questions (and answers) that he poses in this regard are as follows:

- Do I need an architect? (Asked by the client.)
- How do I make sure I get what I want?
- How do we keep on track?
- How do I know that the system has been satisfactorily completed?

Rechtin continues to emphasize heuristics as a major factor in systems architecting, citing heuristics with respect to architecting qualitative change, maintaining system integrity, and systems acceptance. These heuristics and their value apparently came as a "great surprise." Many systems engineers recognize this in terms of "rules of thumb" that they use in order to architect new systems. Such rules are experience-based and are a testament to the extensive domain knowledge of the best systems engineers.

The issue of systems architecting was also examined in some detail at a workshop sponsored by the Navy Department [12.15], with the central question being how to improve architecting for increasingly complex systems and environments. In response, some of the architecting tenets were defined:

- Adherence to fundamental architecting principles
- Recognition of a systems architect
- Client involvement
- Keeping the process creative
- Controlled teamwork

It was also suggested that the systems acquisition process should contain a "systems architecture milestone" that would contain the following:

- A definition of the systems architecture
- System acceptance requirements
- A life-cycle plan
- A rationale that convinces the stakeholders that the system will:
  — satisfy their overall needs
  — satisfy functional, performance, and quality requirements
  — have an acceptable level of risk
  — do the foregoing at least as well as any alternative architecture

Other notions discussed at the workshop were

- The use of an architecture design language (ADL)
- The unique needs of complicated systems
- Methods that enable the architecting of complex systems

- Tools for systems architecting
- Workgroup collaborations
- Taxonomies of system styles
- Training and career paths for systems architects

Another important thrust with respect to systems architecting is represented by the C4ISR (Command, Control, Communications, Computer, Intelligence, Surveillance, and Reconnaissance) Architecture Framework [12.16], as discussed previously in Chapter 9. It will be recalled that this framework was structured around three basic views, namely:

- The operational view
- The systems view, and
- The technical view

This Architecture Framework was confirmed, in part, by the C4ISR Architectures Working Group (AWG) within the Department of Defense (DoD) [12.17]. In its deliberations, the AWG was broken down into four panels to consider:

- The framework
- Interoperability matters
- Data modeling and analysis tools
- Roles and responsibilities

Based upon their report [12.17], the following recommendation areas were considered:

- Establish common architecture terms and definitions
- Implement a common approach for architectures
- Strengthen architecture policy and guidance
- Define and use levels of interoperability
- Build architecture relationships with other DoD processes
- Manage DoD architectures

The reader is urged to examine both the Architecture Framework [12.16] as well as the AWG Final Report [12.17] in order to obtain a full understanding of the significance of these efforts.

Based upon these and other stated interests in this subject, it may be expected that continuing research into and application of various processes of systems architecting will yield results that can be brought into the mainstream of systems engineering practice.

12.2.5   Sustainable Development

In its statement of public policies and priorities, the American Association of Engineering Societies (AAES) officially endorsed the notion of sustainable development, which has been defined as "meeting the needs of the present without compromising the ability of future generations to meet their own needs" [12.18]. It is further stated that "engineers will play a critical role in sustainable development and must acquire the skills, knowledge, and information that are the stepping stones to a sustainable future."

Preserving our resources and our environment are key aspects of sustainable development. Because, as suggested earlier, engineers are and will continue to be deeply involved in such development projects, it is critical that systems engineering be expanded, wherever necessary, to specifically accommodate the principles and practices of sustainable development. As articulated in a position paper from the AAES, "engineers must work with others to adapt existing technologies and create and disseminate new technologies that will facilitate the practice of sustainable engineering and meet societal needs" [12.19]. This unmistakable trend should influence the way projects and systems engineering teams are deployed and managed. Moreover, it is likely that ultimately, the elements of systems engineering will be modified so as to place more emphasis on assuring sustainable development, explicitly including sustainable technologies and processes.

12.2.6   The Structure of Systems Engineering

Fortunately, the overall structure of systems engineering appears to be under constant examination. The Navy workshop [12.15] previously cited did so by exploring the state of the art as well as recommendations for the future in the following key focus group areas:

- Design capture
- Evolutionary systems
- Infrastructure and tools
- Organizational/institutional learning
- Systems architecting
- Reengineering

Results in these areas cannot be reiterated here, but it is clear that the very structure of systems engineering is being analyzed in considerable detail, with participation of many of the best systems engineers in the country.

While mentioning the best systems engineers, it is also necessary to acknowledge the varied, prolific, and significant work, over many years, of Andrew Sage [12.20, 12.21, 12.22, 12.23]. As an example, in 1994, Dr. Sage discussed the "many faces of systems engineering" in the INCOSE inaugural issue of its journal [12.24]. Among the topics examined were

- Systems management
- Systems methodology
- Knowledge types
- Formulation, analysis, and interpretation
- Life-cycle models
- Interactions across life cycles
- Architecture levels

Another examination of the structure of systems engineering was carried out by the Software Productivity Consortium (SPC), expanding its original charter from software considerations to the broader context of systems [12.25]. The result was a tailorable process for systems engineering that was given the name of Generic Systems Engineering Process (GSEP). The basic notion was to view all elements of systems engineering in terms of the detailed processes that were required in order to carry them out. In effect, it was equivalent to taking the thirty elements of systems engineering defined here and answering the question: What process is necessary for the proper execution of each element? In order to give the process descriptions some rigor, ICAM (Integrated Computer-aided Manufacturing) Definition (IDEF) diagrams were used as descriptors. This method is also embodied in several systems/software engineering tools so that the process flows are easily automated.

These are but a few of the many examinations of the basic structure of systems engineering that are under constant scrutiny and reevaluation. It is believed that trends in this direction are helpful and likely to continue indefinitely.

12.2.7 Systems Engineering Environments and Tools

Given the thirty elements of systems engineering and descriptions of the processes required to carry out these elements, two natural questions that follow are:

1. Is there a set of automated tools that the systems engineering team can use?
2. Can these tools be integrated in some fashion to create a "systems engineering environment" (SEE) with which the team can operate in a highly effective and efficient manner?

A book by this author concentrated on answering the first of these [12.26], and also made it clear that a large number of such tools have been developed for other purposes, and that many were focused on software engineering.

Further developments with respect to system of systems (S2) engineering, as discussed earlier in this chapter [12.3, 12.4], depended highly on the availability of systems engineering supporting tools. This situation is true today with tool developers giving emphasis to applications in software engineering and even business process reengineering (BPR) [12.27], in distinction to systems engineering. The need for tools that support systems engineering has now been widely recognized, and groups such as INCOSE and others [12.28] are working on this issue as a continuing activity.

With respect to a systems engineering environment, a leading position was taken by the Air Force's Rome Laboratory in developing the System Engineering Concept Demonstration (SECD). The primary goal of this program was to "increase the productivity and effectiveness of system and specialty engineers involved in the development, maintenance, and enhancement of military computer systems" [12.29]. Within the scope of this effort, the Air Force specifically addressed the automation of the systems engineering role and the various activities that support systems engineering. These activities fell into the categories of engineering, communications, and management.

Ultimately, the program focused on the development of Catalyst, an automated systems engineering environment that targeted the systems engineering team as the user and addressed these three categories. The building blocks of Catalyst were tools that dealt with interface mechanisms and environment frameworks that took the form of:

- Process automation
- Generic engineering and management tools
- Concurrent engineering groupware
- Integration mechanisms
- Environment administration tools

Documentation of SECD was produced in six volumes addressing:

1. Systems engineering needs
2. A process model
3. Interface standards studies
4. Technology assessments
5. Trade studies
6. A security study

From this point, the SECD program transitioned from exploratory to advanced development that involved the building of critical Catalyst components.

Because systems engineering continues to be a process that calls for considerable ingenuity in working with massive amounts of information, it is likely that efforts to build automated systems engineering environments will continue for the indefinite future. One key issue is the extent to which these environments are designed to incorporate tools that are fully integrated, that is, tools that work together so as to minimize rekeying and manual transport of data. Another, of course, is the scope of these tools. In the context of this book, that question may be stated as: To what extent does such an environment cover the full thirty elements of systems engineering? These issues carry forward into constructing a *software* engineering environment, otherwise known as CASE (computer-aided software engineering). Further discussion of these points appear later in this chapter under the subject of software engineering trends.

12.2.8 Education

Systems engineering, in many colleges with engineering programs, has not yet been fully accepted as a discipline of study. It certainly does not have the emphasis given to the old and more widely established fields of electrical, mechanical, chemical, and nuclear engineering. Notable exceptions to this are programs at the University of Virginia, Virginia Tech, the University of Arizona, George Mason University, the University of Maryland, and The George Washington University, among others. In some cases, systems engineering is tied into industrial engineering, operations research, or engineering management. As the demand for more formal training in systems engineering continues, it may be expected that more schools will focus on this need and build more substantial education programs responsive to activities called for in much of our industrial base. This is a relatively slow process, so many firms obtain nonacademic training in systems engineering by contracting with specialty companies that offer short courses that may be adapted to individual needs. Whatever the form of delivery, it can be safely predicted that the development of systems engineering education programs and options will continue for a long time.

12.2.9 Acquisition Practices

Incredible amounts of time and energy have been expended with respect to defining and reforming the processes involved in the acquisition of systems, especially in the government. These efforts, of course, have major impacts on industry and so various companies and industry groups have had significant inputs to the thinking that has gone into acquisition change and reform.

The motivation for acquisition reform has been centered around three key issues—speeding up the process, competition, and fairness. The first of these has been felt rather strongly because the time for the acquisition of largescale systems has become longer and longer. Indeed, many systems are almost obsolete by the time all the preliminary phases are executed and the system is finally fielded. For both government and industry, this is an intolerable condition. With respect to competition and fairness, the number of award protests has increased dramatically in government systems procurements for a variety of reasons. The resolution of these protests has also had a major impact on the time required to acquire a system.

Chapters 2 and 7 discussed the rather large documents known as the 5000 series in the Department of Defense (DoD). This series incorporated a directive, instruction, and documentation requirement and process for the acquisition of military systems. Collectively, they are extensive in their scope and penetrating in their detail.

A number of so-called acquisition reform initiatives have been important to the government. Two such efforts were the Federal Acquisition Reform Act (FARA) and the Federal Acquisition Streamlining Act (FASA) [12.30]. Judging from the extensive dialogue on these matters, reengineering acquisition practices has been and remains a distinctly nontrivial exercise. By looking at the practices over the past thirty years, continuing attention to these matters can be expected to be with us for the indefinite future. These practices impact the way both project managers and systems engineers do their jobs, especially during the proposal development phase.

Another major thrust in the direction of acquisition reform developed internally within the DoD in order to solve some of the problems alluded to earlier. This was basically initiated by the Secretary of Defense in his memorandum of June 29, 1994 [12.31], which called for a move from military specifications and standards to increased use of best commercial practices. The stage was set for this definitive action during the previous year when the Deputy Under Secretary for Acquisition Reform established a Process Action Team (PAT). The charter for this PAT [12.32] was to develop

> (a) a comprehensive plan to ensure that DOD describes its needs in ways that permit maximum reliance on existing commercial items, practices, processes and capabilities, and (b) an assessment of the impact of the recommended actions on the acquisition process.

The PAT produced a "Blueprint for Change" [12.33], which highlighted some thirteen principal recommendations, reproduced in Exhibit 12.2. These activities, along with an implementation plan, defined a significant trend toward acquisition reform that is undeniable. Although some parties suggested more far-reaching approaches [12.32], there is no doubt that reform of the way in which the government acquires large systems is necessary. All of this represents an environment in which both the Project Manager and the systems engineering team must do their jobs.

Exhibit 12.2: Thirteen Principal Recommendations of the Process Action Team (PAT) on Military Specifications and Standards [12.32]

1. All ACAT [Acquisition Category] Programs for new systems, major modifications, technology generation changes, nondevelopmental items, and commercial items shall state needs in terms of performance specifications.

2. Direct that manufacturing and management standards be canceled or converted to performance or nongovernment standards.
3. Direct that all new high value solicitations and ongoing contracts will have a statement encouraging contractors to submit alternative solutions to military specifications and standards.
4. Prohibit the use of military specifications and standards for all ACAT programs except when authorized by the Service Acquisition Executives or designees.
5. Form partnerships with industry associations to develop nongovernment standards for the replacement of military standards where practical.
6. Direct government oversight be reduced by substituting process control and nongovernment standards in place of development/production testing and inspection and military unique quality assurance systems.
7. Direct a goal of reducing the cost of contractor-conducted development and production test and inspection by using simulation, environmental testing, dual-use test facilities, process controls, metrics, and continuous process improvement.
8. Assign Corporate Information Management offices for specifications and standards preparation and use.
9. Direct use of automation to improve the processes associated with the development and application of specifications and standards and Data Item Descriptions (DIDs).
10. Direct the application of automated aids in acquisition.
11. Direct revision of the training and education programs to incorporate specifications and standards reform. Contractor participation in this training effort shall be invited and encouraged.
12. Senior DoD management take a major role in establishing the environment essential for acquisition reform cultural change.
13. Formalize the responsibility and authority of the Standards Improvement Executives, provide the authority and resources necessary to implement the standards improvement program within their service/agency, and assign a senior official with specifications and standards oversight and policy authority.

A particular response to the preceding was the formulation, by the Department of Defense, of an Evolutionary Acquisition Strategy to acquire weapon systems [12.34]. In this approach, the Joint Logistics Commanders gave their formal guidance to Program Managers. They expressed their belief that such an approach provided "a good alternative means to develop and acquire weapon systems while providing for incremental growth over time," and recommended that the Guide be used as a "foundation for effective weapon system acquisition planning." Further, the Evolutionary Acquisition (EA) Process was defined as

> A strategy for use when it is anticipated that achieving the desired overall capability will require the system to evolve during development, manufacture or deployment.

The Guide goes on to indicate that the EA approach was basically the same as Preplanned Product Improvement (P3I), one of the elements of systems engineering, as defined herein. This suggests that new trends may be established by reordering priorities and changing emphasis with respect to elements and processes that already have been well understood. This is an observation that should be kept in mind by Project Managers and systems engineers.

Much of the detail regarding DoD processes and products attendant to the acquisition of systems is contained within the latest version of the DoD 5000 series [12.35]. These documents tend to be updated every several years or so as we gain deeper insights into acquisition matters or wish to change emphasis within overarching goals and objectives. One significant statement of such goals is cited below, as contained within a description of the "Road Ahead" by the then Under Secretary of Defense for Acquisition and Technology, Jacques Gansler [12.36]:

> *Goal One:* Field high-quality defense products quickly; support them responsively
> *Goal Two:* Lower the total ownership cost of defense products
> *Goal Three:* Reduce the overhead cost of the acquisition and logistics infrastructure

Looking down the road, a variety of study teams explored issues and problems in acquisition and logistics, dealing with the following topics:

- Research, development, test, and evaluation (RDT&E) infrastructure
- Product support
- Requirements and acquisition interfaces
- Training and tools for acquisition of services
- Commercial business environment

Integrating the results of these primary studies, as well as other inputs, led to the articulation of some near-term actions that are critically important to the DoD. These are [12.36]:

1. Increased reliance on an integrated civil-military industrial base
2. A new approach to acquisition whereby "price and schedule play a key role in driving design development and systems are reviewed by portfolio"

3. A transformation of the mass logistics system into one that is agile, reliable, and delivers logistics on demand

4. Reduction of acquisition infrastructure and overhead functions and costs

5. A workforce that is adequately trained to "operate efficiently in this new environment and will perpetuate continuous improvement"

6. The institutionalization of continuous improvement as well as change management so as to achieve a virtual learning environment.

12.2.10 Systems Integration

Systems integration is a most interesting term, since many of the largest companies call themselves "systems integrators." If they are asked to describe the business they're in using just a couple of words, they are similarly likely to respond with "the systems integration business." All this, of course, suggests that these special words should be very well defined and that there perhaps should be more books on systems integration than there are on systems engineering. It appears that the opposite is true. With some exceptions [e.g., 12.37], discussions of systems integration tend to be few and far between. However, most companies seem to agree that systems engineering is, in fact, a core competency that they need in order to carry out their primary mission as systems integrators. In addition, these companies appear to document what they mean by systems integration, but for competitive reasons are not anxious to share this information with the rest of the world. They prefer to use their unique approach to systems integration as a means of differentiating themselves in the marketplace. Given this situation, this author would like to suggest his own short-form definition of systems integration:

Systems Integration. The process of bringing together a variety of (possibly disparate) functional elements, subsystems, and components into a larger (meta)system, or system of systems, to provide a highly interoperable and cost-effective solution that satisfies the customer's needs and requirements, while at the same time managing the overall process and delivery of products in a highly effective and efficient manner.

We can make several observations about this definition and what it might imply. First, systems integration looks a lot like the optimal synthesis of systems engineering and program/project management. In that sense, this book, by addressing both project management and systems engineering, might also be called a text about at least a major part of systems integration. Second, in dealing with "systems of systems," systems integration deals also with the topics discussed earlier in this chapter regarding systems of systems. Third, from a technical perspective, systems integration is about searching for ways to find an integrated solution, that is, one that deals as necessary with (1) legacy stovepipe systems, (2) upgrades to legacy stovepipe systems,

(3) commercial-off-the-shelf (COTS) and nondevelopment items (NDI) that need to be part of the solution, (4) reused components, as appropriate, (5) new systems and subsystems that need to be built for the first time. Systems integration also emphasizes the *architecting* of a system with the appropriate *balance* of the above five elements.

In addition, interoperability and compatibility become critical elements in the systems integration process. They also constitute a most difficult part of the problem since (1) it is a nontrivial matter to integrate disparate subsystems so that they interoperate in a harmonious manner, and (2) systems tend to be more failure prone at the interfaces, which is a critical aspect of achieving interoperability and compatibility. Further, the sequence in which parts of systems are integrated is not necessarily obvious, and becomes an important consideration in the systems integration process. Finally, the systems integrator needs to be able to question requirements in order to do his or her job, possibly contrary to some views regarding the subject.

As the last element of commentary here regarding the topic of systems integration, it is necessary to point out two misconceptions about systems integration. The first misconception is that the best solution is one that maximizes the degree of integration of all stovepipe systems, with the goal being 100%. The second misconception is that, if we integrate a set of "best of breed" subsystems, we will necessarily achieve an overall "best" solution. The reader is urged to consider these two propositions, think them through, and use them as necessary in addressing the next systems integration problem that crosses your desk.

12.3 SOFTWARE ENGINEERING TRENDS

Trends in software engineering are at least as extensive as they are in systems engineering. The reason is that software development remains our most significant problem within the context of systems engineering. Put another way, many of the failures in performance, schedule, or cost are traceable to deficiencies in software engineering processes. In this section, we explore trends in a variety of areas. Other more detailed technical trends (e.g., the move toward object-oriented design and programming) are not considered here, but can be found in many texts and documents devoted only to software engineering.

12.3.1 National Software Council

The significance of software as part of our systems is underscored by establishment of the National Software Council (NSC) in 1995. This nonprofit organization was founded to "propel software to the forefront of the national agenda and define the National Software Strategy to preserve U.S. competitiveness and security into the 21st century" [12.38]. Its statement of mission is to "ensure that the U.S. software sector continues to make a strong and growing contribution to national economic prosperity." Thus, the NSC has embraced a rather large vision with potential impacts on a national scale.

The NSC tends to focus on the following three activities:

1. Identify and articulate national software issues
2. Provide a forum for analysis and discussion of software issues
3. Propose policy recommendations to achieve software goals

Clearly, the NSC has to work with people from industry and government to achieve its goals. Its initial prospectus appears to be very thoughtful and well-conceived. Under the able leadership of John Marciniak as the first President, it is likely that the NSC will have a very positive influence on the national agenda with respect to software. Readers are encouraged to track and participate in the future activities of the Council.

12.3.2 Commercial Practices

The previous section on systems engineering discussed the thrust in the DoD toward commercial practices and away from military specifications and standards. This clear trend is reflected as well in the software arena. As an example, in 1994, the Defense Science Board (DSB) examined the issue of the commercial acquisition of defense software [12.39]. This extremely interesting report summarized principal findings and recommendations in the following categories:

- Process credibility
- DoD program management
- DoD personnel
- Use and integration of commercial off-the-shelf (COTS) software
- Software architecture
- Software technology base
- Management control and oversight

The report is strong on the point that the way the DoD currently does business is not compatible with the extensive use of commercial practices. It therefore calls for major changes in current software acquisition and development processes and practices. Exhibit 12.3 [12.39] provides a selected list of some of the findings and recommendations in the DSB report.

Exhibit 12.3: Selected Defense Science Board Findings and Recommendations

Findings

- High life-cycle cost in time and dollars
- Incredibly long (13–15 years) development cycle
- Excessive acquisition agent involvement in design detail and process
- Contractor–government relationship based on mistrust versus mutual trust
- Approach tends to be design it all and then build it
- Little focus on design for reusability
- Requirements and source selection inflexibility
- Complicated regulations
- Program management does not encourage "80% solution for 20% cost"
- Shortage of qualified software personnel in DoD
- Normally no COTS market analysis
- Insufficient advantage taken of commercial research and development (R&D)
- Reasons for trouble on development programs:
  — Poor requirements definition
  — Inadequate process management and control by contractor
  — Lack of integrated product teams (IPTs)
  — Lack of consistent attention to software process
  — Too little attention devoted to software architecture
  — Focus on innovation rather than cost and risk

Recommendations

- Make necessary changes in acquisition regulations
- Establish overarching software life-cycle guidelines
- Renew software program management education and training initiative
- Require trade studies of the use of COTS
- Emphasize use of software architecture
- Strengthen software technology transfer

The DBS report is but one milestone of many that define a significant trend in software engineering and development—a fundamental belief that commercial practices will streamline the process and result in shortened time frames and decreased costs, without any performance penalties.

In November of 2000, another report was produced by the DSB Task Force on Defense Software [12.40]. The Board was asked to explore defense software in relation to the use of commercial practices, and also to develop a strategy that makes appropriate use of these practices. The DSB examined six previous major DoD-wide studies and the 134 recommendations contained in them. They also concluded that only a few of these recommendations had been implemented, an indicator of how difficult it is to get a very large enterprise, such as the DoD, to shift gears and make changes.

As part of their work, the DSB reiterated what they called disturbing statistics that apparently apply to both commercial and government information technology (IT) projects. These numbers, attributed to the Standish Group [12.41], are cited below:

- About 16% of all of the IT projects were completed on time and within budget.
- Some 31% of these projects were canceled before they were completed.
- Adding the above (which is 47%), we are left with about 53% of the projects that are both late and over budget, with the actual expenditure being greater than the budgeted expenditure by more than 89%.
- For those projects that were completed, computed above as 69% of the total, only about 61% had the original set of features that had been specified (39% did not).

The overall recommendations of this DSB Task Force can be summarized by the following, in terms of new actions to be taken by the DoD:

1. More stress should be placed upon past performance and the degree of process maturity.
2. Additional independent expert reviews (IFRs) of programs should be initiated.
3. The software skills of acquisition and program management personnel need to be improved.
4. Best practices in relation to software should be collected, disseminated, and employed.
5. Contract incentives need to be restructured.
6. The technology base that supports software development is in need of strengthening and stabilizing.

We shall see, down the road, if, when, and how these recommendations are implemented.

12.3.3 Reuse

Among the issues cited in the Defense Science Board report is insufficient focus on design for reusability. The matter of software reuse, however, constitutes a significant trend in software development. The basic notion is that many software modules (builds, configuration items, etc.) whose performance has been verified can be reused in at least two contexts:

1. Within a company, utilizing the best software it has developed
2. In a government repository of "certified" software that can be accessed by persons and organizations that have approved reasons to have such access

There are, however, a large number of issues that surround the matter of reuse of software. A Deputy Assistant Secretary of the Air Force, in 1993, distributed a memorandum [12.42] that set forth a software reuse incentive policy. Some of the provisions of that memorandum are listed as follows:

- We should consider designing all new software for reuse.
- Every acquisition strategy panel should explore the full or partial reuse of existing software, including COTS.
- When software reuse may not be feasible, consideration should be given to the design of new software to facilitate reuse in future applications.
- We should identify several specific "test" programs that will require reuse and design for reuse.

A reuse education and training workshop in 1994 [12.43] had the goal of "identifying key reuse concepts and how they can be integrated into an existing curriculum" regarding software engineering. The following subgoals for that workshop reinforce the importance of reuse:

- Identify key software engineering assumptions and concepts relevant to reuse during the software life cycle.
- Explore how to introduce the preceding into various curricula.
- Identify the audience and roadblocks for reuse concepts.
- Compile lists of education resources and references.

There is also some activity regarding the development of a reuse capability model (RCM) [12.44]. This is a "self-assessment and planning aid for improving an organization's reuse capability" and is being addressed by the Software Productivity Consortium (SPC). Studies have also focused on reuse inhibitors, as well as the best of reuse practices [12.45]. Statistics regarding reuse initiatives have been cited as follows [12.46]:

- 14–68% productivity increase
- 20% reduction in customer complaints
- 25% less required time to repair and overall schedule
- 50% reduction in integration time
- 20–35% increases in quality
- 20% less training costs
- 400% return on investment[!]

Clearly, judging from these potential benefits, if organizations can make reuse work, they will increase software development productivity and make them more competitive in the marketplace.

In addition to the initiatives in the Department of Defense and its related contractors, the National Institute of Standards and Technology (NIST) has undertaken an Advanced Technology Program (ATP) in the area of "component-based software." A news release from the U.S. Department of Commerce [12.47] announced the component-based software program as "a five-year, $150 million program to develop the technologies necessary to enable systematically reusable software components." Selected NIST contracts under this ATP program involved:

- Automation of dependable software generation with reusable components
- Component integration: an architecture-driven approach
- Scalable business application development components and tools
- Component-based reengineering technology

From the preceding, it can be seen that there is a great deal of force behind software component reuse and considerable effort is likely to be expended to improve and refine software reuse. This author has also looked in some detail at software reuse and has suggested a reengineered software acquisition process that involves the reuse of entire developer off-the-shelf systems [12.48].

12.3.4 Development Methods

The Defense Science Board report strongly recommends that software be developed more in accordance with commercial practices. This also implies the use of commercial development methods, at least with respect to the use of commercial standards and specifications.

Returning also to Chapter 10, we see the suggestion in Military Standard 498 that there are three well-defined development strategies: (1) grand design, (2) incremental, and (3) evolutionary. The bottom line is that the latter two are preferred approaches, depending on the circumstances. We also note the trend in the direction of evolutionary acquisition of systems (see Section 12.2.9).

A more complete array of software development methods was examined by the Air Force's Software Technology Support Center in Ogden, Utah [12.49], namely:

- Software Development Models:
  — Waterfall
  — Incremental
  — Spiral
- Software Development Techniques:
  — Prototype
  — Cleanroom
  — Object-oriented

Strengths and weaknesses of these various approaches were explicitly cited. It was also indicated that the evolutionary approach was not considered, due to the variation in its meaning in the literature.

By noting the preceding, as well as other investigations of alternative development approaches, it is seen that the situation is in a state of flux, with new ideas and variations on a theme being proposed and analyzed. This investigatory trend will continue as we try to find a process that has the demonstrable productivity increases that we seek. The reader with a special interest in these alternatives that have not been discussed here (e.g., clean-room, object-oriented) is encouraged to examine the extensive literature on these subjects.

12.3.5 Acquisition Practices

Acquisition practices of software systems are related to the following topics already discussed:

- Military Standard 498
- Acquisition trends for systems, including acquisition reform
- The move toward commercial practices
- Various software development approaches
- Reuse of software

The bottom line is that software system acquisition practices are changing, and the expectation is that continuous modifications in current practices are likely to be on our agenda, judging from our history, well into the twenty-first century.

Two of the most recent trends with respect to software acquisition can be identified as

- The emergence of a software acquisition capability maturity model construction
- Simulation and modeling for software acquisition (SAMSA)

For the former, a structure parallel to the Software Engineering Institute (SEI) capability maturity model is being developed, along with key process areas (KPAs) that relate to the software acquisition process. In the latter area, a notable effort is that of Boehm and Scacchi at the University of Southern California [12.50] who indicate that "there are substantial opportunities to rethink how the acquisition of software-intensive systems should occur in ways that address the recurring problems." SAMSA activity is predicated on the notion that it is important to make the acquisition of future software-intensive systems more agile. This can be achieved by reengineering the various software processes across their overall life cycle. SEI's vision for this type of effort is embodied in an approach called VISTA (Virtual Information SysTem Acquisition). VISTA refers to a process whereby "an evolving series of ever more complete and operational systems versions are acquired through a series of short acquisition life cycles." The VISTA approach is based largely on the construction of models and simulations of software development life cycles. Readers with a further interest in this trend should contact the authors cited before.

A significant approach to trying to improve our overall ability to build and field software systems is embodied in GSAM—*Guidelines for Successful Acquisition and Management of Software Intensive Systems* [12.51]. GSAM attempts to articulate success paths, and thereby mitigate against "repeated inappropriate or unsuccessful practices." These paths might also be called "lessons learned" as well as "best practices." We note from a previous discussion of the work of the Defense Science Board in 2000 [12.40] that best practices also represented an area of special focus in that context. GSAM also suggests some key principles, as summarized below:

1. Focus on the real customer.
2. Talk about the program content and not its politics.
3. Understand the full life cycle of the program and its needs.
4. Determine baseline requirements and the scope of the project as soon as possible.
5. Tackle the program as a series of small steps in order to achieve incremental successes.
6. Assume meaningful measurements for cost, schedule, and quality management.
7. Identify and manage key program risks.
8. Capture and utilize data and best practices from earlier programs.
9. Emphasize and sponsor improvement efforts, and assure that managers "walk the talk."

12.3.6 I-CASE (Integrated Computer-Aided Software Engineering)

For a number of years, software vendors have provided tools for software engineers under the general title of computer-aided software engineering (CASE). The purpose of such tools, of course, has been to increase the productivity of software engineering projects. The Air Force's Software Technology Support Center (STSC) has been cataloging and analyzing these tools, for which a representative list is provided in Exhibit 12.4. Formal reports from the STSC on these tools have included the categories of

- Requirements analysis and design
- Software estimation

Exhibit 12.4:   Selected Computer-Aided Software Engineering (CASE) Tools

Name of Tool	Vendor
BACHMAN7Analyst	Bachman Information Systems
CARD tools	CARDTools Systems Corp.
Corvision	Cortex Corp.
Design Aid II	Yourdon/CGI Systems Inc.
EasyCASE Professional	Evergreen CASE Tools, Inc.
Excelerator Series	Intersolv
lEW/Workstations	Knowledgeware Federal Systems
Integrated Systems Engineering Toolset	LBMS, Inc.
Information Engineering Facility	Texas Instruments
ISE Eiffel	Interactive Software Engineering Inc.
Maestro II	Softlab Inc.
MAGEC RAD System	Magec Software
Micro Focus Cobol Workbench	Micro Focus
MicroSTEP	Syscorp International
Natural Engineering Series	Software AG of North America
Objectmaker	Mark V Systems
Objectory	Objective Systems
Oracle CASE	Oracle Corp.
Paradigm Plus	Protosoft, Inc.
POSE	Computer Systems Advisors
ProKappa	Intellicorp, Inc.
Prokit Workbench	McDonnell Douglas Information Systems
Promod CASE Tools	Meridian Software Systems
RDD-100	Ascent Logic Corp.
RTM	Marconi Systems Technology
RTrace	Protocol/Zycad Corp.
Silverrun	Computer Systems Advisors
Softbench	Hewlett-Packard
Software Engineering Toolkit	Caset Corp.
Software Through Pictures	Interactive Development Environments
Statemate	i-Logix, Inc.
Sterling Developer	Sterling Software
System Architect	Popkin Software & Systems, Inc.
System Developer I	Cadware Inc.
superCASE	Advanced Technology International
Synon/2E	Synon, Inc.
TAGS	Teledyne Brown Engineering
Teamwork	Cadre Technologies, Inc.
TreeSoft	+Software
Visible Analyst Workbench	Visible Systems Corp.

- Source code static analysis
- Reengineering
- Documentation
- Project management
- Test preparation, execution, and analysis
- Software engineering environments

Starting around 1990, the DoD formally entered the CASE arena by sponsoring a program known as Integrated Computer-Aided Software Engineering (I-CASE). They issued a request for proposal (RFP) to industry for the purpose of developing a suite of I-CASE tools. The Air Force's Gunter Air Force Base in Alabama became the agent for the I-CASE procurement. As of 1991, the I-CASE development environment was described [12.52] as a central database repository encyclopedia with connectivity to a variety of tools for:

- Requirements analysis and specifications
- Project management
- Quality assurance
- Design
- Testing
- Prototyping
- Code generation
- Configuration management
- Cross development

Other key features of I-CASE included open systems, Ada as a development language, and evolutionary development.

The contract for I-CASE was ultimately won by Logicon in 1994 from the Air Force's Standard Systems Center. The contract had a face value of over $670 million over a ten-year period. The overall notion was to build a standard set of I-CASE tools that could be used throughout the DoD and its contractors. I-CASE has three essential parts:

1. A software engineering environment
2. An operational test environment
3. A run-time environment

Logicon brought its I-CASE product to the marketplace using the name LOGICORE.

I-CASE represents a major investment by the DoD to help solve the software development problem. As of 1995, the DoD issued a statement by the

Assistant Secretary of Defense (Command, Control, Communications, and Intelligence—C3I) reflecting its commitment to I-CASE [12.53]:

> This statement confirms the Department's commitment to the I-CASE initiative and the need to aggressively exploit the advantages available now through the Air Force I-CASE contract. The contract is available to DOD components and any Federal Agency wishing to procure the I-CASE software engineering environment (SEE) or tools. . . .

It appears that I-CASE or some derivative thereof will be an increasingly strong force with respect to how we execute the complex tasks of software engineering.

12.3.7 Architecting

In the Defense Science Board report [12.39] on acquiring defense software commercially, considerable attention was focused on software system architectures. As the report points out, a software architecture consists of:

- Software system components
- The relationship between the components
- Rules for their composition (constraints)

In the same vein, documentation of a software architecture would contain, as a minimum:

- System functionality
- Software system components
- Interfaces, standards, and protocols
- An execution model consisting of:
  — data flows
  — control flow
  — critical timing and throughput considerations
  — error handling

A good software architecture was considered a "prime enabler of flexibility and reuse" and might reduce the costs of changes and upgrades by as much as 30–50% per year. Solid architectures were also seen as a key tool for:

- Evolutionary development (see Section 12.3.4)
- Early involvement of users with functional capability
- Ability to include changing commercial technology
- Reuse
- Assisting in the areas of requirements changes and product line management

Of course, a software architecture must be viewed within the context of an overall systems architecture (see Chapter 9), a point that was not strongly emphasized in the DSB report.

An overview article on software architectures claimed that there is appropriate and renewed interest in this subject [12.54]. This revival of interest has led many researchers to reconsider basic questions, such as the following:

- What is a software architecture?
- Are there generic forms of architectures?
- Are there sets of preferred architectures?
- What is the process whereby an architecture is developed?
- How can we promote the use of good architecting principles in both government and industry?

A variety of cases were cited in which the preceding and other software architecture issues were considered by, for example, the Air Force, (e.g., the CARDS and PRISM programs), industry, and the Software Engineering Institute (SEI) at Carnegie-Mellon University, with the latter apparently taking a lead role in these matters. However, other parts of the government are stepping up to this challenge. An example is the Defense Information Systems Agency (DISA) that set forth a technical architecture framework for information management [12.55]. This framework provided "guidance for the evolution of the DOD technical infrastructure" with respect to information management rather than providing a specific system architecture. The framework, known as TAFIM, was presented in a series of volumes, namely:

Volume 1: Overview
Volume 2: Technical Reference Model
Volume 3: Architecture Concepts and Design Guidance
Volume 4: DoD Standards-Based Architecture Planning Guide
Volume 5: Support Plan
Volume 6: DoD Goal Security Architecture
Volume 7: Information Technology Standards Guidance
Volume 8: DoD Human Computer Interface

Efforts of this type move us all in a positive direction in terms of developing a better understanding of the issue of software and information system architecting.

Finally, with respect to software architecting, we have two topics that have already been discussed in previous sections of this book. The first has to do with the architecting of systems, and the point is made that the same procedures for architecting systems are applicable to software architecting. Chapter 9 addresses these points, as does Section 12.2.4 of this chapter, dealing in part with the C4ISR Architecture Framework [12.16]. In addition, the Appendix illustrates how a software system may be architected. The second point is related to the IEEE standard that is concerned with architectural descriptions [12.56]. Although that standard does not demonstrate how to architect a software system, it does introduce the important notion of architectural *views* of software systems. These views, as per the three important system views as articulated by the C4ISR Framework, provide insight into the ultimate architecture and, by means of reverse engineering, it may be possible to infer a legitimate software architecting process. Until this is demonstrated, however, it is suggested that the reader follow the architecting process defined in Chapter 9.

12.3.8 Reengineering

The motivation for reengineering of software systems lies in the fact that there is a large amount of legacy system software that has to be updated, upgraded, and "reengineered." Reengineering is also associated with the need to maintain these legacy systems. The Air Force's Software Technology Support Center (STSC), discussed before, has had a reengineering initiative [12.57] whose "goal is to encourage reengineering technology transfer and adoption." Part of the STSC's approach is represented by its Software Reengineering Assessment Handbook, which deals with three processes:

1. Reengineering strategy selection
2. Cost analysis
3. Management and priority setting

The STSC has also formulated a nine-step reengineering preparation process consisting of the following activities:

1. Evaluation of needs
2. Formation of the reengineering team
3. Definition of the development/maintenance environment
4. Creation of a set of metrics
5. Analysis of the legacy systems
6. Creation of an implementation plan
7. Ensuring that the test-bed is current and complete
8. Analysis of available reengineering tools
9. Training

Note that this process is preparatory to the main reengineering activities. Readers with a further interest in this, as well as in other software engineering subjects, should get on the mailing list for *CrossTalk,* the STSC's Journal [12.57].

Another important report [12.58] shows the relationship between reengineering and reverse engineering, containing 200 references in these fields. Definitions of important terms are cited by reference [12.59] as:

> **Legacy Systems.** Software systems that are 10–25 years old and often in poor condition
>
> **Reverse Engineering.** A process that identifies components and their interrelationships as well as creating new representations in another form

In effect, the process of reverse engineering takes existing software code and converts it into specifications, design descriptions, and software components. Given this set of results, it is then possible to "forward engineer" the system into a new implementation, to include specific code in another language. The goal is to perform both reverse and forward processes in an automated manner. Conceptually, then, old COBOL code could by such a process be converted into C or Ada code. This type of transformation clearly has many benefits if performed accurately and without too much code redundancy. Six objectives of reverse engineering are cited [12.60] as supporting:

1. Software reuse
2. Documentation
3. Information recovery
4. Maintenance reduction
5. Platform migration
6. Migration to a CASE environment (see Section 12.3.6)

Thus, reengineering and reverse engineering of software are not the same, but reverse engineering is one way to implement reengineering. Research activities in both areas are very active and represent an important trend within the overall topic of software engineering.

12.3.9 Other Trend Areas

In addition to the preceding rather long list of trends in software engineering, we can expect strong and continuing activities with respect to the following:

- The development of new software metrics
- More widespread use and extensions of SEI's basic capability maturity model
- Standards for software development and engineering
- Further illumination of the relationship between systems engineering and software engineering
- Academic programs that blend theory and practice in a more effective manner

The last item suggests that there may not be sufficient integration of the ways in which academia and industry approach or look at the field of software engineering. To be sure, the two have different purposes and goals. But one may argue that their roles may not be well enough related so that each will support the other. In this regard, an article in *Computer* magazine [12.61] asked (and tried to answer) the question: Where is software headed? One conclusion was that there was an apparent chasm between academia and industry and that "the two groups share radically different views on where software is headed." Whatever these differences, some of the topics of interest for the future are:

- Object-oriented programming and design
- Software productivity tools
- Rapid application development
- Algorithms
- Worldwide-Web tools
- Graphical user interfaces (GUIs)
- Multimedia software engineering
- Formal methods
- Group communications

The preceding sections and list are ample testament to the breadth and richness of the field of software engineering. The reader is also referred to an *Encyclopedia of Software Engineering* [12.62] to further explore the subjects discussed here, as well as other related topics.

12.4 PROJECT MANAGEMENT TRENDS

12.4.1 General Management Trends

Every project is executed in some context, whether it be in industry, government, or academia. This organizational setting thus has some influence on the project, establishing an environment of constraint, facilitation, or some combination thereof. In this respect, general management trends can impact what the project personnel are able to do and what they cannot do.

A dozen general management trends that are of note include:

1. Movement toward decentralization
2. Increased span of control
3. Deeper empowerment
4. More training, at all levels
5. Increased use of integrated product teams (IPTs)

6. A focus on business process reengineering (BPR)
7. Continuous improvement (from Total Quality Management)
8. Less loyalty between employer and employee
9. A move toward the "systems approach" [12.63]
10. Recognition of "core competencies"
11. Building of a shared vision for the enterprise
12. Sharper customer focus

Space limitations do not permit the exploration of all these general management trends, although some are revisited in the next chapter. Suffice it to say that the Project Manager and Chief Systems Engineer should be aware of these areas and what is happening with respect to each within their organization. As an example, if training is alive and well in their organization, they have to consider the training needs for their project and schedule their people so as to take advantage of this type of opportunity. In a similar vein, if the enterprise is active in business process reengineering, it might be useful to provide inputs as to the internal processes that have to be streamlined to support the project. This might include processes in accounting, contracts, human resources, and the like. Finally, if the organization subscribes to the notion of deeper empowerment, then one issue might be how to more strongly empower the personnel on the particular project in question. The reader interested in this area should consult the vast literature on the general subject of management.

12.4.2 Project Management Tools

There are numerous software tools available to support the Project Manager and the team. These tools tend to focus on scheduling, and most of them also provide cost aggregation and budgeting capabilities. The Air Force's STSC has compiled a very complete list of such tools [12.64] with considerable information about each tool. The Center has cataloged them in three cost categories: (A) cost less than $300, (B) cost between $300 and $2,500, and (C) cost greater than $2,500. As one would expect, the capabilities of the tools increase as does the price, but a lower-cost tool can be adequate for most smaller projects. A selected list of such tools is provided in Exhibit 12.5, utilizing the cost data in the referenced STSC report. This author has used several tools in category (B); they are extremely cost-effective. The types of features that one might look for in selecting a tool, aside from price, include:

- Gantt charting
- Program evaluation and review technique (PERT) charting
- Operating system/platform

Exhibit 12.5:   Selected List of Project Management Tools

Project Management Tool	Vendor	Cost Category
Advanced ProPath	SoftCorp, Inc.	A
Critical Path Project Manager	Dynacomp, Inc.	A
Fast Track Resource	AEC Management Systems Inc.	A
InstaPlan	MicroPlanning International	A
M2M	MC2 Engineering Software	A
Milestone	Digital Marketing Corp.	A
Harvard Project Manager	Software Publishing Co.	B
Microsoft Project	Microsoft	B
Micro Planner	Micro Planning International	B
Project Scheduler	Scitor Corp.	B
Syzygy	Syzygy Development, Inc.	B
Time Line	Symantec Corp.	B
Artemis Prestige	Lucas Management Systems	C
Open Plan	Welcom Software	C
Promis	Cambridge Management Systems Inc.	C
Qwiknet Professional	Project SW & Development Inc	C
SLIM-Control	Quantitative Software Management	C
Viewpoint	Computer Aided Management Inc.	C

- Work breakdown structure
- Task responsibility matrix
- Resource allocation
- Resource leveling
- Resource limiting
- Resource conflict analysis
- Cost aggregation
- Standard reporting
- Customized reporting
- Ease of use
- Vendor support

In addition to the classical set of project management tools, software has come on the scene under the category of Groupware, which, as the name implies, is intended to support an entire project team. As an example, Lotus Notes was one of the best-known and best-selling packages of this type. It is safe to assume that both project management and Groupware tools will be available indefinitely to support project and systems engineering teams. Both the Project Manager and the Chief Systems Engineer should decide on the support tools that the team requires as soon as possible after the project start date. Considerable efficiencies can be gained by the active use of project management and Groupware tools. As a consequence, they are most highly recommended for a project team.

12.4.3 Two Department of Defense (DoD) Initiatives

The sheer size and complexity of DoD programs and projects make them an excellent showcase for project management issues and trends. Improvements in project management, especially in the software arena, have the potential for improving products and processes and, at the same time, saving literally billions of dollars. In very large organizations, in both government and industry, this is more easily said than done.

One noteworthy initiative within the DoD is its Practical Software Measurement (PSM) program, which is meant to establish a "foundation for objective project management" [12.65]. It also purports to achieve objectivity by focusing in particular upon measurement—what to measure and how. The U.S. Army has played a major role in formulating PSM, which "was developed to help meet today's software management challenges." Here again, we see an emphasis upon software as it plays an increasingly important role in building and fielding our large systems. PSM is applied for the most part at the project level, but it is claimed that the same basic principles can be used to extend PSM to the organizational and enterprise levels.

Since PSM is focused on measurement, it prescribes a set of measurement principles that need to be taken into account in software programs. Some of these principles are:

- Measurement requirements are dependent upon project issues and objectives.
- Software measurement is defined by the software process of the developer.
- It is important to have an independent analysis capability.
- Software measurement needs to occur over the entire life cycle.

The PSM Guide addresses three main topics: (1) tailoring the software measures as appropriate to the project in question, (2) applying the software measures so that the data can be interpreted in terms of useful project information and decisions, and (3) implementing a measurement process that is effective within the specific organization.

A second project management type of initiative within the DoD is related to the "earned value analysis" (EVA) described in Chapter 4. In 1996, the DoD recognized guidelines produced in an industry standard that dealt with Earned Value Management Systems [12.66]. This replaced earlier views of EVA and set forth some thirty-two criteria to be used in EVA. Since EVA has been accepted on many large programs in industry as well as government as a useful way of tracking cost, schedule, and performance (work actually accomplished), the reader with a further interest in this trend should examine the cited reference and standard and the appropriate offices within the DoD.

QUESTIONS EXERCISES

12.1 Define and describe ten categories of software tools that can be used to support
  a. systems engineering
  b. software engineering

12.2 Obtain a copy of the latest IEEE standard on systems engineering and contrast it with Military Standard 499B.

12.3 Obtain a copy of the latest ISO standard on software engineering and contrast it with Military Standard 498.

12.4 Write a five-page report on the current state of practice in software reuse.

12.5 Write a five-page report on the current state of practice in Integrated Computer-Aided Software Engineering (I-CASE).

12.6 Evaluate the relative capabilities of three project management software tools.

12.7 Write a five-page report that explains and evaluates the most recent position of the Department of Defense (DoD) relative to the architecting of systems.

12.8 Write a five-page report that explains and evaluates the most recent status of the Capability Maturity Model(s) for systems and software, including the integrated models.

12.9 Write a three-page report on two new trends in systems engineering.

12.10 Write a three-page report on two new trends in software engineering.

REFERENCES

12.1 *Systems Engineering* (1994). The Journal of the National Council on Systems Engineering (NCOSE) **1**(1), Sunnyvale, CA. *(Note:* NCOSE has changed its name to INCOSE, 2033 Sixth Ave. #804, Seattle, WA 98121–2546.)

12.2 "Systems Engineering in the Global Market Place" (1995). *NCOSE 5th Annual International Symposium*, St. Louis, July 22–26.

12.3 Eisner, H., J. Marciniak, and R. McMillan (1991). "Computer-Aided System of Systems (S2) Engineering." *IEEE International Conference on Systems, Man, and Cybernetics*, Charlottesville, VA., October 13–16.

12.4 Eisner, H., R. McMillan, J. Marciniak, and W. Pragluski (1993). "RCASSE: Rapid Computer-Aided System of Systems (S2) Engineering." *NCOSE 3rd Annual International Symposium*, Arlington, VA, July 26–28.

12.5 Kuhn, D., and S. Garcia (1994). "Developing a Capability Maturity Model for Systems Engineering." *Best Presentations of the NCOSE 4th Annual International Symposium*, San Jose, CA, August 10–12.

12.6 Garcia, S. (1994). *SE-CMM Model Description, Release 2.04*, SEI-94-HB-4. Pittsburgh, PA: Software Engineering Institute, Carnegie-Mellon University.

12.7 Pierson, H. (1995). "Comparison of NCOSE Interim Model & SE-CMM," Virginia Center of Excellence for Software Reuse and Technology Transfer, April 11. Presented as part of NCOSE Notes from the Network 3(4).

12.8 *Systems Engineering Capability Model (SECM)*, EIA/IS 731, (1998). Washington, DC: Electronic Industries Association, Engineering Department.

12.9 L. Ibrahim, et al.,(1997). *The Federal Aviation Administration (FAA) Integrated Capability Maturity Mode (iCMM)l, Version 1.0*, Washington, DC: Federal Aviation Administration, November.

12.10 Software Engineering Institute, Carnegie-Mellon University, Website: www.sei.cmu.edu

12.11 Rechtin, E. (1991). *Systems Architecting*. Englewood Cliffs, NJ: Prentice Hall.

12.12 Rechtin, E., and M. Maier (1997). *The Art of Systems Architecting*. Boca Raton, FL: CRC Press.

12.13 Rechtin, E. (1994). "Foundations of Systems Architecting," *Systems Engineering* **1**(1): 35–42.

12.14 Rechtin, E. (1994). "The Systems Architect: Specialty, Role and Responsibility." *Best Presentations of the NCOSE 4th Annual International Symposium*, San Jose, CA., August 10–12.

12.15 First Annual Workshop on Engineering of Systems in the 21st Century: Facing the Challenge, Focus Group Reports (1994). Sponsored by the Office of Naval Research and Naval Surface Warfare Center, Department of the Navy, Fredericksburg, VA, June 28–30.

12.16 *C4ISR Architecture Framework*, Version 2.0 (1997). Washington, DC: U.S. Department of Defense, Pentagon, December 18; see also Website: www.c3i.osd.mil

12.17 C4ISR Architecure Working Group (AWG) (1998). *Final Report*. Washington, DC: U.S. Department of Defense, Pentagon, April 14; see also Website: www.c3i.osd.mil

12.18 *Public Policies & Priorities: 1995–1996* (1995). Washington, DC: American Association of Engineering Societies.

12.19 *Statement of the AAES on The Role of the Engineer in Sustainable Development: Sustainable Technologies and Processes* (1995). Washington, DC: American Society of Engineering Societies.

12.20 Sage, A. (1992). *Systems Engineering*. New York: John Wiley.

12.21 Sage, A. (1992). *Systems Management for Information Technology and Software Engineering*. New York: John Wiley.

12.22 Armstrong, J., and A. Sage (1995). *An Introduction to Systems Engineering*. New York: John Wiley.

12.23 Sage, A., and W Rouse (1999). *Handbook of Systems Engineering and Management*, New York; John Wiley.

12.24 Sage, A. (1994). "The Many Faces of Systems Engineering," *Systems Engineering* **1**(1): 43–60.

12.25 Cochran, M., et al. (1995). *A Tailorable Process for Systems Engineering*, Software Productivity Consortium Report SPC-94095-CMC. Herndon, VA: Software Productivity Consortium.

12.26 Eisner, H. (1988). *Computer-Aided Systems Engineering*. Englewood Cliffs, NJ: Prentice Hall.

12.27 Conference on Tools & Methods for Business Engineering, (1995). Sponsored by Enterprise Reengineering (7777 Lesburg Pike, Falls Church, VA), Arlington, VA, May 16–17.

12.28 Center of Excellence in Computer-Aided Systems Engineering (CECASE), J. Holtzman, Director, Lawrence, KS.

12.29 Lamonica, F. (1994). "Rome Laboratory System Engineering Research and Development Program," *CrossTalk* (April): 4–7.

12.30 *Federal Computer Week* (1995). July 3; and *Washington Technology*.

12.31 *Secretary of Defense Memorandum on Military Specifications and Standards* (1994). Washington, DC: U.S. Department of Defense.

12.32 *Acquisition Reform: DOD Begins Program to Reform Specifications and Standards* (1994). Report to Congressional Committees, GAO/NSIAD-95-14. Washington, DC: U.S. General Accounting Office.

12.33 *Blueprint for Change: Report of the DoD Process Action Team on Specifications and Standards* (1994). Washington, DC: Office of the Under Secretary of Defense. Reprinted by the U.S. Department of Commerce, National Technical Information Service, AD-A278-102.

12.34 *Joint Logistics Commanders Guidance for Use of Evolutionary Strategy to Acquire Weapon Systems* (1995). Fort Belvoir, VA: Defense Systems Management College Press.

12.35 *The Defense Acquisition System*, DoD Directive 5000.1 (2000); *Operation of the Defense Acquisition System*, DoD Directive 5000.2 (2000); *Mandatory Procedures for Major Defense Acquisition Programs (MDAPS) and Major Automated Information Systems (MAIS) Acquisition Program*, DoD Directive 5000.2-R (2000). Washington, DC: U.S. Department of Defense.

12.36 Gansler, J. (2000). *The Road Ahead*. Washington, DC: Department of Defense, Office of the Under Secretary of Defense for Acquisition and Technology, June.

12.37 Sage, A., and C. Lynch (1998). "Systems Integration and Architecting," *Systems Engineering* **1**(3).

12.38 The National Software Council, John J. Marciniak, President. Kaman Sciences Corporation, Alexandria, VA [(703) 329–7368].

12.39 *Report of the Defense Science Board Task Force on Acquiring Defense Software Commercially* (1994). Washington, DC: Office of the Under Secretary of Defense for Acquisition and Technology.

12.40 Report of the Defense Science Board (DSB) Task Force on Defense Software (2000). Washington, DC: Office of the Under Secretary of Defense for Acquisition and Technology, November.

12.41 The Standish Group. *CHAOS Report*; see also Website: standishgroup.com, 1999.

12.42 Druyun, D. (1994). "Software Reuse Incentive Policy," *CrossTalk* (January): 5. (Druyun was the Deputy Assistant Secretary of the Air Force (Acquisition) at the time.)

12.43 Levine, T. (1994). "Report of the Working Group on Reuse Education," *CrossTalk* (December): 21–25.

12.44 Davis, T. (1994). "The Reuse Capability Model," *CrossTalk* (March): 5–9.

12.45 Hills, F. (1994). "Study Points Way to More Effective Software Reuse," *CrossTalk* (May): 23–24.

12.46 Sodhi, J., and M. Smith (1994). "Marching Toward a Software Reuse Future," *CrossTalk* (September): 20–24.

12.47 "Commerce Department Announces 41 Awards for Advanced R&D in Four Key Technologies," (1994) *U.S. Department of Commerce News* (October): 1–2.

12.48 Eisner, H. (1995). "Reengineering the Software Acquisition Process Using Developer Off-the-Shelf Systems (DOTSS)," *7995 IEEE International Conference on Systems, Man and Cybernetics*, Vancouver, BC, October 22–25.

12.49 Sorensen, R. (1995). "A Comparison of Software Development Methodologies," *CrossTalk* (January): 12–17.

12.50 Boehm, B., and W. Scacchi (1995). *Simulation and Modeling for Software Acquisition (SAMSA)*. Los Angeles: University of Southern California, ATRIUM Laboratory, Information and Operations Management Department, School of Business Administration.

12.51 *Guidelines for Successful Acquisition and Management of Software-Intensive Systems (GSAM)*, Vol. **1**—Version 3.0 (2000), Department of the Air Force, Software Technology Support Center, May; see also Website: web2.deskbook.osd.mil

12.52 Tomlin, B. (1991). "Integrated Computer-Aided Software Engineering." *Software Technology Support Center Conference*, April 16, Hill AFB, Ogden, UT.

12.53 Paige, E. (1995). "DOD Commitment to I-CASE," *CrossTalk* (May): 6.

12.54 Kogut, P., and P. Clements (1994). "The Software Architecture Renaissance," *CrossTalk* (November): 20–23.

12.55 Defense Information Systems Agency (DISA) (1994). *Department of Defense Technical Architecture Framework for Information Management*, volume 1, *Overview, Version 2.0*. Washington, DC: Department of Defense (June 30).

12.56 *Draft Recommended Practice for Architectural Description*, IEEE 1471 (1999). New York: Institute for Electrical and Electronics Engineers (IEEE), December.

12.57 Sittenhauer, C., M. Olsem, and J. Balaban (1995). "Software Reengineering at the STSC," *CrossTalk* (January): 6–8.

12.58 Urban, J., H. Joo, and Y. Wu (1995). "Software Reverse Engineering and Reengineering." Report prepared for Rome Laboratory, RL/COEE, Griffis AFB, New York.

12.59 Chikofsky, E., and J. Cross (1990). "Reverse Engineering and Design Recovery: A Taxonomy," *IEEE Software* **7**(1).

12.60 Frazer, A. (1992). "Reverse Engineering: Hype, Hope or Here?," in P. A. V Hall, ed., *Software Reuse and Reverse Engineering in Practice*. London: Chapman & Hall.

12.61 Lewis, T, et al. (1995). "Where Is Software Headed?" *Computer* **28** (August): 20–32.

12.62 Marciniak, J., ed. (1994). *Encyclopedia of Software Engineering*. New York: John Wiley.

12.63 Senge, P. (1990). *The Fifth Discipline*. New York: Doubleday.

12.64 Berk, K., D. Barrow and T. Steadman (1992). "Project Management Tools Report." Ogden, UT: Software Technology Support Center, Hill AFB.

12.65 *Practical Software Measurement (PSM);* see Website: www.psmsc.com, 1998.

12.66 *Significant Changes Underway in DoD's Earned Value Management Process* (1997), Washington, DC: GAO/NSIAD - 97 - 108, U.S. Government Accounting Office (GAO), May.

____13
SELECTED NEW PERSPECTIVES

13.1 INTRODUCTION

The principal time focus of the new perspectives presented in this chapter is the first decade of the twenty-first century. Rather than trying to systematically cover the full scope of the management of project, program, and systems engineering, subjects have been selected for this chapter that are considered to be of special importance.

13.2 ROLE OF INCOSE

The organization known as INCOSE (International Council on Systems Engineering) has continued to explore and advance the state of the art and practice of systems engineering and its management (see also the citation in Chapter 12 section 12.2.1). INCOSE has continued to move forward during the first decade of this century. Although INCOSE advanced in many areas during this period, only a few will be discussed here. The reader is urged to follow INCOSE closely as it continues to contribute to defining and solving many key issues related to systems engineering.

As the most serious independent organization working on systems engineering matters, INCOSE has been exploring where this discipline might evolve by the year 2020 [13.1]. Indeed, the organization has been developing an "integrating framework" for defining its vision for the future.

Five key areas within that vision include:

1. Global environment
2. Systems
3. Processes
4. Modeling and tools
5. Education and research

In order to proceed with these key areas, INCOSE has been looking at current trends, drivers and inhibitors, changes, and both near-term (2010) and far-term visions (2020). Given the strength and charter of INCOSE, it is likely that the organization will play a major role in how systems engineering evolves between now and the year 2020.

In addition to the above, INCOSE has been a leader in the matter of certification for the systems engineering professional, and has established a set of certification requirements [13.2].

INCOSE's *Systems Engineering Handbook* [13.3] is now in its third version. The organization of this handbook is consistent with a standard known as ISO/IEC 15288, (see section 2.5.6) which deals with systems engineering and its life-cycle processes [13.4]. The four process groups in this standard and in the handbook are:

1. Technical processes
2. Project processes
3. Enterprise processes
4. Agreement processes

Several formal activities support these processes, as discussed briefly in Chapter 2.

Finally, INCOSE produces a quarterly journal called *Systems Engineering* under the pioneering leadership of Andrew P. Sage, its editor-in-chief. This journal carries especially good articles, for both the theorist and practitioner of systems engineering. All members of INCOSE receive the journal.

13.3   ACQUISITION OF SYSTEMS

Matters regarding the acquisition of systems have been discussed at several points in this book. Here we examine new acquisition perspectives that have been documented during the twenty-first century. This documentation includes a directive and an instruction sometimes referred to as the 5000 Series, and it sets the stage for acquisition personnel to consider how to approach the acquisition issue.

13.3.1   5000.1 Directive

The 5000.1 Directive deals with acquisition in the Department of Defense (DoD) and represents its position as of the year 2003. Its primary objective is [13.5]:

> to acquire quality products that satisfy user needs with measurable improvements to mission capability and operational support, in a timely manner, and at a fair and reasonable price.

Key points in this directive are cited and briefly discussed below.

Tailor Program Strategies.   The acquisition agent and program office are authorized to adopt strategies that are considered the best ones for the program particulars.

Streamline and Improve the Process.   This directive gives the acquisition agent the go-ahead to make improvements, where possible, and fits with the tailoring notion.

Adopt Innovative Practices.   This set of practices is focused on the reduction of cycle times and costs, along with encouraging teamwork.

Advanced Technology.   The acquisition authorities are confirming the importance of new and advanced technologies in our systems. They provide superior performance and allow us to deploy these systems in the shortest possible time, given time and cost constraints.

Identify Technology Alternatives.   This suggestion, along with the preceding ones, attempts to make sure that technologies that might be critical to the systems in question have not been overlooked. It is also in consonance with the seven aspects of the systems approach, as defined in this book. The architecting process described in Chapter 9 is also based on the specific definition and evaluation of alternatives.

Consider Multiple Concepts.   Emphasis is placed on the analysis and evaluation of alternative ways to satisfy user needs.

Evolutionary Acquisition.   This approach allows us to build systems in increments and is the preferred procedure to satisfy operational needs. This concept was discussed in Chapter 10 with respect to software systems.

Goals for Cost, Schedule, and Performance.   This area focuses on the top three aspects of all system developments: cost, schedule, and performance (or effectiveness). It is suggested specifically to consider the "minimum" number of parameters that will describe or characterize the program. We need to measure the key parameters rather than all the parameters we can think of. The latter approach increases the cost and schedule and in the past has demonstrated diminishing returns.

Decentralize Acquisition. Decentralization of acquisition will be more efficient and will tend to minimize the introduction of programmatic bottlenecks.

Promote a Competitive Environment. A competitive environment has allowed us to purchase our systems in accord with the primary objective cited above, namely, "at a fair and reasonable price," as well as within required schedules.

Cost-Effective Solution. This key measure defines the overall basis for choosing among the alternatives that have been defined: how they rank on an overall cost-effectiveness basis. This approach is the one defined and recommended in this book (Chapter 9).

Total Systems Approach. In this acquisition directive, the DoD endorses the systems approach in a very specific way, believing that it will yield the best results. Although program offices tend to support this notion, at times, and for various reasons, they are not able to implement this approach.

It is satisfying to see that favored acquisition approaches explicitly consider alternatives and finding the most cost-effective solution. At the top level of systems engineering, doing this involves architecting several systems and selecting a preferred architecture on the basis of cost and effectiveness.

13.3.2 5000.2 Instruction

The 5000.2 Instruction [13.6] is a companion to 5000.1 and addresses the operation of the defense acquisition system. Key points cited in the instruction are described briefly next.

Integrated Architectures, Each with Three Views. As architecting might be within the province of joint programs, architectures are to be integrated as much as possible. Also, explicit support is provided for the operational, systems, and technical views that form the basis for the DoD approach to architecting (see also Chapter 9).

Type of Program. A program should be structured so that it is tailored, responsive, and innovative.

New Acquisition Management Framework. Figure 13.1 shows the new framework that describes the acquisition process. It is especially notable

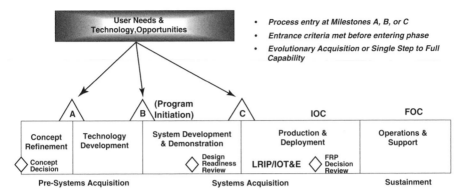

Figure 13.1. Defense Acquisition Management Framework

that technology development and opportunities are stressed. After concept refinement, there is a (new) technology development phase before systems development can be undertaken. Technology readiness and insertion is thus emphasized to assure that our military systems are superior in performance to those of our adversaries. This capability revolves around proper types and levels of technology, which is entirely appropriate.

Integrated Plans and Capability Roadmaps. Although integrated plans have always been stressed, the notions of capabilities and roadmaps might be considered new. For the former, we now have a perspective known as capability-based acquisition. For the latter, roadmaps are a way of showing an explicit pathway from where we are today to where we need to be.

Initial Capabilities Document. As indicated in this instruction, the Initial Capabilities Document (ICD) provides broad, time-phased operational goals, as well as requisite capabilities.

Doctrine, Organization, Training, Material, Leadership, Personnel, and Facilities. The areas of doctrine, organization, training, material, leadership, personnel, and facilities (DOTMLPF) are to be explicitly defined as part of an integrated collaborative process.

Analyses of Alternatives. Guidance for the concept refinement phase (see Figure 13.1) is provided by a strong initial capabilities document (ICD) as well as analyses of alternatives (AoA). One way to interpret the AoA is to say that it is equivalent to the definition and exploration of alternative architectures for a system, a central theme of this book. This AoA is also the basis for a technology development strategy (TDS), which is fully addressed in the next phase.

Technology Development Phase.   "The purpose of this phase is to reduce technology risk and to determine the appropriate set of technologies to be integrated into a full system." This is achieved through a process of iteration and analysis, updating the alternative architectures on the way to being able to choose a preferred architecture. Technology makes a critical difference as we move toward actual system development and demonstration.

System Development Phase.   The system development phase encompasses the more conventional system development, which leads to a system or an "increment of capability" that can be used by the warfighter. Key activities in this phase are systems integration and demonstration. Along the way, test and evaluation processes help to confirm that the system satisfies the overall stated needs and requirements.

Minimum Set of Key Performance Parameters.   There is an appropriate focus on having a *minimum* set of key performance parameters (KPPs) that help to guide the system development process. These parameters must be well defined and measurable. We should be able to determine whether the system actually achieves the required levels for these parameters. Examples of such parameters include the range (of a radar), the detection probability (of a surveillance system), the response time (of an online transaction processor), and the kill probability (of a weapon system). We might also think of these parameters as related to the technical performance measures discussed in some detail in Chapter 7 of this book.

Cost-Effective Operations and Support.   Operations and support (O&S) sustain the system in the field and need to be executed in a cost-effective manner. Here again, cost-effectiveness is a key measure in guiding system selection and behavior. This same perspective was used in formulating alternative architectures and selecting a preferred architecture in the very early days of system development.

Evolutionary Acquisition.   This document clearly confirms that evolutionary acquisition is the preferred approach in which incremental capabilities are delivered. Future capability improvements are recognized, but emphasis is placed on getting a concrete capability into the hands of the warfighter quickly. This incremental approach is distinctly different from the "grand design" approach that did not appropriately meet schedule, cost, and performance goals.

13.3.3   Kadish Report

About two years after the two documents from the 5000 series were published, the acting deputy secretary of defense, Gordon England, requested an integrated acquisition assessment. This was carried out under the leadership of retired General Ronald Kadish, and the report, called the Defense Acquisition Performance Assessment (DAPA) Report, was issued in January 2006 [13.7]. The report gives us further insight into acquisition processes and what might be right, and wrong, with them.

A key player in defense matters over the years, Norman R. Augustine wrote a foreword to the DAPA Report. In it he claimed that there is a lot of agreement regarding the problems that require investigation. But he concluded that "the difficulty resides in having the will to do anything about those problems." He then cited a series of areas in which some progress may be achieved. These areas included:

Seeking experienced capable managers,
Supporting basic research
Starting fewer and finishing more projects
Reducing turbulence
Assigning clear responsibilities
Providing financial reserves
Incrementally budgeting to milestones
Accepting prudent risks
Controlling costs
Disciplining requirements
Using appropriate contractual and competitive instruments
Emphasizing reliability
Creating fast tracks, and
Insisting upon ethical comportment

Augustine was previously the very successful chief executive officer of Lockheed Martin as well as a DoD executive. Thus, it makes considerable sense to pay a great deal of attention to his short list of things to do. Note that most of the items he mentioned can be considered management issues rather than technical matters.

The Kadish report examined 42 key issues in great detail. Integrated assessments of these issues were sorted into six broad areas:

1. Organization
2. Workforce
3. Budget
4. Requirements
5. Acquisition
6. Industry

Some key recommendations made in the Kadish report, in terms of specific suggested changes, are listed in Table 13.1.

In General Kadish's testimony before Congress [13.8], he provided a short summary of the recommendations in each of the preceding categories.

TABLE 13.1   Key Recommendations in the Kadish Acquisition Report [13.7]

Organization
- Realign authority, accountability, and responsibility at the proper levels.
- Streamline the acquisition process.
- Establish four-star acquisition systems commands, at the service level.

Workforce
- Rebuild and value the acquisition workforce.
- Provide leadership with appropriate incentives.

Budget
- Transform the planning, programming and budgeting (PPB) process.
- Establish a distinct and stable program finding account.

Requirements
- Replace the Joint Capability Integration and Development System with the Joint Capabilities Acquisition and Divestment Plan.
- Establish a two-year process to produce the above plan and a continuous Materiel Solutions Plan development process.
- Add an "operationally acceptable" test evaluation category.
- Give program managers authority to defer nonkey performance parameter requirements to later upgrades.

Acquisition
- Adopt a risk-based source selection process.
- Shift to time-certain development procedures.
- Make schedule a key performance parameter.
- Mandate time start and end dates that are clearly defined, and change the acquisition processes to support them.

Industry
- Overcome the consequences of reduced demand by sharing long-range plans and restructuring competitions for new programs.
- Require government insight and favor formal competition for major subsystems when a lead system integrator strategy is pursued.

13.3.4   Capability-Based Acquisition

Some of the acquisition principles may be stated under the overall category of capability-based acquisition (CBA). Under this notion, an initial capability is built as soon as technologically practical. Then this baseline capability is improved through incremental enhancements. This approach to building new systems is considered to be a primary response to grand designs of quite large systems that have not really succeeded. Too many overruns in cost or schedule, or inadequate performance, have resulted in CBA, as well as other changes.

This focus on delivering a capability (versus an overall system) helps to ensure that real and immediate needs of the warfighter are more likely to be met. Incremental enhancements will provide upgrades, and the end user can have a greater input in a more timely and concrete manner. This method also allows us to be more able to respond to changing requirements and uncertainties (e.g., a changing or uncertain threat). Each increment of capability should help in filling gaps and eliminating shortfalls.

New required capabilities can be satisfied by the grouping of legacy systems, new systems, and judicious technology insertion. To make all this work, improved solutions are needed to cross organizational and funding stovepipes.

Although the list of areas for improvement in systems acquisition is rather long, we must continue to remember the advice of Norman Augustine. Even though good solutions are offered, we still need the will and determination to implement these solutions.

13.3.5 Defense Acquisition Guidebook

As a way to assist the acquisition community in the implementation of the new acquisition perspectives, the DoD produced a *Defense Acquisition Guidebook* [13.9]. As documented, the DoD wanted this handbook to be an electronic resource rather than a conventional book. It contains eleven extremely useful chapters:

Chapter 1. DOD Decision Support Systems
Chapter 2. Defense Acquisition Program Goals and Strategy
Chapter 3. Affordability and Lifecycle Resource Estimates
Chapter 4. Systems Engineering
Chapter 5. Lifecycle Logistics
Chapter 6. Human Systems Integration
Chapter 7. Acquiring Information Technology and National Security
        Systems
Chapter 8. Intelligence, Counterintelligence, and Security Support
Chapter 9. Integrated Test and Evaluation
Chapter 10. Decisions, Assessments, and Periodic Reporting
Chapter 11. Program Management Activities

Note the interesting topics in this guidebook, especially those dealing with systems engineering and project management, two central themes of this book. The former is concerned with such topics as:

- Systems Engineering in DoD Acquisition
- Systems Engineering Processes: How Systems Engineering Is Implemented
- Systems Engineering in the System Lifecycle
- Systems Engineering Decisions: Important Design Considerations
- Systems Engineering Execution: Key Systems Engineering Tools and Techniques
- Systems Engineering Resources

Systems engineering clearly plays a central role in the acquisition of DoD systems.

13.4 PROBLEMS IN SYSTEMS AND SOFTWARE

In this section we look briefly at a limited number of systems and software engineering problems that have already been discussed but are of special importance.

13.4.1 Systems Engineering

In January 2003, on behalf of the under secretary of defense (Acquisition, Technology & Logistics [A,T&L]), the National Defense Industrial Association (NDIA) examined 100 separate systems engineering issues and sorted them into the "top five" issues [13.10]:

1. Lack of awareness of the importance of systems engineering in programs
2. Inadequate qualified resources
3. Insufficient tools and environments for systems engineering execution
4. Inadequate requirements engineering
5. Poor initial program formulation

Item 1 basically means that systems engineering is not being applied to programs in an appropriate manner. Regarding item 2, apparently more people trained in the technical and management aspects of systems engineering are needed. Both industry and government, it is believed, can improve backgrounds through formal education (e.g., master's degree) programs at universities. Item 3 looks at tools and environments for systems engineering. This is not a new subject (see earlier chapters and section 13.11), and may simply be a matter of not keeping up with investments needed in industry and government. Item 4 points to requirements matters, which have been a problem area for a long time. Perhaps a more rigorous requirements engineering process would be helpful, but this issue is almost always on the list of why programs fail. For item 5, problems with initial cost and schedule baselines put many programs at risk. We tend to be too optimistic and have to pay the price down the road. At times, that price can be program cutbacks or cancellations.

The reader is advised to look at problem areas discussed elsewhere in this book (e.g., see Chapter 1).

13.4.2 Software Engineering

A defense-oriented workshop [13.12] defined seven key software issues:

1. Requirements engineering
2. Insufficient participation by software engineers in systems engineering

3. Ineffective planning and management by acquirers and suppliers
4. Insufficient quality and quantity of software engineering expertise
5. Verification methods inadequate for large and complex modern systems
6. Cannot assure proper executive of software in distributed environments
7. Inadequate attention to life-cycle issues for COTS/NDI (Commercial-Off-the-Shelf/Non-Development Items) impacts on cost and risk

As with systems engineering, these seven issues are not particularly new. None of them requires technical breakthroughs to address, but all need an infusion of new money, resources, and commitments. A further articulation of problem areas of note can be found in the author's book on managing complex systems [13.13]. Suggestions for improvements relative to the integration of systems can be found at the end of this chapter.

If we broaden our perspective on software issues to the national level, we find an important investigation and contribution by the Center for National Software Studies. Under the most able leadership of Alan Salisbury, they produced a second National Software Summit report [13.14]. Their stated 2015 vision was:

Achieving the ability to routinely develop trustworthy software products and systems, while ensuring the continued competitiveness of the U.S. software industry.

The overall National Software Strategy can be summarized by four imperatives:

1. Improve software trustworthiness.
2. Educate and field the software workforce.
3. Reenergize software research and development.
4. Encourage innovation within the U.S. software industry.

The report provides supporting rationales and details [13.14]. Note that these results were obtained by looking at our software issues and problems from a national, and at times international, perspective.

These and other studies of software indicate that many of our system vulnerabilities lie, significantly, in how we produce and maintain software.

13.5 INTEGRATION OF SYSTEMS

In this section, we briefly explore several issues that pertain to the art and science of systems integration. Some of these issues relate to the systems and software problems discussed in the previous section of this chapter and also in Chapter 12. It is hoped that the reader will gain some new and useful perspectives with respect to this expansive and important subject.

13.5.1   Systems of Systems Revisited

In section 12.2.2 of Chapter 12, we explored the matter of system of systems engineering (SoSE). This was an attempt to examine the meaning of systems engineering when we are dealing with *systems* of systems. Emphasis was placed on disciplines of particular importance in such a case, with a resultant top-level structure of:

1. Integration engineering
2. Integration management
3. Transition management

Section 12.2.2 provided a substructure for the top-level structure.

Chapter 12 paid special attention to matters of *integration* and *management*. In engineering *systems of systems*, we must pay even more attention to these two areas, given that the individual systems are or have been subject to a basic systems approach.

Other investigators have made important contributions in the domain of systems of systems and the engineering thereof. For example, Sage and Cuppan examined systems of systems and federations of systems, concluding that both exhibited the "behaviors of complex adaptive systems" [13.15]. Related concepts, such as a "new federalism" and evolutionary acquisition, are discussed in some detail. Charles Keating and several colleagues at Old Dominion University examined the "emerging discipline" of SoSE to point the direction toward future investigation [13.16]. Mark Maier has set forth a set of "architecting principles for systems of systems" and also emphasized the role of communications relative to the overall architecture [13.17]. The Defense Acquisition University (DAU), in the *Defense Acquisition Guidebook*, advises that the factors of greatest interest include:

- Greater scope and complexity (of integration)
- Dynamic as well as collaborative engineering activities
- Consideration of design optimization
- Reconfiguration of the overall architecture
- Extensive modeling and simulation (M&S) to ascertain overall behavior
- Design and engineering under greater uncertainty
- Rigorous interface definition and control

It is clear that the systems engineering of systems of systems is here to stay, and will need and experience continuing exploration during the coming years. Groups like INCOSE and the System of Systems Engineering Center of Excellence [13.18] are sure to be involved.

13.5.2 Integration of Stovepipes

The topic of the integration of stovepipes can be directly related to building systems of systems, *or* it can turn out to be basically unrelated, depending on one's basic definition of what constitutes a system of systems.

In many federal government programs, there is a tendency to want to "fully" integrate stovepipes. This is usually based on the premise (actually, an assumption) that the integration of stovepipes will lead to a superior system. This may, *or may not*, be the case, depending on the functionality and internal nature of the stovepipes as well as any related constraints (e.g., budget, schedule). A simple example may serve to illustrate the point.

Suppose the acquisition agent for a system defines the top-level requirements as:

1. The overall system shall carry out these functions:
   - Spreadsheet
   - Database management
   - Presentation management
   - Word processing
2. The overall system shall be the synthesis of the "best of breed" systems that execute the above functions.

Assume further that the "best of breed" analysis yields these results [13.13]:

Stovepipes	Best of Breed Selection	Manufacturer
Spreadsheet	Lotus 1-2-3	Lotus
DBMS	Oracle	Oracle
Presentation Manager	Powerpoint	Microsoft
Word Processor	Word Perfect	Corel

Under these conditions, the integration of stovepipes would lead us to try somehow to make systems from four different manufacturers play together. This is a horrendous prospect, and clearly much inferior to simply going with either Microsoft Office or Lotus Smartsuite. This example demonstrates how much trouble one can get into under the overall task of integrating stovepipes. This author, as part of an advisory group for a system several years ago, saw a stovepipe integration program fail. The stovepipes were not easily integrable, and the project manager just ran out of dollars and time.

Is there a plausible answer to this knotty problem? The answer is yes, and lies in approaching the integration of stovepipes problem in two steps:

1. Architecting alternatives as per the approach in Chapter 9
2. Selecting the preferred architecture, and its attendant level of integration, on the basis of cost-effectiveness calculations and considerations, within all program constraints, including schedule and budget

13.5.3   System Complexity

Broadly speaking, we can assume that systems integration (SI) becomes more difficult as the subordinate systems, subsystems, and/or stovepipes become more complex, under more or less most definitions of complexity. At some point down the road, as with our measurement of software complexity (see Chapter 10 and cyclomatic complexity), we will be in a position to measure *system* complexity in an agreed-upon manner. That will give us some guidance on to how to evaluate the ease or difficulty of system integration efforts and will be an enormous step forward.

From an engineering perspective, it is not difficult to enumerate factors that are highly correlated with systems complexity. These factors include [13.13]:

- System size
- Number of functions to be instantiated
- Parallel versus serial operation
- Number of modes of operation
- Duty cycle (static versus dynamic)
- Real-time operations
- Very high performance
- Number of interfaces
- Different types of interfaces
- Degree of integration
- Nonlinear behavior
- Human–machine interactions

Researchers in systems engineering are making progress in this area, trying to develop algorithms that will yield specific and quantitative measures of complexity, at the systems level. Many of listed factors likely will be key variables in such measures.

13.5.4   Horizontal Fusion and Netcentric Notions

Are the trends suggesting more or less systems integration as we move farther into the twenty-first century? Looking at what is happening in various executive agencies of the government (e.g., Department of Homeland Security, Department of Defense), we see moves that are clearly in the direction of more, rather than less, systems integration. This will result in more integration of stovepipes and also more systems of systems.

As an example, we look here a bit more closely at the DoD's approach to information generation, handling, and use. In effect, the DoD is attempting to assure that all warfighters obtain the right information, in the right place, and at the right time. A major focus is the Horizontal Fusion Initiative that will enable netcentric operations (NCO) [13.19]. This portfolio of programs is designed to effect a transformation in how information is generated and used, both strategically and tactically [13.20]. These programs emphasize:

- High connectivity with security
- Trusted massive-bandwidth network systems
- A web-enabled environment
- Real-time situation awareness across the battle space
- Information that is tailored for posting to shared spaces
- Information that is fused and exchanged
- In-depth information assurance (IA)
- Collaborative efforts between communities of interest
- Quick and agile adaptation to deal with uncertainty
- Joint integrated capabilities
- Better and more timely decision making, at all appropriate levels
- Building many facilitating systems, such as the Global Information Grid (GIG) and the Transformational Satellite Communications System (TSAT)

It is believed that stovepipe systems will *not* lead to agile information sharing. We will need to have people, processes, and technologies working together in new and more integrated ways that are not achievable with old platform-oriented programs. We will be able to leverage information so that it will streamline business and battlefield processes and serve as a true force multiplier. All of these net-centric notions are designed to deal with the threat profiles of the new century. The supporting plans and systems represent leading-edge transformations of current capabilities.

13.5.5 Joint Capabilities Integration and Development System

In moving from fully integrated multifunctional systems to increments of capability, it becomes important to be able to identify, with some precision, what gaps in capability exist so that a gap-filler increment can be designed and built. This determination can be quite complex, involving the answers to at least five questions in the defense arena:

1. What are our current threats?
2. Which of these threats are fully addressed and which are not?
3. How can we prioritize the threats?
4. For threats not being addressed, and of medium to high priority, do we have any concrete plans to develop gap fillers?
5. If not, what are the precise functions and features that we need to address the gaps?

All of these questions involve both a broad and deep analysis of current and planned capabilities and systems. The questions are simple, but getting to the answers with the required specificity is a major activity.

The Joint Capabilities Integration and Development System (JCIDS) is designed to deal with this matter. A brief look at this system will make more concrete some of the acquisition issues discussed in this chapter.

JCIDS is intended to "allow joint forces to meet the full range of military challenges of the future" [13.21]. It emphasizes:

- Highly networked operations
- Interoperability
- Coordination among components
- Collaborative environments
- Assessment of existing and proposed capabilities
- Discovering potential redundancies
- Identification of capability gaps
- Technologically sound, sustainable, and affordable increments of capability
- Supportable, innovative solutions (including business areas)
- Knowledge management/decision support tools

Systems like JCIDS demonstrate some of the complexities that are attendant to the problem of assuring and implementing the incremental approach. The systems reject the earlier "grand design" notions since they have not satisfied the rapid response needs of our warfighters, with the required superior use of the most up-to-date technologies.

13.5.6 "Top Dozen" Integration List

Systems integration is without doubt one of the most important aspects of building and fielding large-scale systems. As discussed in Chapter 12, many of the most capable and largest companies in the country call themselves SIs and are properly proud of their capabilities in bringing our most important systems into being. But, at the same time, we have major problems in this arena that do not seem to go away.

In that context, the author offers a top dozen integration list, a set of suggestions for how to improve the manner in which we carry out integration activities.

Suggestion One. *When attempting to integrate stovepipes, do not accept 100 percent integration as an a priori goal.*

The degree to which two or more stovepipes are integrable depends on many factors, including at least their functionality, the specific ways in which the functions (and subfunctions) have been instantiated, and the use and nature of the software. Some stovepipes can be integrated relatively easily; others basically not at all, without an enormous effort in time and dollars. Integrating stovepipes, as a discipline, should follow precisely the methods of architecting that are delineated in this book. Formulate at least three alternative architectures, evaluate all three on a cost-effectiveness basis, and then choose the preferred alternative. The level or degree of integration is *not* the criterion by which this is accomplished. Focus on the cost-effectiveness of the systems being built. Cost-effectiveness of alternatives should be the primary basis for integrating stovepipes.

Suggestion Two. *Always architect a set of alternatives from which to select the preferred architecture.*

This suggestion confirms the approaches cited earlier and in Chapter 9. The acquisition principles discussed earlier in this chapter support this key notion in at least two very specific ways:

1. The explicit use of an analysis of alternatives (AoA)
2. The top-level criterion of system cost-effectiveness

By following this suggestion, the likelihood of basing our detailed design on the best architecture is increased.

Suggestion Three. *Insist that all team members have specific skills in at least one of the two major elements of systems integration, namely, systems engineering and project management.*

Systems engineering and project management have been accepted as the most important aspects of systems integration, assuming that subject matter technical expertise is sufficiently represented. Therefore, at this point people with the requisite skills are brought on to the project. If personnel do not have the necessary skills, the likelihood of project success tends to diminish. In addition, it is a good idea to choose project personnel who have already demonstrated that they know how to operate as a team.

Suggestion Four. *Think of requirements as tentative needs that, at times, can be treated as variables subject to tradeoffs.*

Requirements that are questionable or potentially incorrect (see Chapter 8) should be scrutinized and challenged since they often are major drivers of the system architecting and design processes. A key phrase, cited in Chapter 9, was taken from a DoD document in the context of trade studies:

Desirable and practical trade-offs among stated *user requirements*, design, program schedule, functional and performance *requirements*, and life cycle costs shall be identified and executed.

Chapter 8 discussed how a program was able to save many millions of dollars by changing a single requirement from a one-second response time to a four-second response time [13.22].

Some of the design principles cited by Eb Rechtin, one of our premier systems engineers [13.23], reinforce these points. One example is:

[E]xtreme requirements should remain under challenge throughout system design, implementation and operation.

Setting poor or extreme requirements in concrete has given us no end of trouble over the years.

Suggestion Five. *Accept technology insertion as a key driver for architecture and design.*

The acquisition management framework, shown in Figure 13.1, makes it clear that technology insertion is a critical aspect of building today's systems. Note the dominant position of "User Needs and Technology Opportunities" as well as the specific "Technology Development" phase. We cannot initiate system development until we have confirmed the role that we wish to have technology play in the system we are building. This position is also reinforced by the requirements of both the Systems Engineering Management Plan (SEMP) and the Systems Engineering Plan (SEP). Being on the forward edge of technology leads to superior systems that are crucial in the defense world. Without this type of explicit attention to technology, we may wind up at the trailing edge, and fall behind in both military and commercial domains.

Suggestion Six. *Assure that all systems are subjected to a risk analysis and mitigation discipline.*

It is this author's firm belief that it is both possible and highly desirable for key senior members of the architecting and design team to sit down together, at the beginning of a program, and define the top dozen risks that the team faces in trying to build a new system. This process can be reenacted every three months to ensure that changes are taken into account on a continuing basis. Besides stating the risk areas, the team can begin risk mitigation activities early enough so that mitigation can be an integral part of the architecting and design processes. In other words, our best systems engineers know very early where the high-risk areas are. We need the discipline and the will to incorporate fixes before the risks seriously compromise or degrade the program.

Suggestion Seven. *Accept evolutionary/incremental "chunking" of capabilities to shorten timelines, live within budgets, and improve performance.*

According to current acquisition and systems engineering principles, the evolutionary approach is the one most likely to lead to success. This belief is confirmed by a capability-based acquisition concept, which means that we build increments of the system in a particular and well-defined sequence. For example, if ten specific top-level functions are part of the system, the (overlapping) sequence in which we build the system is:

- First build and integrate to functions 3 and 7 (where function 7 contains an embedded database management system) (increment A)
- Then build to functions 8 and 10, and integrate them (increment B)
- Then build to functions 1, 2, and 6, and integrate them (increment C)
- Then build to functions 4, 5, and 9, and integrate them (increment D)
- Then integrate, as per the overall SEP, increments A, B, C, and D

Under this notion, distinct increments are built in an overlapping sequence, with each increment representing a workable capability that can be delivered to the end user (for the DoD, the warfighter) and used in the real world as soon as possible.

We believe that this evolutionary approach will avoid many of the pitfalls and problems of years past. Variations on this theme are within the purview of the acquisition team, based on the "tailoring" aspects of the acquisition plan.

Suggestion Eight. *Confirm that schedules and budgets are sufficient to build, test, and deliver a system that satisfies the ultimate requirements.*

Too often, schedules and budgets are accepted that all know cannot realistically be achieved. In other words, we start out under enormous pressures to build systems with low success probabilities and then somehow are surprised that we are failing. This is clearly not a satisfactory approach.

The systems integrator (SI) enterprise must have the expertise to obtain solid estimates of what it should take to build a system with the stated requirements, within so-called time and dollar constraints. These estimates, as a minimum, should be provided by no less than three independent internal teams that have been assigned to look at this issue in detail. If the bottom line is that there is a less than 50 percent chance of meeting time, cost, and performance requirements, the development plan needs to be changed, perhaps radically. It may be that an incremental approach can be found that has a high probability of success by delivering smaller interoperable capabilities. We need to get away from a pattern of behavior that leads us to "overpromising and underdelivering" on a continuous basis. Facing the facts is a good place to start.

Suggestion Nine. *Gain efficiency and leverage through reuse methods, while avoiding the pitfalls of reinventing the wheel.*

After many years of developing large-scale systems, we still tend to approach each new system as one of a kind, especially in the federal government arena. This clean sheet of paper approach is often selected for two reasons:

1. It is the way we have done business in the past.
2. We tend to believe that each new system has a unique and inviolate set of requirements.

If we wish to gain major efficiencies in the future, we need to change that mind-set. Real-time high-performance weapon systems will likely stay with the current approach. Non–real-time systems can be developed under more of a cookie-cutter approach, which is likely to lead to major reductions in cost and time to develop without serious performance compromises. These types of systems include personnel tracking systems, inventory management systems, configuration control systems, and various types of financial management systems.

Some enterprises have figured out the way to achieve an important level of standardization in the latter types of systems and have flourished as a consequence. Related concepts whereby we reuse entire systems have been suggested but not necessarily adopted [13.13]. Under these notions, requirements are more flexible, leading to major time and cost improvements.

The area of software reuse remains promising but it has not yet achieved its full potential. Many software engineers would prefer, and believe it is more efficient, to write new code rather than search for and integrate already existing code. If we are to truly transcend the one-of-a-kind barrier in systems integration, we need to pay a lot more attention to reuse notions across the board. Reuse has the potential for order-of-magnitude improvements in several important domains, as demonstrated when and where COTS (Commercial-Off-The-Shelf) solutions have been selected.

Suggestion Ten. *Reduce complexity and implement the K.I.S.S. principle whenever possible.*

Related to suggestion 9, but not identical, is the notion of systematically eschewing complexity and moving more toward simplification. Although we seem to claim that we understand perfectly well the K.I.S.S. principle, we very likely do not use it in a systematic way. This leads to interactions and failure modes that threaten our missions.

An example might be drawn from the manned satellite systems that we have already flown. We had a requirement for a very high availability of, in effect, a laptop computer. Two obvious but clearly different approaches were at hand. One was to build a new and ultra-reliable laptop (perhaps one with an individual 0.99999 reliability). Another approach was to introduce n-fold redundancy, using a set of identical COTS laptops. If each laptop had a reliability of, say, 0.8, then we would need, with no new developments, a set of some eight laptops from one of many COTS suppliers. NASA made the correct choice of employing simple redundancy to achieve its mission goal. This is a clear case of trying to simplify the design and avoid single-point catastrophic failures.

Finally, with respect to this tenth suggestion, we come to some ideas emphasized by Eb Rechtin in his groundbreaking book on systems architecting [13.23]. On the same page, he discussed the K.I.S.S. principle and cited the Occam's Razor notion: "The simplest solution is usually the correct one."

Suggestion Eleven. *Understand and accept the acquisition concepts and constraints under which your system is being acquired.*

We have spent considerable space in this chapter on the topic of the acquisition of systems. The 5000 series and supporting materials provide ground rules and guidance for use by the people whose job it is to acquire the systems that we need. The systems integration firms and personnel can develop immediate insights, from these materials, as to what is driving the acquisition personnel. In some cases, the acquisition framework is shown with considerable clarity (e.g., see Figure 13.1). In others, various approaches are spelled out that ultimately need to be appreciated and followed by the systems integrator. Indeed, the existence of several possibilities is the primary reason why so much attention is paid in this chapter to recent acquisition perspectives. For example, the evolutionary approach is defined and insisted on with respect to building our systems. Choosing a different approach basically violates this guidance, and is not recommended, especially for the systems integrator.

> **Suggestion Twelve.** *Have the will, determination, discipline, and accountability to utilize all preferred practices and processes.*

We need to make substantive improvements in building and fielding new systems, whether they are constructed on a clean sheet of paper or represent the results of integrating existing stovepipes, or are a combination of the two. In most cases, the nature of these improvements is well known, having been studied and accepted by various review panels and boards over the years. The DAPA Report discussed earlier in this chapter is one such example.

With all this advice, are we in possession of coherent and recommended ways to proceed from where we are today, in order to solve our problems? The answer seems clear, and might be expressed as Augustine's admonition, from earlier in this chapter:

> The difficulty resides in having the will to do anything about these problems.

If there is a silver bullet to our systems and software problems, Norman Augustine has pointed us in the right direction.

QUESTIONS/EXERCISES

13.1 Write a three-page overview of the key points in the *Systems Engineering Handbook,* version 3, published by INCOSE.

13.2 Obtain a copy of 5000.1 and develop your list of the twenty most important aspects of this document.

13.3 Obtain a copy of 5000.2 and develop your list of the twenty most important aspects of this document.

13.4 What ten points are emphasized in *both* 5000.1 *and* 5000.2?

13.5 What is a "capability roadmap"? Construct an illustrative example.

13.6 Consider a system with eight major functional areas. Develop a graphic showing how to formulate an evolutionary acquisition sequencing diagram.

13.7 Contrast the Joint Capabilities Integration and Development System (JCIDS) with the Joint Capabilities Acquisition and Divestment Plan.

13.8 Write a three-page overview of how you might "solve" each of the systems engineering problems described in this chapter.

13.9 Write a three-page overview of how you might "solve" each of the software engineering problems described in this chapter.

13.10 Define the ten most important aspects of systems integration, with a brief explanation as to why you selected each item.

REFERENCES

13.1 "Systems Engineering Vision 2020," Version 2.0, INCOSE, October 6, 2006

13.2 See the INCOSE Web site: www.incose.org.

13.3 *Systems Engineering Handbook—A Guide for System Life Cycle Processes and Activities*, Version 3, INCOSE-TP-2003-002-03, June 2006.

13.4 International Standard ISO/IEC 15288, "Systems Engineering—System Life Cycle Processes," 2002-11-01

13.5 U.S. Department of Defense (2003). *The Defense Acquisition System*, Directive 5000.1. Washington, DC: DoD, May 12.

13.6 U.S. Department of Defense (2003). *Operation of the Defense Acquisition System*, Instruction 5000.2. Washington, DC: DoD, May 12.

13.7 U.S. Department of Defense (2006). "Defense Acquisition Performance Assessment," Kadish Report. Washington, DC: DoD, January.

13.8 "Testimony of Lt. General R. T. Kadish" (2006). The Defense Acquisition Performance Review Project, testimony before the 109th Congress, House Armed Services Committee, March 29.

13.9 *Defense Acquisition Guidebook*; see http://akss.dau.mil/dag.

13.10 "Top Five Systems Engineering Issues in Defense Industry" (2003). National Defense Industrial Association, Systems Engineering Division, Task Group Report, January.

13.11 Eisner, H. (1988). *Computer-Aided Systems Engineering*. Upper Saddle River, NJ: Prentice-Hall.

13.12 Schaeffer, M. D. (2006). "DoD Systems and Software Engineering—Taking It to the Next Level." Systems and Software Engineering, Office of the Deputy Under Secretary of Defense (A&T), October 25.

13.13 Eisner, H. (1995). *Managing Complex Systems—Thinking Outside the Box*. New York: John Wiley & Sons.

13.14 "Software 2015: A National Software Strategy to Ensure U.S. Security and Competitiveness" (2005). Report of the 2nd National Software Summit, Center for National Software Studies, April 29.

13.15 Sage, A., and C. Cuppan (2001). "On the Systems Engineering and Management of Systems of Systems and Federations of Systems." *Information, Knowledge, and Systems Management* **2**(4): 325–345.

13.16 Keating, C., et al. (2003). "System of Systems Engineering." *Engineering Management Journal* **15**(3).

13.17 Maier, M. "Architecting Principles for Systems of Systems," www.infoed.com/Open/PAPERS/systems.htm.

13.18 See www.sosece.org.

13.19 Stenbit, J. (2004). "Horizontal Fusion: Enabling Net-Centric Operations and Warfare." *Crosstalk, The Journal of Defense Software Engineering* (January), pp. 4–6.

13.20 Grimes, J. (2006). "From the DoD CIO: The Net-Centric Information Enterprise." *Crosstalk, The Journal of Defense Software Engineering* **19**(7), pp. 4–6.

13.21 Chairman of the Joint Chiefs of Staff, Instruction CJCSI 3170.01D (2004). "Joint Capabilities Integration and Development System," U.S. Department of Defense, March 12.

13.22 Boehm, B. (2002). "Unifying Software Engineering and Systems Engineering." *COMPUTER Magazine* (March), pp. 114–116

13.23 Rechtin, E. (1991). *Systems Architecting*. Upper Saddle River, NJ: Prentice-Hall.

____14
INTEGRATIVE
MANAGEMENT

14.1 INTRODUCTION

The previous chapters have been organized so as to explore the two primary areas of project management, and systems engineering and its management. Interrelationships between these two primary areas have been mentioned in a variety of places in this book but perhaps have not been sufficiently emphasized. In fact, essentially all systems engineering activities occur within some type of program or project context. So the real world connects them whether or not any commentary about them does so. And many commentaries, particularly the collection of extant books, tend to deal with project management *or* systems engineering, but not both. Indeed, one of the key purposes of this book is to bring together these two interrelated subjects under one cover. This chapter moves even further in terms of the interconnectedness between project management and systems engineering management. This is done by explicitly considering the matter of integrative management in this final chapter.

Integrative management is defined as a set of practices whereby people, processes, tools, and systems are brought together into harmonious interoperation so as to maximize their efficiency and effectiveness. The latter, in turn, are further defined as follows [14.1]:

Efficiency: The capability of producing desired results with a minimum of energy, time, money, materials, or other costly inputs

Effectiveness: The capability of bringing about an effect or accomplishing a purpose, sometimes without regard to the quantity of resources consumed in the process

It is a fundamental responsibility of management to seek these conditions and to act as a prime facilitator. The key word is "integrated," so we are looking for interoperability, wholeness, and organic strength. This means paying attention to an overall strategy for integrative management, as well as developing detailed tactics to implement such a strategy. As a management activity, above all, it means paying attention to the interactions between people, the systems that they are in, and the systems that they are trying to build.

We see integrative management at work in an organization when we are able to observe the following:

- Strong and effective teams
- Commitment to "getting the job done"
- Deep interest in the technical issues
- Constructive problem solving
- Corporate support for the needs of the project
- Little or no complaining
- Short and productive meetings
- Rapid flow of information
- Effective computer support
- Involved and happy people

Although these do not guarantee integrative management, they are strong indicators. The following sections explore in greater detail the various domains and contexts of integrative management, including:

- Managers as integrators
- Teams as integrators
- Plans as integrators
- The systems approach as integrator
- Methods and standards as integrators
- Information systems as integrators
- Enterprises as integrators

14.2 MANAGERS AS INTEGRATORS

In Chapter 1, we introduced the project organization and the project triumvirate—the Project Manager (PM), the Chief Systems Engineer (CSE), and the Project Controller (PC). The project organization chart, shown in

Figure 1.2, indicated that all three members of the triumvirate have significant responsibilities and, in fact, are usually managers. This management team serves as the primary integrative factor for the project. If the three persons are working together in an integrative fashion, then the project members will usually follow their example. If there is less than complete trust and information exchange, this, too, will be conveyed to project staff, and a different model of behavior may be established.

A key attribute of the triumvirate is that of being able to communicate. A dozen essentials of an effective communicator have been described in Exhibit 6.1 in Chapter 6. Among the items listed is to be able to "synthesize and integrate." In this context, integration refers mainly to thinking in terms of how pieces of information, as well as human behavior interact and might be brought together in a harmonious way. Appropriate communications itself serves as an integrator. It conveys information in a "holistic" manner so that people on the project can see and experience the interrelationships between information and people. Because one of the principal forums for communications is in the context of a team, the matter of teams as integrators is considered in the next section.

14.3 TEAMS AS INTEGRATORS

The management team, represented by the Project Manager, Chief Systems Engineer, and Project Controller, although critical to the notion of integrative management, is not the only team on a project. Examining the project organization (Figure 1.2) more closely reveals that at least eight(!) additional teams are suggested by the chart. These include the following:

- The team of Chief Systems Engineer and the lead engineers in each of the six supporting disciplines
- The six teams represented by the lead engineers in the six disciplines together with the personnel working in those areas
- The team consisting of the Project Controller as team leader and the personnel heading up the work teams in the areas of scheduling, costing, personnel assignments, facilities, and contract liaison

These teams represent opportunities for adopting an integrative style such that all members of the project know, understand, and sign up to the goals and objectives of the project. In short, all of these subteams have to behave in the manner described in some detail in Chapter 6. If conscientious team behavior patterns are the norm, then a major step toward integrative management has been taken.

Suggestions for building a team that will support and facilitate integrative management were set forth in Exhibit 6.2. Especially key items in terms of integrative management include:

- Encouraging participative, possibly consensual, behavior
- Integrating
- Coordinating
- Facilitating
- Communicating

Adherence to these notions will go a long way toward building an overall integrative management team.

Another important idea with respect to team behavior and integrative management is that of "integrated product teams" (IPTs). IPTs bring together (in a team) all the people who have an influence on how a product is to be developed and sold. Often, the context for an IPT is a new product for a company so that it is clear that, as a minimum, the following functional areas have to be represented in the IPT:

- Management
- Marketing and sales
- Research and development
- Engineering
- Production
- Finance and accounting
- Contracts

Because all relevant functions are part of the process, IPTs are also at times referred to as multidisciplinary teams, multifunctional teams, cross-functional teams, or work groups. The latter have been examined in detail, including their characteristics and beneficial consequences [14.2].

Another aspect of the IPT is that the preceding representation occurs at the very outset of a project. This is in distinction to earlier practices that tended to operate serially. The latter often led to reengineering a product when, for example, the production folks threw the design back over the wall to engineering with the claim that the design was not producible! The IPT, operating over the full life cycle of a product, was able to identify and reconcile possible problems very early and obviate the need for excessive rework and iterations. The IPT notion is also considered a part of concurrent engineering, which is addressed later in this chapter and is included as one of the thirty elements of systems engineering (see Exhibit 7.3). It also has been reported on quite extensively in the literature, particularly with respect to manufacturing processes.

14.4 PLANS AS INTEGRATORS

Plans are descriptions of perspectives and agreed-on future activities. As such, they are integrative because they convey the project agenda to all project personnel as well as other corporate staff people. A documented plan indicates the project intent and gets everyone tuned to the same set of goals, objectives, and approach. A plan gets everyone focused in the same direction.

It is of interest to cite three types of plans here, namely:

1. The project plan
2. The systems engineering management plan (SEMP)
3. The software development plan (SDP)

The project plan, from Chapters 2 and 3, contains seven elements:

1. Needs, goals, objectives, and requirements
2. Task statements, a statement of work, and a work breakdown structure
3. The technical approach to the project
4. A project schedule
5. Organization, staffing, and a task responsibility matrix
6. The project budget
7. A risk analysis

This plan should be conveyed to all project personnel, although there is often resistance to such an action. By doing so, however, everyone has an opportunity to see the "game plan" and to understand how he or she fits into it. By being explicit about what is understood about the project, including budgets and perceived risks, there is an opportunity for people to try to find solutions to problems and to possibly bring new ideas to the attention of project management. The project plan also has to be updated from time to time so that it represents the best and most recent view of the project.

Unlike the project plan, the SEMP focuses exclusively on systems engineering activities. An outline of the contents of a SEMP, within the framework of Military Standard 499B, was provided in Exhibit 2.4. Although the project plan and the SEMP have some elements in common, that is, cost and schedule information, the SEMP also addresses the key issues of:

- The overall systems engineering process
- How technical performance measurement is to be achieved
- Reviews that are scheduled to occur
- The role of technology, with emphasis on transition
- Technical integration teams

The last item, in particular, fosters integrative thinking and action and is described in detail in what follows [14.3]:

Technical Integration Teams. The contractor shall describe how the various inputs into the systems engineering effort will be integrated and how multi-disciplinary teaming will be implemented to integrate appropriate disciplines into a coordinated systems engineering effort that meets cost, schedule and performance objectives. The contractor shall include: (a) how the contractor's organizational structure will support team formation; (b) the composition of functional and subsystem teams; and (c) the products each subsystem and higher level teams will support (e.g., teams organized to support a specific product in the work breakdown structure (WBS), "team of teams" utilized for upper level WBS elements).

We note the emphasis on the key concepts of integration, multidisciplinary teaming, and coordinated effort. The notion of teams here reinforces the discussion of teams as an integrative element in the preceding section of this chapter.

With respect to the Software Development Plan (SDP), Military Standard 498 [14.4] contains a Data Item Description (DID) that describes the purpose of an SDP as follows:

The Software Development Plan (SDP) describes a developer's plans for conducting a software development effort. The term "software developer" in this DID is meant to include new development, modification, reuse, reengineering, maintenance, and all other activities resulting in software products.

The SDP provides the acquirer insight into, and a tool for monitoring, the processes to be followed for software development, the methods to be used, the approach to be followed for each activity, and project schedules, organization, and resources.

Clearly, the SDP is a key document with respect to the software development effort, itself one of the thirty elements of systems engineering. As noted before, this plan conveys to the team, management, and the customer how the software development is to be accomplished. As such, it is an integrative point of reference for all software development activities and how they fit into the overall systems engineering effort.

The three plans, although extremely important, are not the only plans that are called for on many projects. Examples of other plans, called for in Military Standard 498, include

- A software installation plan
- A software test plan
- A software transition plan

as well as various manuals, descriptions, and reports. Here again, project management is encouraged to distribute these documents as widely as possible to gain the integrative effect that is produced when the entire project team knows what is happening and what progress is being made.

14.5 THE SYSTEMS APPROACH AS INTEGRATOR

The systems approach as well as the thirty elements of systems engineering (see Chapter 7), also have an integrative influence, both explicitly and implicitly. By identifying the thirty elements, all activities can be viewed in the context of how they affect, and are affected by, the other elements. For example, when carrying out the reliability-maintainability-availability (RMA) and integrated logistics support (ILS) elements, the engineers must be cognizant of the architecture design and synthesis element as well as several others.

Three of the thirty elements of systems engineering, however, are more sharply related to integrative management, namely:

- Integration
- Interface control
- Concurrent engineering

Integration, in a systems engineering context, refers to the connecting of pieces of hardware and software to make larger builds. In this framework, we are integrating structures, whether they be chunks of hardware or program code. Successful integration requires that all the interfaces be correct so that the pieces, in fact, are able to interoperate in the intended manner. Many projects and systems engineering efforts have failed because of major integration problems.

A related element is that of interface control. In this case, we make explicit the need to examine and control all interfaces as a singular element of systems engineering. By doing so, we are also assisting in the process of integrative management because we are systematically examining all interfaces and attempting to assure that successful integration is achieved. It is also a means whereby potential problems are uncovered before they create havoc with the integrated system, including the backtracking and reengineering of major parts of the system.

Concurrent engineering, another of the thirty elements of systems engineering and discussed in Chapters 7 and 9, has been formally defined as [14.5]:

A systematic approach to the integrated, concurrent design of products and their related processes, including manufacture and support. . . intended to cause the developers, from the outset, to consider all elements of the product life cycle from concept through disposal, including quality, cost, schedule, and user requirements.

The emphasis for concurrent engineering is on the word "concurrent," strongly supporting parallel activities that bring all the required disciplines together at all stages of the life cycle of a system. It is also clearly related to the notions of IPTs, cross-, multifunctional, and multidisciplinary teams. The reader with a further interest in concurrent engineering is referred to the many books as well as a bibliography [14.6] on this subject.

An overview of the systems approach was presented in Section 1.4 of Chapter 1. This approach, embedded in the processes of systems engineering, is a major factor in integrative management.

14.6   METHODS AND STANDARDS AS INTEGRATORS

Both methods and standards tend to support the achievement of integrative management. Because they normally set forth well-considered ways of doing business and are widely distributed, they are read by many researchers and practitioners. This leads to their incorporation into the processes and procedures of a large number of enterprises. This, in turn, promotes a common understanding and vocabulary regarding best practices. This, of course, is an ever-evolving set of activities as we discover new and better ways of building systems.

A few of the methods that have had such effects include

- Structured analysis for systems
- Capability maturity modeling (CMM)
- Total Quality Management (TQM)
- Business process reengineering (BPR)

Of course, by formalizing the processes and elements of project as well as systems engineering management, as represented here, the notion of integrative management is also supported.

The standards that we have discussed in this book also tend to enhance integrative management. These have included military standards (e.g., Military Standards 499B, 498, and others) as well as commercial standards (e.g., those produced and promulgated by the international standards organization [ISO]; the Electronic Industries Association [EIA]; and the Institute of Electrical and Electronics Engineers [IEEE]). Often, the requirements and specifications for new systems cite these standards; it therefore becomes important for industry to be aware of their provisions and meaning. Again, this builds common ways of looking at the systems development process across the industry. In turn, this supports and facilitates integrative management.

14.7 INFORMATION SYSTEMS AS INTEGRATORS

Information systems, aside from serving as critical elements in any enterprise in such domains as accounting, inventory control, proposal preparation, marketing and sales, have had the salutary effects of bringing people closer together and reducing the time it takes for a broad range of communications. Communication networks, in particular, interconnect computers that previously were in stand-alone configurations. This interconnection capability is itself integrative by its very nature.

Networks used by organizations include local-area networks (LANs), wide-area networks (WANs), and metropolitan-area networks (MANs). With the introduction of the Internet, exchanges of information with the world at large has been greatly enhanced. Being able to sit at home or in the office and have access to vast amounts of information has, indeed, constituted a revolution in human and organizational behavior.

A related example is the relatively simple technology associated with e-mail. Previous early morning behavior within an organization included getting the mandatory cup of coffee and sitting down to plan the activities of the day. In today's interconnected world, the early hours are often taken up with going through one's e-mail messages and responding to them. To that extent, we are now all "on call" most of the time between the capabilities of our voice mail and our electronic mail systems.

Integrative management has also been supported by a class of software known as Groupware, as referred to earlier in Chapter 12. This type of software is used in meetings between groups of people so as to assist in a variety of group processes ranging from tactical problem solving to strategic planning. Systems of this type have been available from Ventana Systems, IBM, and others. Groupware is sometimes also referred to under the general category of group decision support systems (GDSSs).

Integrative behavior and exchange of information have also been facilitated by previously discussed domain-specific information systems such as systems and software engineering environments. An example is the I-CASE (integrated computed-aided software engineering) initiative presented in Chapter 12. In this situation, the integration is specifically focused on the interoperability of several software tools that support software engineering. Through this type of technical information integration, it is presumed that the inter-connectedness and productivity of software teams will be increased.

Clearly, entry into the new "information age" has great potential for enhancing integrative behavior and management. It appears likely that we will continue to receive great benefits from our information systems in terms of the rapid conveyance of large amounts of information in extremely short periods of time. It is important, however, to assure that the human side of our organizations is not trampled in the process.

14.8   ENTERPRISES AS INTEGRATORS

The project and its systems engineering activities are both embedded in some type of enterprise, whether it be in industry, government, or academia. If that enterprise behaves in an integrative fashion, then a conducive environment is established within which the project is executed. If not, there is usually a negative flow-down to the project and its internal operations. A negative environment puts additional pressures on the managers within the project. Some of the issues with respect to the overall project environment have been discussed in Section 1.6 of Chapter 1.

Experience has shown that a critical factor in terms of integrative management is top management. In particular, it must encourage, support, and exhibit team behavior. As suggested by a President and Chief Executive in industry [14.7], "teamwork starts at the top." Some of the points made in this regard include:

- The single most important quality is to build effective teams.
- Teamwork is a critical success factor.
- Important contributions are made by individuals in a variety of functional areas.
- The example for integrative teamwork must come from the top.

Although the top of an organization appears to be somewhat removed from a particular project, it has its definite effects, either for the good or otherwise.

Another example that exemplifies integrative management may be drawn from management literature, as previously alluded to in Chapter 1. This is Senge's "learning organization" [14.8], wherein five disciplines are emphasized:

1. Building of a shared vision
2. Personal mastery
3. Mental models
4. Team learning
5. Systems thinking (the fifth discipline)

The last two are of special importance in terms of integrative management. Team learning carries with it at least two important ideas. The first is the significance of teams as a critical element of successful project execution. This point has been reiterated many times in this book. The second idea is that of learning. The organization itself must adopt a position of continuous learning to keep up with its business and remain current (or ahead of the pack) in its critical technology areas. A company that is not learning soon falls behind its competition and begins to lose its key people. The trouble often is that organizations "don't know what they don't know." This is a situation to be guarded against through the use of seminars, workshops, colloquia, and bringing "new blood" into the company. A particularly cogent exploration of the roles of dialogue and culture in organizations [14.9] is suggested reading in addition to Senge's book.

Systems thinking is basically the same as the systems approach referred to in this chapter as well as other parts of this book. According to Senge, this fifth discipline serves as an integrator of the other disciplines. As Senge indicates, "by enhancing each of the other disciplines, it continually reminds us that the whole can exceed the sum of its parts." Integrative management seeks that same goal.

As we look at the environment external to the project itself, we must also not lose sight of the customer. Integrative management means that the customer is a crucial part of the equation and that all project and systems engineering tasks are ultimately traceable to the needs and requirements of the customer. More than that, the customer is not a static concept; rather, customer focus brings a vital and dynamic force to bear on the project. Interactions on a day-to-day basis with the customer usually yield a better process and a better product. The "customer is king" slogan brings accountability to each and every project member, often in ways that transcend internal management oversight.

Finally, integrative management implies a connectedness to the community within which the enterprise operates. As "no man is an island," no company should keep itself removed and isolated from its community. Involvements with the community foster a condition of well-being and also serve as a positive example to everyone in the company. Such involvements may include any or all of the following:

- Contributions to charitable causes and organizations
- Support for educational institutions
- Science project sponsorships
- Co-op programs
- Memberships in local professional organizations
- Chamber of commerce support
- Sponsorship of special events and sports teams

Community connectedness supports a "world waiting to be born," as suggested by M. Scott Peck [14.10]. It is a basic truth for the individual and also a fundamental value for the organization.

14.9 THINKING OUTSIDE THE BOX

As in many fields, at least some of the conventional wisdom of management and systems engineering needs to be challenged. In the context of this book, such challenges might well originate through the project manager (PM) and/or the chief systems engineer (CSE). If they do, there is some chance that next steps will be taken, such as coming up with a different approach and following that with new actions. Such approaches can "move the ball down the field," a good way to make some progress. One of the best systems engineers, Eberhardt Rechtin, did exactly that by documenting a set of heuristics that he found were of particular interest in building systems [14.11]. Here are several of them that are especially noteworthy:

No complex system can be optimized for all parties;
Don't slice through regions of high information flow;
A model is not reality;
System structure should resemble functional structure.

Many others can be found in Rechtin's book, which is most highly recommended.

This author suggested a generic set of potentially new approaches that were called "thinking outside the box" [14.12]. Nine such notions were documented as:

1. Broaden and generalize
2. Crossover
3. Question conventional wisdom
4. Back of the envelope
5. Expand the dimensions
6. Obversity
7. Remove constraints
8. Thinking with pictures
9. Systems approach

Although there is not space here to examine each of these ideas in detail, we briefly cite below some areas in which thinking outside the box (TOTB) may differ from conventional wisdom.

System Requirements. Perhaps the conventional wisdom (CW) here is that system requirements are not and should not be subject to change. On the other hand, TOTB challenges that idea by recognizing that (a) poor requirements *need* to be challenged, (b) we often "lock into" a set of requirements when we know the least about a system, (c) some government approaches tell us that requirements tradeoffs can be important in a program, and (d) some practitioners have demonstrated that changes can be quite constructive [14.13].

Processes and Products. The CW here tends to be that if the process is right, the product will necessarily also be right. We agree that process is indeed critically important, but unfortunately it may not guarantee the correct product. The missing ingredient could well be the appropriate subject matter expertise. For example, an excellent hardware engineer whose expertise is in building heavy machinery is likely not to be very productive when asked to build operating system software for a computer, even if he or she follows precisely the accepted software engineering processes.

Integrating Stovepipes. CW in this domain tends to be the more integration, the better. The challenge to that is simply that this could easily lead to systems that violate schedule, cost and performance requirements or constraints. Depending upon the nature and details of the stovepipes, the "best" integration approach could range from almost none, to a fair amount. *Assuming* high levels of integration as a goal can well lead to disastrous results. A better solution? Use the architecting procedure illustrated in chapter nine.

Measurements. The CW here might be "measure everything that you can." The TOTB approach tends to be to measure those "key performance parameters" (KPPs) that give us enough information to manage a program properly. Measure what you need to vs. what you can is a way to focus on the most important aspects of a system and its development.

Do It Right the First Time. This suggestion can be attributed to Philip Crosby as he approaches the issue of quality management [14.14]. If first-time correctness were possible to achieve for all systems, it might be easy to embrace. However, we have discovered that this imperative can often fail, leading us to early decisions that are incorrect. We have also learned that a systematic iterative process helps us through this issue and can avoid errors. We should also keep in mind yet another of Rechtin's heuristics, namely, "build in and maintain options as long as possible for complex systems" [14.11]. This includes using "TBD" (to be determined), to be filled in when we have the information needed to finalize a requirement or specification.

Dollar and Time Reserves. At least some of the CW in many enterprises calls for "management" reserves at several levels of management. If three such layers each take ten percent dollar reserves, then the PM has less than three-quarters of the money originally budgeted. This may well be enough only to assure failure at the PM position. Is there a better way? How about giving the PM enough money and time to get the job done, with a probability of at least eighty percent?

"Interchangeable" People. The CW here might well be that people with the same nominal title and pay grade will have about the same level of productivity and performance. A closer look, however, reveals that there are many more variables to consider. It's the job of management and the chief engineer to find precisely the right person for the job, and be able to explain why. Good management is able to do this on a consistent and continuing basis.

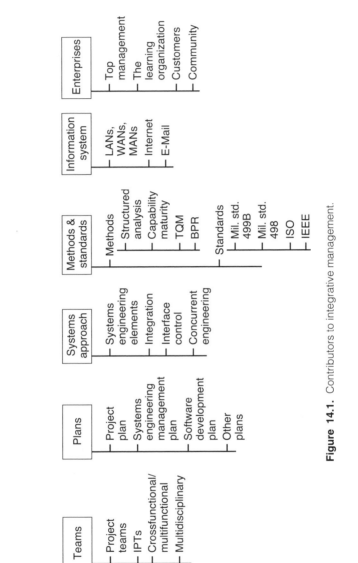

Figure 14.1. Contributors to integrative management.

The Customer is Always Right. This is an "inside the box" platitude that can lead an organization a long way in the wrong direction. Real-world experiences have shown that customers can be wrong, and when they are, their instructions should be questioned (in an appropriate manner). Deliberately doing the wrong thing cannot be excused or tolerated, even if that action is traceable to a direct customer request.

14.10 SUMMARY

This final chapter has provided some insight into integrative management whereby project management and systems engineering management are brought together into a harmonious and mutually supportive framework. The major contributors to integrative management are summarized in Figure 14.1.

Our final summary is a set of some dozen suggestions as to how to achieve integrative management. In particular, one should attempt to ensure that:

1. Managers have and exhibit appropriate integration and synthesis skills.
2. There is effective communications at all levels.
3. Divisive behavior is not tolerated.
4. Integrated teams are established and listened to.
5. Plans are coherent, informative, updated, and widely disseminated.
6. The systems approach is encouraged and used.
7. Appropriate methods and standards are employed.
8. Internal and external information systems are up to date and in good operating condition.
9. The enterprise adopts the principles and practices of a "learning organization."
10. Strong customer contact, dialogue, and commitment is maintained.
11. The enterprise is connected to the community at large.
12. Top management actively supports all of the preceding.

QUESTIONS/EXERCISES

14.1 Identify and discuss six indicators of integrative management, other than those listed at the beginning of this chapter.

14.2 Identify and discuss six actions that an organization can take to achieve integrative management, other than those listed at the end of this chapter.

14.3 Cite three consistent patterns of behavior of your boss that
  a. support integrative management
  b. do not support integrative management

14.4 Investigate and write a three-page paper regarding the documented operation and benefits of integration product, cross-functional or multi-disciplinary teams.

14.5 Discuss, in four pages, how each of the following might support integrative management:
  a. structured analysis for systems
  b. capability maturity modeling (CMM)
  c. Total Quality Management (TQM)
  d. business process reengineering (BPR)

14.6 Describe five actions that top management should take to encourage integrative management in a company.

14.7 If you served as president of a large company, identify and explain ten specific steps you would consider taking in order to achieve integrative management.

14.8 Write a five-page report elaborating upon the topic of systems integration (see Chapter 12), as well as its relationship to integrative management.

14.9 Does your organization tend to "think outside the box"? If yes, how is that achieved? If no, what five new actions would you suggest to encourage this type of thinking?

14.10 Identify and discuss three ways in which your organization might avoid negative aspects of GroupThink.

REFERENCES

14.1 Mondy, R. W., and S. R. Premeaux (1993). *Management: Concepts, Practices, and Skills*. Boston: Allyn and Bacon.

14.2 Bowen, D. (1995). " Work Group Research: Past Strategies and Future Opportunities," *IEEE Transactions on Engineering Management* 42(1).

14.3 *Systems Engineering*, Military Standard 499B (1991). Washington, DC: U.S. Department of Defense.

14.4 *Software Development and Documentation*, Military Standard 498 (1994). Washington, DC: U.S. Department of Defense.

14.5 *The Role of Concurrent Engineering in Weapons System Acquisition*, Report R-338. Alexandria, VA: Institute for Defense Analyses.

14.6 Pennell, J., and M. Slusarczuk (1989). *An Annotated Reading List for Concurrent Engineering*, Document D-571. Alexandria, VA: Institute for Defense Analyses.

14.7 Serpa, R. (1991). " Teamwork Starts at the Top," *Chief Executive* **66**: 30–33.

14.8 Senge, P. (1990). *The Fifth Discipline*. New York: Doubleday.

14.9 Schein, E., " On Dialogue, Culture and Organizational Learning," *Engineering Management Review* **23**(1):23–29.

14.10 Peck, M. Scott. (1993). *A World Waiting to Be Bom*. New York: Bantam Books.

14.11 Rechtin, E., (1991). *Systems Architecting*. Englewood Cliffs, NJ: Prentice-Hall.

14.12 Eisner, H., (2005). *Managing Complex Systems—Thinking Outside the Box*. Hoboken, NJ: John Wiley.

14.13 Boehm, B., "Unifying Software Engineering and Systems Engineering," Computer Magazine, March 2000 pp. 114–116.

14.14 Crosby, P., (1984). *Quality without Tears*. New York: New American Library, Penguin Books.

APPENDIX: SYSTEMS ARCHITECTING—CASES

A.1 INTRODUCTION

Chapter 9 presented the basic principles of systems architecting with the following key elements, a subset of the thirty elements of systems engineering:

- Requirements analysis/allocation
- Functional analysis/allocation
- Architecture design/synthesis
- Alternatives analysis/evaluation
- Technical performance measurement
- Life-cycle costing
- Risk analysis
- Concurrent engineering

In this appendix, we show some short case studies of systems architecting to more clearly demonstrate the process and some of the requisite outputs. These cases are drawn from assignments given in graduate courses in systems engineering at The George Washington University. This author gratefully acknowledges the efforts and insights of a variety of students in the graduate program in systems engineering [A.1].

The cases presented simulate the architecting process rather than going through all elements in detail. In particular, the technical performance measurement element is limited because it is usually a long and difficult procedure to truly assess all important elements of performance, including trade-off studies and sensitivity analyses. In one of the cases, sensitivities were analyzed, utilizing a decision support system that is particularly well-suited to an examination of how parameters vary and cause changes in the overall architecture evaluations.

The instructions given for the development and assessment of architectures were rather broad. The first assignments started with the definition of requirements, which was left for the students to formulate. After considerable discussion, these requirements were responded to with the synthesis of alternative architectures. The procedure (presented in Section 9.5.3) recommended was to develop three architectures:

1. A low-cost, minimum-effectiveness alternative
2. A "baseline" alternative
3. A high-performance (high-cost) alternative

Not all alternatives mapped precisely into these three situations, but in all cases, at least three alternatives were defined. From this point, the alternatives were evaluated based on a well-defined set of evaluation criteria. Weighting and ranking "systems" were most often set forth as a means of carrying out the evaluations. In some cases, systems engineering elements beyond the scope of architecting were also considered.

Some systems turned out to be software-based, with varying functionalities and host computers. The four cases described in the remainder of this appendix deal with:

1. A logistics support system
2. A decision support system for software defects
3. A systems engineering environment (SEE)
4. An anemometry system

A.2 A LOGISTICS SUPPORT SYSTEM (CASE 1)

The system being architected is LEAPS, a Logistics Evaluation, Analysis, and Planning System. The Level 1 functional decomposition of the system is shown in Figure A.1 This decomposition places in evidence the eight major functional areas for LEAPS. The general requirements for the system, sometimes also called system requirements, are:

- DOS PC-Based (target machine: 386, 25-MHz processor, 80-Meg hard drive, mouse)
- Event-driven (activities initiated based on completion of prior activities)
- Windows-compatible

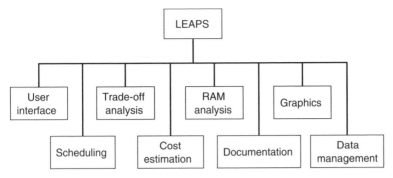

Figure A.1. Level 1 functional decomposition.

- "Transparent" access to commercially available packages
- Flexibility required for growth to meet expanding user demands

The requirements specific to the individual functional areas are listed in Exhibit A.I.

The alternative architectures to meet the stated requirements are shown in Table A.1. We note that as we move from Alternative 1 to 3, we increasingly rely on commercial software to implement the stated functional capabilities.

The evaluation of the three alternatives is depicted in Table A.2. This figure shows the following weighted scores:

Alternative 1: Score = 4.71
Alternative 2: Score = 6.20
Alternative 3: Score = 6.10

Thus, there is a preference for Alternative 2, but the composite score is not a great deal higher than that for Alternative 3.

This evaluation process was accomplished through pairwise comparisons, followed by a manual consistency check. Schedule considerations were developed in relative terms under the following ground rules:

- Using existing commercial packages = 1 time unit
- Modifying commercial packages = 3 time units
- Developing new software components = 8 time units

A preliminary costing was carried out to develop the numbers in Table A.2 under the two categories of acquisition and maintenance cost. The four key performance evaluation criteria were:

1. Ease of use
2. Speed

3. Data sharing
4. Consistency

Exhibit A.I: Functional Requirements—LEAPS

- User interface:
  — Provide a uniform look and feel for all functional operations
  — Provide ready access to all functional capabilities
- Scheduling:
  — Handle all life-cycle phases (20–30 years)
  — Generate project start and completion dates based on the completion of program milestones
- Trade-off analyses:
  — Handle unique aspects of integrated logistics support elements
  — Address other supportability-related factors (i.e., environmental impacts/limitations)
  — Support development of alternative courses of action
- Cost estimation:
  — Address all life-cycle phases for all system components (ground, space, air)
  — Accommodate proven cost-estimating experience, cost-estimating relationships, and new cost-estimating requirements
  — Support formalized cost formats (i.e., Big 6)
- RAM analysis:
  — Include "canned" desk-top RAM models and access to external models and simulations
  — Accept user-developed RAM models
- Documentation:
  — Include templates for standard logistics reports
  — Provide user capability to develop templates for special report and study formats
- Graphics:
  — Provide graphical representation of analysis results (RAM, trade-off, cost estimating)
  — Provide user capability to develop specialized graphics as desired
- Data management:
  — Maintain milestones data with the linkages to program activities
  — Provide access to logistics support analysis records
  — Link cost data to system components (through work breakdown structures and logistics control numbers)

TABLE A.1 Alternative to Function Mapping—LEAPS

	User Interface	Scheduling	Trade-Off Analysis	Cost Estimation	RAM Analysis	Documentation	Graphics	Data Management
Alternative 1	Build	Build	Build	Build	Build	MS Word	MS Power Point	Build
Alternative 2	Build	Modify MS Project	MS Excel	Build	MS Excel	Power Word	MS Power Point	Paradox for Windows
Alternative 3	Build	MS Project	MS Excel	Modify ground and space	MS Excel	MS Word	MS Power Point	MS Access

Note: Ms = Microsoft.

TABLE A.2 Alternative Evaluation—LEAPS

	Alternative 1							Alternative 2							Alternative 3						
Cost 0.55																					
Acquisition 0.44	9	1	3	1	3	7	7	5	7	1	7	7	7	7	7	5	7	7	7	7	7
Maintenance 0.11	9	3	3	3	3	7	7	5	7	1	7	7	7	7	7	5	7	7	7	7	7
Schedule 1.12																					
Acquisition 0.09	7	1	3	1	3	9	9	5	9	1	9	9	7	7	9	3	9	9	9	9	7
Maintenance 0.03	7	3	3	3	3	9	9	5	9	1	9	9	7	7	9	3	9	9	9	9	7
Performance 0.33																					
Ease of Use 0.16	7	9	9	9	9	7	9	7	9	9	9	7	5	5	9	5	9	9	9	7	7
Speed 0.02	5	9	9	9	9	5	9	7	9	9	9	5	5	7	9	3	9	9	9	5	7
Data sharing 0.05	5	9	9	9	9	7	9	5	7	9	7	7	5	5	7	1	7	7	7	7	5
Consistency 0.10	7	9	9	9	9	9	9	5	9	9	9	9	3	3	9	9	9	9	9	9	7
Alternative score				4.71							6.20							6.10			

Raw scores are measures of goodness: least good = 1; most good = 9.
Total score reflects weightings indicated.

In this case, a separate risk analysis and definition of possible future improvements (preplanned product improvements—P3I) were also part of the case. These were supplemented by an analysis that included a market assessment, a calculation of internal rate of return (IROR) for the product, the formulation of schedules, and a first-order look at sales (revenues), costs, and the estimated break-even point.

A.3 A SOFTWARE DEFECTS ASSESSMENT SYSTEM (CASE 2)

The software defects assessment system provides the software manager with an automated decision-support tool to assist in the software maintenance process by prioritizing software defects both accurately and efficiently. The system is called SD-DSS (Software Defects—Decision Support System).

The top-level functional identification for this system is shown in Figure A.2, which shows the various requirement areas under the following four major functional categories:

1. Data management
2. Model management
3. Dialog management
4. System requirements

Exhibit A.2 provides an overview of requirements for each of these functional areas.

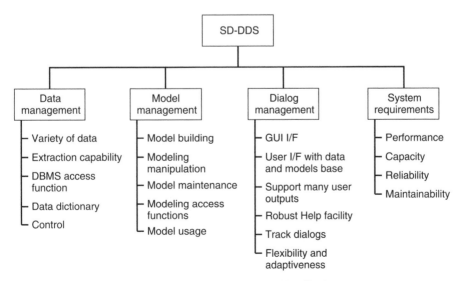

Figure A.2. Top-level functional identification.

Given these functional areas and requirements in each area, the next step is to structure alternative architectures for such a system. Three alternatives are depicted in Exhibit A.3. In broad terms, the alternatives reflect different choices in terms of host hardware and the associated software that would normally be attendant to different computer configurations.

The evaluation of the three alternatives is represented in Table A.3. This "comparison matrix" lists the evaluation in three categories: (1) cost, (2) functionality, and (3) system. Each of these categories is broken down so that it is then possible to assess each alternative at a lower level of detail. Expansion of the subcriteria in terms of ideas as to how to make the calculations is shown in Table A.4.

With respect to the category of cost, the following points are relevant:

- Total life-cycle cost was based on the following components over a ten-year period:
  - Research, Development, Test, and Evaluation (RDT&E): engineering design, software development, and documentation
  - Production and Construction: Commercial off-the-shelf (COTS) hardware and software and facility modifications
  - Operations & Maintenance (O&M): system operations, maintenance personnel, activation, and materiel support
  - Retirement and Disposal: retirement personnel, transportation, and handling
- Parametric cost estimation:
  - Constructive Cost Model (COCOMO) used for RDT&E and O&M costs
- Bottom-up unit cost estimation used for production, construction, retirement, and disposal costs

A summary of life-cycle costs (LCC) over a ten-year period is provided in Table A.5. The percent of costs by cost category is also listed in the far-right column.

The system architect assessed all evaluation categories with the assistance of a commercial software package known as Expert Choice [A.2], as illustrated in Figure A.3 Functionality measures were analyzed by performing

Exhibit A.2: Functional Requirements—SD-DSS

Data Management Requirements
- The system shall collect/extract software-defect-related data from the following sources for inclusion into the DSS database management system (DBMS):

  — Automated configuration management system
  — Financial/accounting system
  — Personnel system
  — Software reliability information (i.e., metrics)

- The system shall provide the basic data manipulation functions such as select, update, delete, and insert. The query capability shall support ad hoc query and report generation capability.
- The system shall provide a catalog of all data with definitions and the ability to provide usage information on those data.
- The system shall provide comprehensive data security, providing protection from unauthorized access, recovery capabilities, archival capabilities, etc.

Model Management Requirements

- The system shall allow users to create models that will structure and evaluate the qualitative and quantitative factors involved in prioritizing software defects.
- The system shall create models easily and quickly, either from scratch or from existing models contained in the model base.
- The system shall allow users to manipulate models to include sensitivity analysis, "what if" analysis, goal seeking, and cost-benefit analysis.
- The system shall generate a directory of models for use by different individuals of the organization.
- The system shall manage and maintain the model base with management functions analogous to database management: store, access, run, update, link, catalog, and query.
- The system shall track models, data, and application usage.

Dialog Management Requirements

- The system shall have a user-friendly graphical user interface.
- The system shall be able to interact with several different dialog styles.
- The system shall be able to accommodate a variety of input devices.
- The system shall be able to present data with a variety of formats and output devices (i.e., color graphics, three-dimensional presentations, and data plotting).
- The system shall provide "help" capabilities, diagnostics, training, and suggestion routines to assist to user.
- The system shall be able to capture, store, and analyze dialog usage to facilitate improving the dialog system.
- The system must provide flexibility and adaptiveness to accommodate different problems and technologies.

Exhibit A.2: (*Continued*)

System Requirements

- Performance:
  - The system must have an overall end-user response time of five seconds.
- Capacity:
  - The system must be able to accommodate up to fifteen concurrent users at three geographic locations.
  - The database must be able to store summary defect information for up to 1M defects.
  - The model base must be able to support up to 1500 models.
- Reliability:
  - The system availability must be greater than 90%.
- Maintainability:
  - Preventive maintenance must be scheduled during weekday off-hours and weekends.

pairwise comparisons. The results in terms of overall scores for the three alternatives were as follows:

Alternative 1: Score = 0.361
Alternative 2: Score = 0.334
Alternative 3: Score = 0.305

An illustrative risk analysis was carried out for SD-DSS with the results as shown in Exhibit A.4. Note that risks were considered in the categories of overall funding, cost, schedule, administrative, and technical.

System enhancements were analyzed under the systems engineering element of preplanned product improvement (P3I), with the following results:

- Enhance SD-DSS with a knowledge acquisition system (i.e., an expert system) consisting of:
  - A knowledge base
  - An inference engine
  - A blackboard (workboard)
  - A user interface
- Enhance SD-DSS with an expert critiquing subsystem consisting of:
  - An executive
  - An evaluator
  - A transformer
  - An elicitor

Exhibit A.3: SD-DSS Alternatives

Alternative 1
Classical Time-Sharing Monolithic Architecture
- Software:
  - — Centralization of all three functional areas onto monolithic platform
  - — Functional management areas designed various COTS application packages and customization
    - — Data management: DB2 with Query Management Facility
    - — Model management: Application System, Lotus/M
    - — Dialog management: Data Interpretation System
- Hardware:
  - — Partitioned IBM mainframe or dedicated AS400 series machine
  - — Graphics terminals

Alternative 2
Local Area Network (LAN)-Based Workstation System Architecture
- Personal workstations connected to a minicomputer in client-server arrangement:
  - — Workstations handle the interface, high-speed graphics display devices, and some of the data-reduction function
  - — File server handles the data management and the bulk of the analysis
- Software:
  - — Functional management areas designed with various COTS application packages and customization
    - — Data management: Oracle
    - — Model management: Criterium, Oracle, Lotus/M
    - — Dialog management: Sun Solaris 2.0
  - — UNIX OS and Novelle UNIXware NOS
- Hardware:
  - — Sun SPARCserver 10 File/Application Server
  - — Sun SPARCstation Workstation

Alternative 3
LAN-Based PC Network with File Server Architecture
- Total PC solution where information from the corporate database is downloaded to the PC LAN environment and then delivered to the PCs
- Software:
  - — Functional management areas designed with various COTS application packages and customization:
    - — Data management: Borland's Paradox 4.5
    - — Model management: Expert Choice, MS Office for Windows
    - — Dialog management: MS Visual Basic 3.0
  - — DOS 5.0 OS with Windows 3.1 and Novelle Netware 3.11 NOS

Exhibit A.3: (*Continued*)

- Hardware
  — Pentium, 66-MHz PC File Server
  — 486DX2, 33-MHz PC Workstation
- Enhance SD-DSS to handle the overall defect resolution process by adding configuration management functionality.
- Add technology upgrades with new COTS hardware and software as they become available.

This case study of a software defects tracking and assessment system thus focused on the elements of requirements definition, alternatives evaluation, life-cycle costing, risk, and P3I. Special attention was paid to the sensitivity of the alternatives selection by utilizing the Expert Choice commercial software package and displaying these sensitivities in real time. This simulated system architecting process provided fertile ground for exploring both variations and extensions of the basic process. Exercises of this type are necessarily limited, but provide an excellent overview of key steps that are critical to systems architecting.

A.4   A SYSTEMS ENGINEERING ENVIRONMENT (CASE 3)

A systems engineering environment (SEE) basically provides an automated information center in which a team can carry out the thirty elements of systems engineering. Developing such an environment is considerably more than an academic exercise. As an example, the Rome Laboratory of the Air Force has a program to structure an SEE known as Catalyst [A.3, A.4]. This is being formulated as a comprehensive environment for a systems engineering team. The Navy also has been interested in systems engineering, having requested information and proposals from industry on the broad subject of the engineering of complex systems [A.5]. Subordinate areas of interest to the Navy under this program have included:

- System information capture
- System understanding, guidance, and synthesis
- System reengineering
- Process
- Integration

Thus, there is strong and continuing interest in developing systems engineering as well as an automated environment that facilitates the practice of systems engineering. One approach to the formulation of an SEE follows.

TABLE A.3 Comparison Matrix—SD-DSS

Criteria	Subcriteria: Measures	System		
		Alternative 1	Alternative 2	Alternative 3
Cost	RDT&E ($K)	200	380	287
Cost	Prod/Con ($K)	1563	330	103
Cost	O&M ($K)	780	1500	1080
Cost	Retire ($K)	50	50	50
Functionality	Data: collection/extraction	Subjective	Subjective	Subjective
Functionality	Data: data manipulation	Subjective	Subjective	Subjective
Functionality	Data: data dictionary	Subjective	Subjective	Subjective
Functionality	Data: security	Subjective	Subjective	Subjective
Functionality	Model: model creation	Subjective	Subjective	Subjective
Functionality	Model: sensitivity	Subjective	Subjective	Subjective
Functionality	Model: management	Subjective	Subjective	Subjective
Functionality	Dialog: consistency	Subjective	Subjective	Subjective
Functionality	Dialog: styles	Subjective	Subjective	Subjective
Functionality	Dialog: flexibility/adaptability	Subjective	Subjective	Subjective
Functionality	Dialog: help	Subjective	Subjective	Subjective
Functionality	Dialog: I/O devices supported	Subjective	Subjective	Subjective
System	RMA: availability (MTBF/MTBF + MTTR)	0.92	0.98	0.9
System	RMA: schedule maintenance (mean time in hours)	2	6	10
System	Performance (seconds)	1.2	0.76	1.9
System	Capacity: Saturation (no. of users)	46	38	22
System	Capacity: DBMS (defect records in millions)	10	6	5
System	Capacity: Models (models in thousands)	2500	2000	1500

TABLE A.4 Ideas on Calculating Evaluation Criteria— SD-DSS

Criteria	Subcriteria: Measures	Ideas on Calculating Evaluation Criteria	Alternative 1	Alternative 2	Alternative 3
Cost	RDT&E ($K)	Prefer lower RDT&E costs over life cycle	200	380	287
Cost	Prod /Con ($K)	Prefer lower prod /con costs over life cycle	1563	330	103
Cost	O&M ($K)	Prefer lower O&M costs over life cycle	780	1500	1080
Cost	Retire ($K)	Prefer lower retirement costs over life cycle	50	50	50
Functionality*	Data: collection/ extraction	Prefer ability to handle more information sources	0.114	0.481	0.405
Functionality*	Data: data manipulation	Prefer robust ad hoc query & report gen. facility	0.126	0.458	0.416
Functionality*	Data: data dictionary	Prefer robust DD facility	0.33	0.33	0.334
Functionality*	Data: security	Prefer design with comprehensive security	0.652	0.235	0.113
Functionality*	Model: model creation	Prefer robust model creation facility	0.126	0.458	0.416
Functionality*	Model: sensitivity	Prefer sophisticated sensitivity analysis	0.169	0.387	0.444
Functionality*	Model: management	Prefer robust model management	0.169	0.387	0.444
Functionality*	Dialog: consistency	Prefer less variation in menu design	0.5	0.25	0.25
Functionality*	Dialog: styles	Prefer support higher number of dialog styles	0.126	0.458	0.416
Functionality*	Dialog: flexability/ adaptability	Prefer open systems architecture design	0.126	0.458	0.416
Functionality*	Dialog: help	Prefer accurate and informative help	0.169	0.387	0.444
Functionality*	Dialog: I/O devices supported	Prefer higher number of different I/O devices supported	0.139	0.435	0.426
System	RMA: availability (MTBF/MTBF+MTTR)	Prefer higher system availability	0.92	0.98	0.9
System	RMA: schedule maintenance (mean time in hours)	Prefer shorter maintenance time	2	6	10
System	Performance (seconds)	Prefer shorter end-user response time	1.2	0.76	1.9
System	Capacity: saturation (no. of users)	Prefer larger number of users	46	38	22
System	Capacity: DBMS (defect records in millions)	Prefer higher number of records	10	6	5
System	Capacity: models	Prefer higher number of models	2500	2000	1500

¹Indicates pairwise comparisons performed.

TABLE A.5 LCC Matrix—SD-DSS

Design	Cost Category	FY93	FY94	FY95	FY96	FY97	FY98	FY99	FY00	FY01	FY02	Costs (K)	%LCC
Alt 1	1. Research & development	$100	$100									$200	7.71
	2. Production & construction	$185	$178	$150	$150	$150	$150	$150	$150	$150	$150	$1,563	60.28
	3. Operations & maintenance			$180	$120	$80	$80	$80	$80	$80	$80	$780	30.08
	4. Retirement and disposal										$50	$50	1.93
	Total (actual K$)	$285	$278	$330	$270	$230	$230	$230	$230	$230	$280	$2,593	100.00
	Total (present K$)	$259	$230	$248	$184	$143	$130	$118	$107	$98	$108	$1,625	
Alt 2	1. Research & development	$180	$200									$380	16.81
	2. Production & construction	$255	$75									$330	14.60
	3. Operations & maintenance			$240	$180	$180	$180	$180	$180	$180	$180	$1,500	66.37
	4. Retirement and disposal										$50	$50	2.21
	Total (actual K$)	$435	$275	$240	$180	$180	$180	$180	$180	$180	$230	$2,260	100.00
	Total (present K$)	$395	$227	$180	$123	$112	$102	$92	$84	$76	$89	$1,481	
Alt3	1. Research & development	$127	$160									$287	18.88
	2. Production & construction	$50	$53									$103	6.78
	3. Operations & maintenance			$240	$120	$120	$120	$120	$120	$120	$120	$1,080	71.05
	4. Retirement and disposal										$50	$50	3.29
	Total (actual K$)	$177	$213	$240	$120	$120	$120	$120	$120	$120	$170	$1,520	100.00
	Total (present K$)	$161	$176	$180	$82	$75	$68	$62	$56	$51	$66	$975	100.00

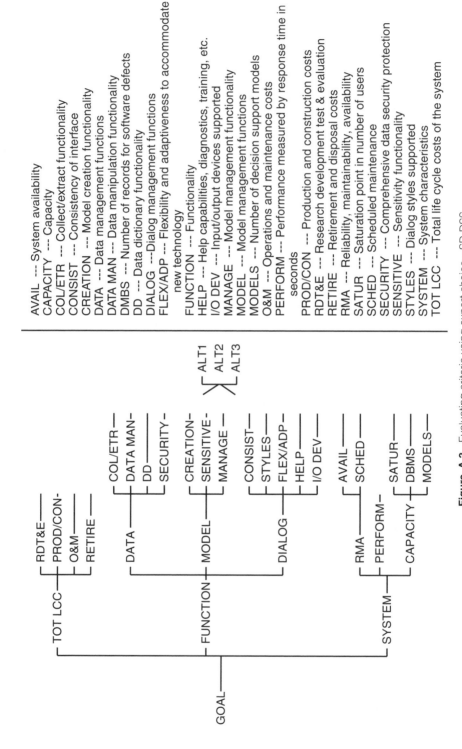

AVAIL --- System availability
CAPACITY --- Capacity
COL/ETR --- Collect/extract functionality
CONSIST --- Consistency of interface
CREATION --- Model creation functionality
DATA --- Data management functions
DATA MAN --- Data manipulation functionality
DMBS --- Number of records for software defects
DD --- Data dictionary functionality
DIALOG ---Dialog management functions
FLEX/ADP --- Flexibility and adaptiveness to accommodate
    new technology
FUNCTION --- Functionality
HELP --- Help capabilities, diagnostics, training, etc.
I/O DEV --- Input/output devices supported
MANAGE --- Model management functionality
MODEL --- Model management functions
MODELS --- Number of decision support models
O&M --- Operations and maintenance costs
PERFORM --- Performance measured by response time in
    seconds
PROD/CON --- Production and construction costs
RDT&E --- Research development test & evaluation
RETIRE --- Retirement and disposal costs
RMA --- Reliability, maintainability, availability
SATUR --- Saturation point in number of users
SCHED --- Scheduled maintenance
SECURITY --- Comprehensive data security protection
SENSITIVE --- Sensitivity functionality
STYLES --- Dialog styles supported
SYSTEM --- System characteristics
TOT LCC --- Total life cycle costs of the system

Figure A.3. Evaluating criteria using expert choice—SD-DSS.

Exhibit A.4: SD-DSS Risk Assessment

Program Risks

- Funding: Medium
  - Reduction of corporate fiscal resources
  - Risk mitigation: business process improvement effort to reduce costs
- Cost/Schedule Risk: Medium
  - Similar corporate programs realized 10% cost and schedule growth
  - Risk mitigation: prototyping, user engineering during design, and utilization of corporate metrics program
- Administrative: Low
  - Experienced and stable development team in place

Technical Risks

- Code & Unit Testing: Medium
  - Development of three software modules (builds)
  - Risk mitigation: maximal use of COTS SW, structured programming techniques, and early implementation of CASE Tools
- Integration Testing: Medium
  - Integration of three software modules (builds)
  - Risk mitigation: CASE tools and early introduction of test plans
- Transition/Activation: Medium
  - Uploading of software defected data from external sources
  - Risk mitigation: early introduction of real defect data into the code & unit test and integration test phases

The objectives set forth for the SEE were:

- To enable a more effective and efficient design of large-scale systems
- To provide a facility/center with CASE expertise
- To provide tools for managing complex representations and data
- To allow different teams to simultaneously work on the various elements of system design

The high-level functional requirements of the SEE are shown in Figure A.4, with the key requirements under each in Exhibit A.5. Three architectures were set forth, as represented in Exhibit A.6. The mapping of the three architectures against the functional areas is shown in Table A.6.

Two different evaluation methods were employed to assess the merits of the three architectures. The evaluation matrices for these methods are shown in Table A.7. The methods differ in the weights assigned to the evaluation criteria; the second method led to the calculation of a "cost-benefit ratio." A

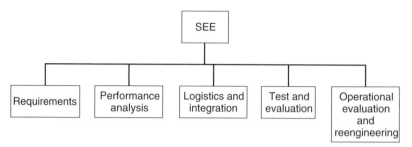

Figure A.4. High-level functional requirements of a SEE.

1–3–5–7–9 scale was utilized for the technical criteria. Results for the three alternatives are as follows:

Alternative 1: All Macintosh computers
- Technical score = 6.2 (Method 1); 5.7 (Method 2)
- Cost = 190 units
- Benefit-to-cost ratio = 0.03

Alternative 2: All PC (DOS) computers
- Technical score = 7.0 (Method 1); 6.0 (Method 2)
- Cost = 170 units
- Benefit-to-cost ratio = 0.0353

Alternative 3: Macintosh + DOS computers
- Technical score = 5.9 (Method 1); 7.5 (Method 2)
- Cost = 340 units
- Benefit-to-cost ratio = 0.0221

By examining the two methods of evaluation, it was possible to see how the final evaluations change as different procedures are followed. This examination of sensitivities provided a broader perspective regarding how final architectural alternatives might be evaluated to derive a preferred architecture.

Exhibit A.5: SEE Requirements

Requirements
- Needs/goals/objectives
- Mission engineering
- Requirements analysis
- Functional analysis
- Functional allocation
- Specification development

Performance Analysis
- Alternative evaluation
- System design/analysis
- Scheduling
- Life-cycle costing
- Technical performance measurement
- Program/decision analysis
- Risk analysis
- RMA

Logistics and Integration
- Interface definition and control
- Integration logistics support
- Integration
- Configuration management

Test and Evaluation
- Quality assurance
- Requirements traceability

Operational Evaluation and Reengineering
- Operations
- Operational evaluation
- Modeling:
  — Prototyping
  — Mathematical models
  — Simulation models

Exhibit A.6: Alternative Architectures—SEE

Architecture 1
- All Macs
- Microsoft Office (Word, Excel, Powerpoint)
- SE tools:
- IDEE
  — System Architect
  — Oracle
  — Expert Choice
  — Microsoft Project
  — SLIC
  — RTM
  — Extend

Exhibit A.6: (*Continued*)

Architecture 2

- All PCs
- Microsoft Office (Word, Excel, Powerpoint)
- SE tools:
  — CORE
  — Design IDEE
  — System Architect
  — DBMS: Oracle
  — Expert Choice
  — @Risk
  — Foresight
  — RTM
  — GPSS/H
  — MATLAB

Architect 3

- Hybrid (MACs, PCs, and workstations)
- Microsoft Office
- SE tools:
  — CORE
  — ROD-100
  — Design IDEE
  — System Architect
  — DBMS: Oracle
  — Expert Choice
  — @Risk
  — NETSIM
  — Foresight
  — Microsoft Project
  — SLIC
  — Extend
  — RTM
  — GPSS/H
  — MATLAB
  — Demo

A.5 AN ANEMOMETRY SYSTEM (CASE 4)

This system architecting dealt with a severe climate anemometry system (SCAS) and was the only case of the four that was hardware-based. A top-level

TABLE A.6 Requirements Traceability Matrix—SEE

	Architecture 1	Architecture 2	Architecture 3
Requirements			
Needs/goals/objectives	IDEF	CORE, IDEF	RDD-100, CORE, IDEF
Mission engineering	IDEF	CORE, IDEF	RDD-100, CORE, IDEF
Requirements analysis	IDEF	CORE, IDEF	RDD-100, CORE, IDEF
Functional analysis	IDEF	CORE, IDEF	RDD-100, CORE, IDEF
Functional allocation	IDEF	CORE, IDEF	RDD-100, CORE, IDEF
Specification development	IDEF	CORE, IDEF	RDD-100, CORE, IDEF
Performance analysis			
Alternative evaluation	LOW	Expert Choice	Expert Choice
System design analysis	Object-Time	System-Developer	TeamWork
Scheduling	Fast Track	Microsoft Project	Microsoft Project
Life-cycle costing	COCOMO	COCOMO	COCOMO, PRICE
Technical performance	Lisa-2B	Lisa-2B	Lisa-28
Measurement			
Program/decision analysis	Decide	Expert Choice	Expert Choice
Risk analysis	@Risk	@Risk	@Risk
RMA	RELAX	RPP	SLIC
Logistics and integration			
Interface definition and control	OMEGA	SLIC	SLIC
Integrated logistics support	OMEGA	SLIC	SLIC
Integration	OMEGA	SLIC	SLIC
Configuration management	OMEGA	SLIC	SLIC
Test and evaluation			
Quality assurance	TRACE	MATRIX	RTM
Requirements traceability	TRACE	MATRIX	RTM
Operational evaluation & reengineering			
Operation	MODSIM II	Demo	Demo
Operational evaluation	MODSIM II	Demo	Demo
Modeling			
Prototyping	Demo	Demo	Demo
Mathematical models	IThink	MATLAB	MATLAB
Simulation models	Extend+BPR	GPSS	NETSIM

TABLE A.7 Architecture Selection—SEE

Evaluation Criteria	Weight	Architecture 1 (All Macintosh)					Architecture 2 (All PC)					Architecture 3 (MAC & PC)				
		Require.	Perf. Anal.	Log. & Int.	T&E	OE & Reeng.	Require.	Perf. Anal.	Log. & Int.	T&E	OE & Reeng.	Require.	Perf. Anal.	Log. & Int.	T&E	OE & Reeng.
							Method 1									
Cost	35%	9	7	7	7	7	9	9	9	9	9	3	3	3	3	3
Speed	5%	5	5	5	5	5	5	7	5	5	7	7	7	5	7	9
Ability to handle large problems	10%	5	5	5	7	7	5	7	5	7	7	7	7	7	7	9
Ability to handle growth in users	10%	5	5	5	5	5	5	5	5	5	5	7	9	7	7	7
Ability to meet new requirements	10%	5	5	5	5	5	5	5	5	5	5	7	9	7	9	7
Capability to meet new applications	5%	7	6	5	5	5	7	7	7	7	7	9	9	9	9	9
Operability	15%	5	7	7	7	7	5	5	7	7	5	5	7	5	7	5
Currency of software	10%	7	7	5	5	7	5	7	5	7	7	5	5	5	7	5
Total	100%															
Total score (benefit)			6.2					7					5.9			
Total cost			190					170					340			

TABLE A.7 Architecture Selection—SEE (continued)

Evaluation Criteria	Weight	Architecture 1 (All Macintosh)					Architecture 2 (All PC)					Architecture 3 (MAC & PC)				
		Require.	Perf. Anal.	Log. & Int.	T&E	OE & Reeng.	Require.	Perf. Anal.	Log. & Int.	T&E	OE & Reeng.	Require.	Perf. Anal.	Log. & Int.	T&E	OE & Reeng.
							Method 2									
Speed	10%	3	5	5	5	5	5	7	5	5	7	7	7	5	7	9
Ability to handle large problems	15%	5	5	5	7	7	5	7	5	7	7	7	7	7	7	9
Ability to handle growth in users	15%	5	5	5	5	5	5	5	5	5	5	7	9	7	7	7
Ability to meet new requirements	15%	5	5	5	5	5	5	5	5	5	5	7	9	7	7	7
Capability to meet new applications	10%	5	5	5	5	5	7	7	7	7	7	9	9	9	9	9
Operability	20%	7	7	7	7	7	5	5	7	7	5	5	7	5	7	5
Currency of software	15%	5	7	5	5	7	5	7	5	7	7	5	5	5	7	5
Total	100%															
Total score (benefit)			5.7					6					7.5			
Total cost			190					170					340			
Benefit-to-Cost Ratio			0.03					0.035294					0.022059			

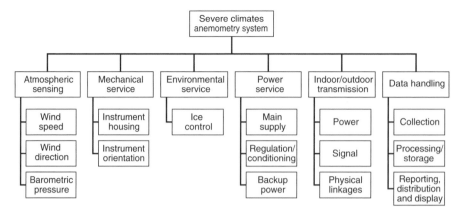

Figure A.5. Top-level SCAS functional decomposition.

functional decomposition for this system is shown in Figure A.5 The requirements for each of the decomposed elements were defined in considerable detail, and are not presented here because the definition and evaluation of alternatives were particularly well-executed. Table A.8 shows three alternative designs, represented as (1) an as-is design, (2) a moderate upgrade, and (3) a major upgrade. We note that each element of these alternatives is set forth at the second level of functional decomposition, all shown in one figure.

The evaluation framework for this system is presented in Table A.9. Here we see the three system alternatives mapped against a set of evaluation criteria listed in the following six categories:

1. Performance
2. Human factors
3. Maintainability
4. Cost
5. Risk
6. Other

The evaluation was carried out at the subcriterion level, with both a weighting and rating set of numerics. Values were normalized to the "as-is" case. As shown in the table, the total scores for the three alternatives are:

1. As-is: score $=100$
2. Moderate upgrade: score $= 128.44$
3. Major upgrade: score $= 125.67$

We thus see that both upgrades represent significant improvements over the as-is alternative, according to the evaluator.

The architect then displayed some of the results in graphical form. Figure A.6 shows a cost-effectiveness plot using an effectiveness metric based on

TABLE A.8 Alternative System Architectures—SCAS

Functional Areas		System Alternatives Described by Functional Area		
		AS-IS	Moderate Upgrade	Major Upgrade
1. Atmospheric sensing	1.1 Wind speed sensing	COTS pitot tube	COTS pitot tube, transducer	COTS pitot, radio transducer
	1.2 Wind direction sensing	Shaft drive	Shaft drive	Shaft drive
	1.3 Pressure sensing	COTS pitot tube	COTS pitot tube, transducer	COTS pitot, radio transducer
2. Mechanical service	2.1 Instrument housing	Machined aluminum assembly	Add molded composite outer coating	Lightweight, more compact, formed-machined composite body
	2.2 Orientation positioning	Wind-vaned / COTS bearings	Reduction in tail boom length	High precision bearing & balancing
3. Environmental service	3.1 Ice control	Calrod htrs./thermocouples/ analog feedback temp. control	Add digitized computer temp. control	Onboard microprocessor control heat temp. sens. units/heat pipes

TABLE A.8 *(continued)*

4. Power service			
4.1 Main power supply	Commercial feed—220/110 V	Commercial feed—220/110 V	Commercial feed—220/110 V
4.2 Power regulation/cond.	COTS conditioners/lightning rods/fuses	Add ground-fault interrupters	Add interrupters and custom lightning arrest system
4.3 Backup power	Battery—for instruments only	Gas generator w/interruption sensor/switcher (COTS)	High-power, high-reliability diesel generator with autoswitching
5. I/O transmission			
5.1 Power transmission	Stranded wire harnesses	Stranded wire harness	Custom slip rings
5.2 Signal transmission	Foil-shielded wire harnesses	Coaxial wire w/slip rings	2-way radio only—no wiring
5.3 Physical linkages	Shaft/conduit, press. tubes	Shaft/smaller conduit/shielded transducer wire	Minimal shaft for physical support & power only
6. Data handling			
6.1 Data collection	Potentiometer/indoor pneum. diff'l. press. cell & barograph	Magnetic position sensors/in-head electr. press. cell & X-ducer	Optical position sensors/ In-head electrical press. Cell & transducer
6.2 Data processing/storage	Manual data-base entry	Automatic computer-controlled	Automatic computer control
6.3 Reporting, dist., display	Physical meters/ devices/manual reporting	GUI/modem access to data views COTS DBMS	Advanced customized DBMS/packet radio network

TABLE A.9 Evaluation Framework—SCAS

Evaluation Criteria	Weight (%)	As-Is			Moderate Upgrade			Major Upgrade		
		Raw	Norm	$W \times N$	Raw	Norm	$W \times N$	Raw	Norm	$W \times N$
Performance	33									
Vaning function/stability	6	—	1.00	6.00	—	1.50	9.00	—	2.25	13.50
Avg. power consumption (kW)	5	3	1.00	5.00	2.5	1.20	6.00	2	1.50	7.50
Impact resistance/robustness	3		1.00	3.00		0.95	2.85		0.85	2.55
Speed of data processing	3		1.00	3.00		4.00	12.00		4.00	12.00
Data availability	3		1.00	3.00		3.00	9.00		4.00	12.00
System availability	5	.995	1.00	5.00	.997	1.00	5.01	.997	1.00	5.01
System reliability	5	.986	1.00	5.00	.998	1.01	5.06	.990	1.00	5.02
Useful service life (yr)	3	5	1.00	3.00	7	1.40	4.20	6	1.20	3.60
Subtotals				33.00			53.12			61.18
Human Factors	14									
Ease of use	4	—	1.00	4.00	—	2.00	8.00	—	2.50	10.00
Operator safety	5		1.00	5.00		1.50	7.50		1.50	7.50
Bystander safety	5		1.00	5.00		2.00	10.00		2.00	10.00
Subtotals				14.00			25.50			27.50
Maintainability	10									
Freq. of sch. Maintenance (/yr)	3	2	1.00	3.00	1	2.00	6.00	1	2.00	6.00
Ease of maintenance	4		1.00	4.00		1.20	4.80		2.00	8.00
Complexity of assembly	3		1.00	3.00		1.00	3.00		1.50	4.50
Subtotals				10.00			13.80			18.50

TABLE A.9 *(continued)*

| | | System Alternative Scores | | | | | | | | |
| | | As-Is | | | Moderate Upgrade | | | Major Upgrade | | |
Evaluation Criteria	Weight (%)	Raw	Norm	W × N	Raw	Norm	W × N	Raw	Norm	W × N
Cost	17	—	—	—	—	—	—	—	—	—
Development cost ($k)	8	60	1.00	8.00	100	0.60	4.80	250	0.24	1.92
Production cost ($k, 50 units)	7	250	1.00	7.00	300	0.83	5.83	500	0.50	3.50
Disposal/decomm. cost ($k)	2	10	1.00	2.00	15	0.67	1.33	20	0.50	1.00
Subtotals				17.00			11.97			6.42
Risk	16	—			—			—		
Cost risk	5		1.00	5.00		0.50	2.50		0.33	1.65
Schedule risk (Development)	2		1.00	2.00		0.50	1.00		0.25	0.50
Performance risk	3		1.00	3.00		0.85	2.55		0.33	0.99
Technological risk	6		1.00	6.00		0.90	5.40		0.10	0.60
Subtotals				16.00			11.45			3.74
Other	10	—			—			—		
Manufacturability	4		1.00	4.00		0.90	3.60		0.50	2.00
Market potential/demand	4		1.00	4.00		1.50	6.00		0.75	3.00
Appearance/aesthetic quality	1		1.00	1.00		2.00	2.00		3.00	3.00
Expandability/upgradability	1		1.00	1.00		1.00	1.00		0.33	0.33
Subtotals				10.00			12.60			8.33
Total Score				100.00			128.44			125.67
Rank				3			1			2

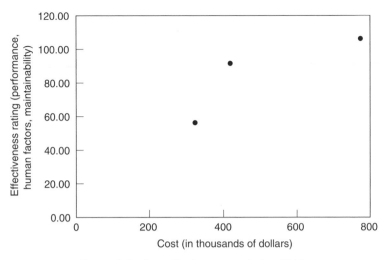

Figure A.6. Cost-effectiveness analysis—SCAS.

the criteria of performance, human factors, and maintainability. This graph distinctly indicates a "knee-of-the-curve" phenomenon. There is a large gain in effectiveness for relatively little cost increase in moving from the as-is alternative to the moderate upgrade. The major upgrade carries with it a large cost increase for only a modest increase in effectiveness. A rather imaginative next step for the architect was to examine the relationship between effectiveness and risk, as illustrated in Figure A.7. Although risk increases monotonically in moving from the as-is to the major upgrade, there is almost no "knee-in-the-curve" as compared with the previous figure.

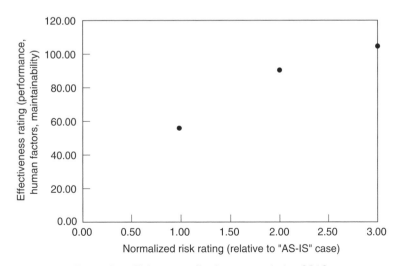

Figure A.7. Risk versus effectiveness analysis—SCAS.

A.6   SUMMARY

This appendix illustrated the systems architecting process by briefly presenting four basic case studies of

1. An automated logistics evaluation, analysis, and planning system
2. A software defects decision support system
3. A systems engineering environment (SEE)
4. An anemometry system

In all cases, students elected to develop the system requirements, define three alternative architectures that would satisfy those requirements, evaluate the alternatives by a weighting and rating system, and examine other system aspects such as risk and preplanned product improvement. The architecting was carried out both by individuals and in teams. The process of defining and evaluating alternatives allowed the system architects to range far beyond a point solution to explore many alternatives. Ultimately, this led to the selection of a preferred system. In a classroom environment, the process was executed over a fourteen-week period. In the real world, the process often extends over several years, including the time required to fully develop system requirements.

REFERENCES

A.1 Students who developed the case studies presented in this chapter are Richard C. Anderson (Case 4), S. Gulu Gambhir (Case 3), David A. Grover (Case 1), Robert H. Laurine, Jr. (Case 2), and Hassan Shahidi (Case 3). This author thanks them again for their diligence, inventiveness, and participation, as well as their permission to include this material here.

A.2 *Expert Choice. Contact Professor Ernest Forman*, The George Washington University, Washington, DC.

A.3 *Catalyst Requirements, Annex II, System Specification for the Catalyst System* (1992). Rome, NY: U.S. Air Force, Rome Laboratory.

A.4 Comer, E. (1992). " Catalyst: Automating Systems Engineering in the 21st Century." *Proceedings of the Second Annual International Symposium of the National Council on Systems Engineering*, Seattle, July 20–22.

A.5 Broad Agency Announcement (BAA) (1993). *Engineering of Complex Systems*. Silver Spring, MD: Naval Surface Warfare Center, Dahlgren Division Detachment, White Oak.

INDEX